Managing Wastewater
In Coastal Urban Areas

Committee on Wastewater Management for Coastal Urban Areas
Water Science and Technology Board
Commission on Engineering and Technical Systems
National Research Council

NATIONAL ACADEMY PRESS
Washington, D.C. 1993

NATIONAL ACADEMY PRESS • 2101 Constitution Ave., N.W. • Washington, DC 20418

NOTICE: The project that is the subject of this report was approved by the Governing Board of the National Research Council, whose members are drawn from the councils of the National Academy of Sciences, the National Academy of Engineering, and the Institute of Medicine. The members of the committee responsible for the report were chosen for their special competencies and with regard for appropriate balance.

This report has been reviewed by a group other than the authors according to procedures approved by a Report Review Committee consisting of members of the National Academy of Sciences, the National Academy of Engineering, and the Institute of Medicine.

Support for this project was provided by the National Academy of Engineering, National Science Foundation Grant No. BCS-9002867, U.S. Environmental Protection Agency Contract No. X-817001-01-0, the City of San Diego, the Freeman Fund of the Boston Society of Civil Engineers, and National Oceanic and Atmospheric Administration Contract No. 50-DGNC-9-00139.

Library of Congress Cataloging-in-Publication Data

Managing wastewater in coastal urban areas / Committee on Wastewater Management for Coastal Urban Areas, Water Science and Technology Board, Commission on Engineering and Technical Systems, National Research Council.
p. cm.
ISBN 0-309-04826-5
1. Sewage disposal. 2. Runoff—Environmental aspects. 3. Coastal zone management. I. National Research Council (U.S.). Committee on Wastewater Management for Coastal Urban Areas.
TD653.M34 1993
628.1'682—dc20 93-1845
CIP

Printed in the United States of America

COMMITTEE ON WASTEWATER MANAGEMENT FOR COASTAL URBAN AREAS

JOHN J. BOLAND, *Chair*, Johns Hopkins University, Baltimore, Maryland
BLAKE P. ANDERSON, County Sanitation Districts of Orange County, Fountain Valley, California
NORMAN H. BROOKS, California Institute of Technology, Pasadena
WILLIAM M. EICHBAUM, The World Wildlife Fund, Washington, D.C.
LYNN R. GOLDMAN, California Department of Health Services, Emeryville
DONALD R. F. HARLEMAN, Massachusetts Institute of Technology, Cambridge
ROBERT W. HOWARTH, Cornell University, Ithaca, New York
ROBERT J. HUGGETT, Virginia Institute of Marine Sciences, Gloucester Point
THOMAS M. KEINATH, Clemson University, Clemson, South Carolina
ALAN J. MEARNS, National Oceanic and Atmospheric Administration, Seattle, Washington
CHARLES R. O'MELIA, Johns Hopkins University, Baltimore, Maryland
LARRY A. ROESNER, Camp Dresser & McKee Inc., Orlando, Florida
JOAN B. ROSE, University of South Florida, Tampa
JERRY R. SCHUBEL, State University of New York at Stony Brook

WSTB LIAISON

RICHARD A. CONWAY, Union Carbide Corporation, South Charleston, West Virginia

PANEL ON ENVIRONMENTAL PROCESSES

NORMAN H. BROOKS, *Chair*, California Institute of Technology, Pasadena
WAYNE R. GEYER, Woods Hole Oceanographic Institution, Woods Hole, Massachusetts
ROBERT J. HUGGETT, Virginia Institute of Marine Sciences, Gloucester Point
GEORGE A. JACKSON, Texas A & M University, College Station
ALAN J. MEARNS, National Oceanic and Atmospheric Administration, Seattle, Washington
CHARLES R. O'MELIA, Johns Hopkins University, Baltimore, Maryland
DONALD W. PRITCHARD, Professor Emeritus, The University at Stony Brook, New York (resigned 9/6/91)
JERRY R. SCHUBEL, State University of New York at Stony Brook

PANEL ON HEALTH, ECOSYSTEMS, AND AESTHETICS

LYNN R. GOLDMAN, *Chair*, California Department of Health Services, Emeryville
WILLIAM M. EICHBAUM, The World Wildlife Fund, Washington, D.C.
ROBERT W. HOWARTH, Cornell University, Ithaca, New York
ROBERT J. HUGGETT, Virginia Institute of Marine Sciences, Gloucester Point
ALAN J. MEARNS, National Oceanic and Atmospheric Administration, Seattle, Washington
JOAN B. ROSE, University of South Florida, Tampa

PANEL ON POLICIES, INSTITUTIONS, AND ECONOMICS

JOHN J. BOLAND, *Chair*, Johns Hopkins University, Baltimore, Maryland
JOHN M. DeGROVE, Florida Atlantic University/Florida International University, Fort Lauderdale
WILLIAM M. EICHBAUM, The World Wildlife Fund, Washington, D.C.
KATHERINE FLETCHER, Campaign for Puget Sound, Seattle, Washington

PANEL ON SOURCE CONTROL AND PUBLICLY OWNED TREATMENT WORKS TECHNOLOGY

THOMAS M. KEINATH, *Chair*, Clemson University, Clemson, South Carolina
BLAKE P. ANDERSON, County Sanitation Districts of Orange County, Fountain Valley, California
TAKASHI ASANO, University of California, Davis
GLEN T. DAIGGER, CH_2M Hill, Denver, Colorado
DONALD R. F. HARLEMAN, Massachusetts Institute of Technology, Cambridge
BILLY H. KORNEGAY, Engineering-Science, Inc., Fairfax, Virginia
JAMES F. KREISSL, U.S. Environmental Protection Agency, Cincinnati, Ohio
JOSEPH T. LING, 3M Company (retired), St. Paul, Minnesota
P. AARNE VESILIND, Duke University, Durham, North Carolina

PANEL ON SOURCES

LARRY A. ROESNER, *Chair*, Camp Dresser & McKee Inc., Orlando, Florida
JAMES P. HEANEY, University of Colorado, Boulder
VLADIMIR NOVOTNY, Marquette University, Milwaukee, Wisconsin
WILLIAM C. PISANO, Havens and Emerson, Inc., Boston, Massachusetts

Staff

SARAH CONNICK, *Study Director*
PATRICIA L. CICERO, *Senior Project Assistant*
JACQUELINE MACDONALD, *Staff Officer*
LYNN KASPER, *Editorial Assistant*

Interns

BETH C. LAMBERT, *Summer Intern*
SUSAN MURCOTT, Marblehead, Massachusetts

WATER SCIENCE AND TECHNOLOGY BOARD

The National Academy of Sciences is a private, nonprofit, self-perpetuating society of distinguished scholars engaged in scientific and engineering research, dedicated to the furtherance of science and technology and to their use for the general welfare. Upon the authority of the charter granted to it by the Congress in 1863, the Academy has a mandate that requires it to advise the federal government on scientific and technical matters. Dr. Bruce M. Alberts is president of the National Academy of Sciences.

The National Academy of Engineering was established in 1964, under the charter of the National Academy of Sciences, as a parallel organization of outstanding engineers. It is autonomous in its administration and in the selection of its members, sharing with the National Academy of Sciences the responsibility for advising the federal government. The National Academy of Engineering also sponsors engineering programs aimed at meeting national needs, encourages education and research, and recognizes the superior achievements of engineers. Dr. Robert M. White is president of the National Academy of Engineering.

The Institute of Medicine was established in 1970 by the National Academy of Sciences to secure the services of eminent members of appropriate professions in the examination of policy matters pertaining to the health of the public. The Institute acts under the responsibility given to the National Academy of Sciences by its congressional charter to be an adviser to the federal government and, upon its own initiative, to identify issues of medical care, research, and education. Dr. Kenneth I. Shine is president of the Institute of Medicine.

The National Research Council was organized by the National Academy of Sciences in 1916 to associate the broad community of science and technology with the Academy's purposes of furthering knowledge and advising the federal government. Functioning in accordance with general policies determined by the Academy, the Council has become the principal operating agency of both the National Academy of Sciences and the National Academy of Engineering in providing services to the government, the public, and the scientific and engineering communities. The Council is administered jointly by both Academies and the Institute of Medicine. Dr. Bruce M. Alberts and Dr. Robert M. White are chairman and vice chairman, respectively, of the National Research Council.

Preface

At least 37 percent of the U.S. population is located in counties adjacent to the oceans or to major estuaries, most of them in relatively concentrated urban areas. The waste from this population and its associated activities is a major contributor to the widely documented deterioration of ocean and coastal waters. Beaches are closed, fisheries and shellfish beds are quarantined, and, in many areas, harbor sediment has become so contaminated that dredging cannot be accomplished safely. The U.S. Environmental Protection Agency (EPA) and others have expressed concern over the relative lack of progress in improving the quality of estuarine and coastal waters. However, two of the most highly publicized coastal wastewater policy disputes in the late 1980s involved cases of alleged *overcontrol.* Both Boston and San Diego complained that they were being asked to provide upgraded wastewater treatment facilities that would produce no significant improvement in ocean water quality at great costs.

A number of factors combined to focus the attention of the Water Science and Technology Board (WSTB) on this issue. First was the conspicuous paradox of complaints of overcontrol in the midst of widespread concern over deteriorating water quality. Another notable feature was the large amount of money involved: both Boston and San Diego face secondary treatment construction costs on the order of several billion dollars. Finally, there had been no outside review of the policies laid down by the Clean Water Act since the National Commission on Water Quality report in 1975. Consequently, in 1989, at the direction of Congress, the EPA requested that the WSTB advise the agency on opportunities to improve wastewater management policy for coastal urban areas in the future.

The WSTB then initiated the study that led to this report. Financial support was provided by the EPA, the National Science Foundation, the National Oceanic and Atmospheric Administration, the City of San Diego, the Freeman Fund of the Boston Society of Civil Engineers, and the National Academy of Engineering. The Committee on Wastewater Management for Coastal Urban Areas was formed and charged with the completion of various aspects of this study. The Statement of Task required the Committee to examine issues relevant to wastewater management in urban coastal areas. Among other things, the Committee was directed to consider:

- environmental objectives, policies, and regulations;
- technology and management techniques; and
- systems analysis and design, including environmental modeling.

The Committee was not asked to review past decisions. Instead, it was directed to identify opportunities for improving the current system through which coastal urban wastewater and stormwater are managed.

The Committee consisted of fifteen engineers, scientists, and environmental policy specialists. The first meeting was held in Washington, D.C. in May 1990. As a result of its initial assessment of the problem, the Committee formed five panels with a combined membership of 30, including the Committee members. The Committee met six times and heard from a wide range of local, state, and federal officials, as well as independent engineers and scientists and other concerned individuals. The panels met more than 20 times, and additional meetings were held by an executive subcommittee and an editorial subcommittee.

It is fair to say that the Committee experienced more than the usual amount of difficulty in preparing this report. The problems arose, not from any significant disagreement among the members, but from uncertainty about how to present the Committee's findings. On the one hand, the Committee has assembled much information of immediate use to wastewater management practitioners, much of it never before published: comparisons of treatment processes, design procedures, a guide to risk management, and evaluations of alternative regulatory instruments. On the other hand, the report provides a wide-ranging critique of existing wastewater management policy and proposes a new and fundamentally different paradigm: integrated coastal management.

The Committee's problem was to present the practical information without obscuring the policy recommendations and to highlight the policy recommendations without crowding out the practical data. This report represents a set of compromises reached after several major changes of outline and countless revisions. It expresses the consensus of the Committee, but it fails to capture everything that every member would have wished.

The issues addressed by the Committee take place within complex and diverse institutional and political structures. Also, as may be expected, data on the progress and status of water quality improvement efforts are less than complete in most cases, and because of site-specific and methodological differences difficult to compare from one case to another. As much of the available information on experiences with existing policies is less than complete, and therefore perhaps subject to broad interpretation, the recommendations in this report reflect the collective judgement of the Committee and are based on a substantial examination of experiences around the nation.

In reaching its conclusions, the Committee consulted with more than 150 distinguished scientists, engineers, public officials, regulators, and citizens (see Appendix G). The Committee is grateful to them for sharing their knowledge, insights, and accumulated experience in these matters. The quality and usefulness of this report has been improved immeasurably by this assistance, so generously offered. Any errors or shortcomings are, of course, the responsibility of the Committee.

In the course of this study, a large volume of material was transmitted to, among, and from Committee and panel members. The reports and papers reviewed at various times in the life of the Committee now amount to more than one meter of shelf space. Another meter would be needed for all of the meeting notebooks, panel reports, and Committee report drafts. The management of this flood of material, along with the arrangement of meetings and monitoring of the comings and goings of Committee and panel members, adds up to a formidable workload interspersed with numerous deadlines. Despite the size of this task, everything was done as needed calmly and efficiently by Senior Project Assistant Patricia Cicero and her colleagues in the office of the Water Science and Technology Board.

Several other WSTB staff members must be mentioned here. WSTB Director Stephen D. Parker was one of the first to understand the need for this study. He guided and shaped the concept through nearly two years of preliminary discussions and was instrumental in securing adequate financial support. He continued to provide advice and perspective throughout the life of the study. Additional assistance was provided by Research Associate Jacqueline MacDonald, who served as staff to one of the panels.

Special thanks are due, as well, to Susan Murcott, a graduate student at the Massachusetts Institute of Technology, who began working with the Committee as a 1990 Summer Intern. Following that summer, Susan continued her association with the Panel on Source Control. She served as a researcher and technical writer, making many important contributions to the panel's report.

To the extent that this report serves its intended purpose, the major credit belongs to Study Director Sarah Connick. Sarah had the overall responsibility for managing the numerous activities that made up the study and for insuring that the final product met the expectations of the Board and the Committee. It would be accurate to say that she carried out this responsibility fully and with complete professionalism, but that would miss the point. Sarah was, in a very real sense, a contributor to this project as well as its staff director. Her knowledge, judgment, and sense of balance were always present, even when the Chair and other Committee members temporarily lost their way. Whatever usefulness this report may have in the public policy arena is largely due to Sarah's consistent attention to purpose and priorities.

Finally, we return to the subject of Boston, San Diego, and other situations where waivers from the Clean Water Act's secondary treatment requirement were requested. Amid much controversy, San Diego's application was withdrawn and Boston's was not approved. The provision of the Act under which waivers are administered, Section 301(h), continues; however, the opportunity to enter the program has since expired. Meanwhile these and other relevant controversies live on. This report will doubtless be scrutinized by persons on all sides of these controversies, seeking evidence of the Committee's views.

The Committee makes no explicit recommendations on these or any other specific cases. Rather, the report offers a detailed proposal for the way in which coastal wastewater systems should be planned and such issues should be considered in the future. The approach the Committee advocates, integrated coastal management, is more demanding in many ways than existing wastewater management policy, but it is inherently flexible in application. The key decisions in the Boston and San Diego cases were made within the more rigid context of existing policy. Whether these decisions were correct at the time is an issue that the Committee did not address. Whether findings in this report will permit any of these decisions to be revisited is essentially a legal matter that is outside the purview of a National Research Council committee. The Committee does believe the report contains technical information and analysis that should be immediately useful to coastal areas faced with environmental problems. In addition, the wastewater management policy proposed here will greatly improve the ability to resolve future conflicts in the best interests of the environment and the community.

JOHN J. BOLAND, *Chair*
Committee on Wastewater Management
for Coastal Urban Areas

Contents

Managing Wastewater

In Coastal Urban Areas

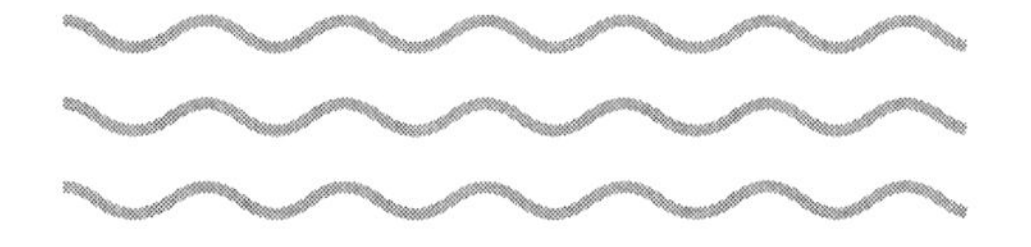

Executive Summary

Although significant progress has been made in improving the nation's water quality over the past 20 years, many coastal areas continue to suffer from persistent environmental problems and can expect to encounter new problems in the future. Today's coastal water-quality management practices do not provide adequate protection from some types of problems and in some cases are overprotective of other types of problems. Much of the debate over how to protect and improve coastal water-quality has focused on urban wastewater and stormwater management.

This report, as requested of the National Research Council by the U.S. Environmental Protection Agency (EPA) at the direction of Congress, examines issues relevant to wastewater management in coastal urban areas. These issues include environmental objectives, policies, and regulations; technology; management techniques; systems analysis and design; and environmental modeling. The National Research Council was not asked to review past policies or decisions. Instead, it was directed to identify opportunities for improving the current system through which coastal urban wastewater and stormwater are managed. The report identifies several key areas in which specific progress could be made, and recommends a new framework for coastal management toward which current management practices should evolve. It addresses the management of marine and estuarine areas in particular and does not consider the Great Lakes.

The scope of activities involved in the management of wastewater and stormwater in coastal cities is large and complex. In the broadest terms, the purpose of managing these wastes is to protect the environment while using it for waste disposal. At least 37 percent of the United States' population resides along the coast, mostly in urban areas. More than 1,400 municipal wastewater treatment plants provide service to the coastal population, discharging 10 billion gallons of treated effluent per day. During the period from 1972 to 1992, about $76 billion were spent in constructing or expanding publicly owned treatment works; $50 billion of this total came from federal grants. At an estimated operating cost ranging from $300 to $500 per million gallons of treated effluent, the national expenditure for operating these plants is between $1.1 billion and $1.8 billion per year.

The management of wastewater and stormwater in coastal urban areas takes place in the context of a multitude of other human activities and natural processes within the coastal zone. Some major factors that cause perturbations in the coastal zone include, in no special order, municipal wastewater and stormwater discharges; combined sewer overflows; other urban runoff; direct industrial wastewater discharges; agricultural runoff; atmospheric deposition; ground water flow; boating traffic; shipping; dredging and filling; leaching of contaminated sediments; oil and gas production; introduction of nonindigenous species; harvesting of fish and shellfish; freshwater impoundment and diversion; and land-use changes in coastal drainage basins.

THE CURRENT APPROACH TO WASTEWATER MANAGEMENT IN COASTAL AREAS

While treatment plant and outfall technologies often dominate discussions of wastewater issues, they are only two of many important pieces that together make up a coastal wastewater management strategy. Other, less visible components of a management strategy include source control efforts to discourage the production of undesirable wastes and prevent their introduction into wastewater and stormwater drainage systems; education to encourage changes in behavior such as appropriate methods for disposal of automobile oil; monitoring to assure compliance and ascertain the effectiveness of management strategies; and environmental studies to improve understanding of the impact of wastewater management strategies and point toward opportunities for improvement. Water conservation and reclamation programs also can be important components of an integrated strategy.

Federal Legislation

Current wastewater and stormwater management policies are rooted in the 1972 amendments to the Federal Water Pollution Control Act, reauthorized in 1977 and 1987 as the Clean Water Act. The 1972 act set the nation on a fundamentally new course for protecting its waters. It asserted federal authority over the quality of navigable waters, required the establishment of uniform minimum federal standards for municipal and industrial wastewater treatment, set strict deadlines for compliance, established a national discharge permit system, and provided substantial amounts of federal grant money to help pay for the newly required projects. The 1972 act resulted in a tremendous effort to control water pollution and produced notable water-quality improvements around the country, particularly in rivers and lakes.

While the approach laid out in the 1972 act produced rapid and effective improvements in many areas, it has not always allowed a process that adequately addresses regional variations in environmental systems around the country or responds well to changing needs, improved science, and more complete information. A provision in the 1977 Clean Water Act attempted to recognize the differences in how municipal wastewater discharges affect marine waters versus freshwater rivers, lakes, and streams. For a limited time, coastal publicly owned treatment works (POTWs) were allowed to apply for waivers which would exempt them from the federal minimum requirement of secondary treatment if they were able to demonstrate that their treatment and disposal practices provided adequate protection of the environment.[1] Dischargers who were granted waivers also had to institute source control and monitoring programs that went far beyond those required for dischargers that met the technology-based requirement.[2] The opportunity to apply for an initial waiver expired at the end of 1982; some applications are still pending. Approximately 40 dischargers are currently operating under waivers; they range from small community systems in the continental United States, Alaska, Hawaii, and the trust territories, to large municipal systems such as the County Sanitation Districts of Orange County, California.

Progress and Emerging Concerns

As improvements have been made in the quality of point source discharges, the impacts of other sources of pollution, diffuse or nonpoint sources, have become more apparent. In some areas, even if pollution from all point sources were controlled, nonpoint contributions would still cause significant environmental problems. Thus, any solution to coastal environmental

[1]Section 301(h) of the Clean Water Act requires that applicants demonstrate "the attainment or maintenance of water quality which assures . . . protection and propagation of a balanced, indigenous population of shellfish, fish and wildlife, and allows recreation activities, in and on the water", in addition to source control, monitoring, and other requirements.

[2]Technology-based requirements are performance standards based on the capability of an existing technology, as opposed to performance standards based on the receiving water requirements.

problems must address the entire range of sources of disruption that causes adverse impacts.

Since 1972, important changes have taken place in government, science, engineering, and the expectations of the public in regard to wastewater and stormwater management and environmental protection. Budget limitations at all levels of government point to the need to spend public money more efficiently. Much has been learned from experience in managing coastal environments. Advances in science have greatly improved understanding of coastal environmental processes, and advances in engineering have led to the development and use of improved technologies for managing coastal resources. As a result of the 1972 act, there is now a well developed permitting system for point source dischargers coupled with a federal enforcement authority.

Constituents of Concern

Wastewater and stormwater management strategies focus on controlling the release of potentially harmful constituents to the environment. As with any activity that affects the environment, the potential for harm depends on the magnitude of the insult, where it occurs, and the characteristics of the stress. In general, a wastewater constituent may be considered to be of high concern if it poses significant risk to human health or ecosystems well beyond points of discharge and is not under demonstrable control. A wastewater constituent may be generally considered to be of lower concern if it causes only local impact or is under demonstrable control.

In the collective judgement of the Committee, in general, it may be anticipated that national level priorities for wastewater constituents in coastal urban areas over the next several decades will be as described below and summarized in Table ES.1. It is noted, however, that priorities may differ at the local and regional level depending on site-specific circumstances.

High Priority

Nutrients. Many estuaries of the Atlantic and Gulf coasts currently experience widespread eutrophication from excess inputs of nutrients, usually nitrogen, and more are vulnerable to excess nutrient enrichment. Secondary treatment does not remove significant amounts of nitrogen from wastewater. Nutrients come from a variety of point and diffuse sources. To adequately address their effects on coastal water bodies, all relevant sources need to be identified and compared, and the most important inputs reduced or otherwise diverted.

Pathogens. Over 100 pathogenic viruses and bacteria have been identified in runoff and sewage. Numerous shellfish beds and bathing beaches

TABLE ES.1 Anticipated National-Level Priorities for Constituents of Concern

Priority	Pollutant Groups	Examples
High	Nutrients	Nitrogen
	Pathogens	Enteric viruses
	Toxic organic chemicals	PAHs
Intermediate	Selected trace metals	Lead
	Other hazardous materials	Oil, chlorine
	Plastics and floatables	Beach trash, oil, and grease
Low	Biochemical oxygen demand (BOD)	
	Solids	

NOTE: Within each priority group the order of listing does not indicate further ranking.

are closed due to unacceptable levels of coliform bacteria each year. However, neither the true extent of contamination by actual human pathogens nor the dominant sources of contamination are adequately known in most regions.

Toxic Organic Chemicals. Chronic industrial and wastewater point sources of toxic chemicals such as chlorinated dioxins, polynuclear aromatic hydrocarbons (PAHs), and solvents have been identified and controlled or are readily subject to control with existing technology. In fish and shellfish, levels of some toxic organics (including chemicals no longer produced in the United States, such as PCBs and DDT) are dropping nationwide, while others such as petroleum hydrocarbons are apparently not declining. Urban runoff, combined sewer overflows and contaminated sediments due to past uncontrolled discharges are major continuing sources of toxic organic chemicals in many coastal urban areas. Although the original source of contamination may have been controlled, contaminated sediments may continue to be secondary sources of contamination to fish, shellfish, and seabirds for many years or decades.

Intermediate Priority

Metals. Elevated concentrations of potentially toxic metals such as mercury, cadmium, and tin are still found in shellfish in localized urban areas, but these problems are not large-scale or region-wide concerns. Dissolved metals may affect species distributions in coastal ecosystems. Most metals do not biomagnify through marine food webs. Source control has been effective in several areas in reducing concentrations. Future problems can be expected to be with lead and localized cases of contamination by organometals. As with toxic organic chemicals, metals from past uncontrolled discharges still contaminate sediments especially near harbors, and can be a significant source of contamination to overlying waters and local aquatic life.

Oil and Other Hazardous Materials. The probability of major oil spills is low, but their immediate impacts on coastal ecosystems and local industries (e.g., fishing, tourism) can be devastating. Of greater consequence, however, are the thousands of unpublicized small spills and leaks (e.g., illegal disposal of used automobile crankcase oil in storm drains) which occur daily in coastal urban areas and may add up to large chronic inputs of petroleum hydrocarbons.

Toxic chemicals used in wastewater treatment (e.g., chlorine compounds) and industrial and commercial settings (e.g., solvents, arsenicals) are transported across urban coastal areas and subject to accidental release. Though not a central part of the wastewater management issue, spills need to be accounted for in addressing coastal quality issues.

Floatables and Plastics. Beaches continue to be fouled by trash from land-based sources, especially following episodic weather conditions such as storms and unusual changes in coastal currents. Marine debris poses hazards to wildlife as well as people, and is aesthetically displeasing. There is considerable opportunity for use of predictive simulation models, such as oil spill trajectory models, to identify sources of marine debris and develop control strategies.

Low Priority

Biochemical Oxygen Demand (BOD). In open coastal waters and well-flushed estuaries, oxygen depletion due to BOD from wastewater discharged through a well-designed outfall is generally not of ecological concern. In these situations, organic material from wastewater is a minor, localized cause of oxygen depletion, especially relative to that due to nutrients. In most coastal urban areas in which BOD from wastewater is of significant concern, it is being controlled under existing requirements.

Solids. Settleable and suspended solids from large wastewater outfalls were once the major cause of localized accumulations of anaerobic sediments and damaged seafloor ecosystems. Where they were significant in the United States (e.g., large municipal outfalls, pulp mills), these conditions have been controlled with primary or advanced treatment, and high dilution outfalls. Today, the degree of solids removal required is driven by the need to protect sediments from accumulations of particle-associated pollutants. Heavy urban runoff, including combined sewer overflows (CSOs), in some areas may still be a source of localized solids accumulations and warrant control.

KEY ISSUES RELATING TO WASTEWATER AND STORMWATER MANAGEMENT

The Committee identified seven specific areas which present opportunities for improving wastewater management in coastal urban areas. Then, based on its analysis of these and other issues, the Committee proposes a new framework for managing coastal waters, integrated coastal management.

Regional Differences

Finding: Because of the wide variations encountered in coastal systems, it is not possible to prescribe a particular technology or approach at the national level that will address all water quality issues at all locations satisfactorily. Any such approach would necessarily fail to protect resources in some coastal regions and would place excessive and unnecessary requirements on others.

Recommendation: *Coastal wastewater and stormwater management strategies should be tailored to the characteristics, values, and uses of the particular receiving environment based on a determination of what combination of control measures can effectively achieve water and sediment quality objectives.*

Discussion: The environmental effects of a POTW discharge from an outfall or urban stormwater from a shoreline outlet depend strongly on the physical, chemical, and biological nature of the receiving water body, and its geography and bathymetry. The degree of flushing of the receiving water with relatively uncontaminated ocean water is a major factor in determining the concentration of nutrients or persistent contaminants in coastal or estuarine waters. In general, this coastal exchange is much slower for the estuaries and shallow coastal shelf waters along the East and Gulf coasts than for the deeper narrow shelf waters of the Pacific coast.

The opportunity for accumulation of wastewater particles and any associated pollutants in bottom sediments also depends greatly on receiving water characteristics. Estuaries may trap sediments and pollutants because flocculation is enhanced where fresh water mixes with salt water. Along the open coasts, deposition is more likely to occur in areas with slow currents and limited exchange with deep water.

Finally the resources to be protected, and water and sediment quality objectives may be quite different among various regions and discharge sites.

The engineering and scientific capability needed to account for these variations has developed significantly over the past 20 years.

Nutrients in Coastal Waters

Finding: Nutrient enrichment, primarily due to nitrogen, is an important problem in many estuarine and some coastal marine systems.

Recommendation: *Greater attention should be focused on preventing excess regional enrichment of nitrogen and other nutrients at levels that are harmful to ecosystems.*

Discussion: Nutrient enrichment can cause oxygen depletion, reduced fish and shellfish populations, nuisance algal blooms, and dieback of seagrasses and corals. While not known to be a problem along much of the open Pacific coast, excess nutrient enrichment, or eutrophication, is a persistent problem in many estuaries, bays, and semi-enclosed waterbodies along the Atlantic and Gulf coasts, and may even be of concern over a large scale in some more open areas along these coasts. Nitrogen controls primary production and eutrophication in most temperate estuaries and coastal waters, although phosphorus can be of concern in many tropical waters and perhaps in some temperate estuaries. By contrast, in freshwater systems, phosphorus is almost always the nutrient limiting growth. It may be important to keep nitrogen and phosphorus concentrations low relative to silicon to avoid causing nuisance algal blooms such as red and brown tides.

Both the sources of nutrients to coastal waters and the associated effects occur at the regional scale making them difficult to measure, assess, and manage. Nutrient inputs to coastal waters come from both point and diffuse sources including wastewater treatment plants, agricultural runoff, urban runoff, ground water seepage, atmospheric deposition, and release of previously accumulated nutrients from bottom sediments.

Source Control and Water Conservation

Findings:

1. Reduction or elimination of pollutants at their sources is an effective tool for managing both point and diffuse sources. For example, for trace metals and toxic organics, source control is more efficient than removal at central plants, which may then have problems of safe disposal of large volumes of contaminated sludge.
2. Water conservation reduces the volume of sewage requiring collection and treatment, however, it does not change the total mass of wastewater pollutants; in fact, pollutant concentrations may actually be increased. The benefits of water conser-

vation include reduced cost of facilities for water supply and wastewater treatment, and reduced impacts in the region from which surface or ground water supplies are extracted.

Recommendation: *Source control of pollutants should be strongly encouraged by incentives and regulation.*

Discussion: Many toxic substances are difficult and/or expensive to remove from wastewater. Often, however, these materials can be prevented from entering the wastestream or significantly reduced in amount through pollution prevention programs. For example, industrial pretreatment and source control programs have already achieved significant reductions of trace metals, toxic organics, and oil and grease in the influent and thus in the effluent and sludge products from municipal wastewater treatment plants (AMSA 1990). In the case of urban runoff, erosion controls at construction sites, street sweeping, storm drain warning signs, and public education efforts have led to improvements in some areas. In new developments, stormwater designs can significantly slow runoff and increase infiltration into the ground and improve stormwater quality.

Levels of Treatment

Findings:

1. Important water and sediment quality problems in the coastal zone include excessive levels of nutrients, pathogens, and toxic substances.
2. Toxic pollutants are often associated with particles in wastewater discharges. Particle removal is therefore a very important treatment step for protecting sediments from excessive carbon enrichment and accumulation of toxic substances.
3. Chemically-enhanced primary treatment has been used successfully to increase the removal of suspended solids in POTWs. Removals of 80 to 85 percent have been achieved with low doses of chemicals; higher removals are possible with higher doses. This level of removal for suspended solids is nearly equivalent to the EPA performance standard for secondary treatment. EPA requires that 30-day averages for removal of suspended solids be at least 85 percent, with effluent concentrations of less than 30 mg/l.
4. The depletion of dissolved oxygen (DO) is generally not of ecological concern in the ocean or in open coastal waters. Where low DO levels are of concern, as in some estuaries, they are more likely to result from eutrophication by nutrients rather than from point source inputs of BOD. In these situations, secondary or any other treatment implemented

solely for BOD removal produces little improvement in receiving water quality.

5. Implementation of an environmental quality-based approach in coastal areas would require levels of treatment in POTWs that, depending on regional needs and receiving-water characteristics, will often be different, either higher or lower, than current requirements.

Recommendation: *Coastal municipal wastewater treatment requirements should be established through an integrated process on the basis of environmental quality as described, for example, by water and sediment quality criteria and standards, rather than by technology-based regulations.*

Discussion: A wide array of wastewater treatment processes is available, however, the costs of treatment and volumes of waste sludges produced tend to increase with increasing removal capabilities. Generally, it is simplest to remove large solids, oil, and grease, then BOD, and then nutrients. Some removal of fine solids, toxic metals and organic substances, and pathogens can be expected with most treatment systems.

Environmental and human health concerns associated with wastewater contaminants differ depending on the location and mechanism of their introduction into coastal waters. Accordingly, wastewater treatment, sludge disposal practices, and other management controls should be guided by water and sediment quality requirements of the receiving waters. Wastewater solids are of concern in most environments because of the possible toxicity of associated heavy metals, organic substances, and pathogens. BOD is of interest in most bays, estuaries, and semi-enclosed waterbodies because of the effects of oxygen depletion on aquatic life. Widespread problems of oxygen depletion in estuaries and coastal waters are much more likely to result from excess nutrient enrichment than from BOD originating directly from wastewater flows. BOD from wastewater flows is generally not important in the open ocean.

Chemically enhanced primary treatment is an effective technology for removing suspended solids and associated contaminants. It has potential application in situations where BOD is not a significant concern. It can also be combined with biological treatment for BOD and/or nutrient removal.

Stormwater and Combined Sewer Overflows

Finding: Urban runoff and CSOs are major contributors to water quality problems in coastal urban areas.

Recommendation: *Stormwater and CSO abatement requirements should be based to the greatest extent possible on an understanding of regional* *and local hydrology and coastal oceanography. They should be designed in conjunction with other regional environmental protection programs to produce the most cost-effective program for achieving the desired level of protection for receiving waters.*

Discussion: Many older cities, primarily in, but not limited to, the northeastern United States, have combined collection systems that carry both stormwater and municipal sewage. During even small rainstorms or if improperly maintained, these systems can overflow, discharging untreated sewage, industrial wastewater, and urban runoff into nearby waterways.

The way in which urban runoff and CSOs affect receiving waters is significantly different from continuous, point loadings. Rainfall induced loads are not constant, but intermittent, pulsed loads. In general, the greatest concentration of pollutants is contained in the first flush of stormwater, with concentrations decreasing as a storm continues.

Reducing pollutant loads from urban runoff and CSOs is significantly more challenging and potentially more costly than removing pollutants from municipal and industrial wastewaters. Wastewater treatment processes are designed to treat relatively constant and continuous flows, and perform poorly when subjected to the extreme variations in flow that are characteristic of stormwater flows.

Currently, pollutant removal efficiencies of treatment facilities for CSOs and urban runoff cannot be stated with sufficient confidence to design a facility plan that will limit pollutant loads from these sources to a prescribed level. Given the cost of constructing these facilities on a large scale in urban areas ($20 to $60 million per square mile for combined sewer areas and $6 thousand to $3.8 million per square mile for stormwater facilities (APWA 1992)), a serious, well-funded research program is needed.

In the absence of the ability to predict pollutant discharge concentrations accurately, there have been proposals to legislate technology-based requirements mandating the capture and treatment of precipitation from all storms up to a certain size and frequency. The difficulty with such requirements is the same as that for other technology-based treatment requirements: in some cases they are likely to result in costly overcontrol; in others, undercontrol with continued adverse environmental effects; and in relatively few cases will they likely meet the environmental protection requirements of a particular region in a cost-effective way.

Detecting Human Pathogens

Finding: Although concentrations of coliform bacteria higher than conventional standards indicate unacceptably high risk of exposure to human pathogens through water contact sports or consumption of shellfish, the opposite is not true—concentrations of coliform bacteria below the standards do not reliably predict that waters and shellfish have safe levels of pathogens.

Recommendation: *The EPA, public health agencies, and wastewater treatment agencies should vigorously pursue the development and implementation of techniques appropriate for routine monitoring to measure more directly the presence of pathogens, particularly in marine and estuarine waters.*

Human pathogens (e.g., enteroviruses associated with diabetes, diarrhea and meningitis, and protozoa such as *Giardia*) can be detected routinely in untreated wastewater. Levels of such pathogens present in treatment plant discharges vary as a function of the level of infection in the community that produces the wastewater and the type of treatment processes used.

The traditional method for assessing the presence or potential presence of human pathogens in wastewater effluent, stormwater, and the ambient environment has been to use coliform bacteria as an indicator of disease-causing organisms. However, coliforms are not predictors of the presence or survival of pathogens, such as viruses or parasites. For example, in the United States, outbreaks of illnesses due to enteric pathogens such as the hepatitis A virus continue to occur and are associated with consumption of shellfish from areas contaminated by nearby wastewater discharges although coliform standards were being met. The risk of disease transmission related to wastewater management practices needs to be better understood.

Evaluation and Feedback

Finding: The effectiveness of management systems and approaches can only be determined and corrected when necessary, if there is adequate monitoring, research, evaluation, and feedback.

Recommendation: *Management systems should be flexible so that they may be changed as needed to respond to new information about environmental quality and the performance of existing management systems.*

Discussion: It is important that management systems be sufficiently flexible to allow for changes and improvements in response to new information. Evaluations should take into account both the effectiveness of specific components and that of the whole integrated management effort. In order to improve future decisions and control measures, evaluations should focus on lessons learned through implementation experiences. For example, the ability to use mathematical models to predict the behavior of sewage effluent in coastal systems has advanced dramatically over the past 20 years. However, comparatively little effort has been put toward the prototype verification of model predictions. In order to make good use of and improve these tools, it is important that follow-up studies be carried out.

INTEGRATED COASTAL MANAGEMENT

The Committee believes that whether because of any of the seven issues identified above or because of other concerns, most coastal cities now, or in the near future, will face the need to address complex coastal quality management issues. With increasing population pressures, increasing recognition of the importance of nonpoint sources in coastal waters, and the decreasing availability of public funding at the federal, state, and local levels, coastal cities face the need to establish objectives and set priorities for protecting coastal resources. The Committee therefore proposes a framework for managing coastal resources toward which coastal environmental quality management should evolve. This framework, integrated coastal management, should provide the opportunity to consider regional differences, multiple sources of perturbations, costs, and benefits in the development of management strategies.

Finding: Wastewater and stormwater management strategies should be developed in the context of each other and other important sources of perturbation in the coastal zone. Successful management strategies must take into account the multiple sources and identify approaches for controlling inputs in the most cost-effective manner. For example, there may be cases where it is more effective and/or efficient to control other sources rather than upgrading wastewater treatment systems or where cooperative efforts among stormwater agencies, wastewater agencies, water supply agencies, and other institutions charged with managing and/or protecting the region's resources can arrive at mutually beneficial solutions in a more cost-effective manner. Urban coastal wastewater and stormwater management should be based on the following concepts:

- Water quality and sediment quality criteria and standards should be established taking account of risk, uncertainty, and variability among regions and sites.
- A water- and sediment-quality driven approach should be used, with appropriate modeling, to design pollution control systems.
- All sources of pollutants should be considered in the development of regional strategies.
- Policies should be integrated across all media, taking account of environmental impacts on water, air, and land.
- Management options should reflect societal goals and priorities, incorporate public inputs, be cost effective, consider relative risks, and achieve benefits at least commensurate with the costs of controls.
- Management actions should be implemented incrementally so as to provide the flexibility to respond to feedback from monitoring the effect of operating systems, new research results, new technologies, and changing objectives.

Recommendation: *Wastewater and stormwater management and other protection strategies for coastal urban areas should be developed and implemented within a system of integrated coastal management (ICM). ICM is an ecologically based, iterative process for identifying and implementing, at the regional scale, environmental objectives and cost-effective strategies for achieving them.*

Integrated coastal management, as proposed here, is an approach that holds much promise for effective and efficient systematic management of the coastal environment. ICM is an ecologically based, iterative process for identifying, at a regional scale, environmental objectives and cost-effective strategies for achieving them. Through ICM, environmental and human resources that require protection can be identified, the multitude of factors that may contribute to adverse impacts can be considered, and the relative importance of various impacts and contributors can be weighed. The ICM process is flexible so that problems can be addressed at a variety of levels of integration depending on their complexity, and priorities can be set in an explicit manner. ICM, as proposed here, has two fundamental objectives: (1) to restore and maintain the ecological integrity of coastal ecosystems, and (2) to maintain important human values and uses associated with those areas.

Six key principles underlie the implementation of ICM. (1) Management actions need to be developed on the basis of the best scientific knowledge available about ecological functions as well as on a comprehensive understanding of human needs and expectations. (2) Management objectives should be expressed as water- and sediment-quality based, and other environmentally and health based goals. (3) Comparative assessment of both risk scenarios and management options should drive the selection of management strategies. (4) A trans-disciplinary perspective is critical in coastal problem solving. (5) The system should function in a context that is responsive to scientific uncertainty about functions of coastal ecosystems. (6) The system should be driven by science and engineering together with public expectations.

The Process

ICM is a three-part process which should be implemented on a continuing basis. It is iterative with the aim of making incremental improvements in coastal environmental quality over time. The three principal components of the process are (1) dynamic planning, (2) selection and implementation, and (3) research and monitoring. The relationships among these components are shown in Figure ES.1. Of the three, the dynamic planning process is perhaps the most complex. It is within this component that ICM objec-

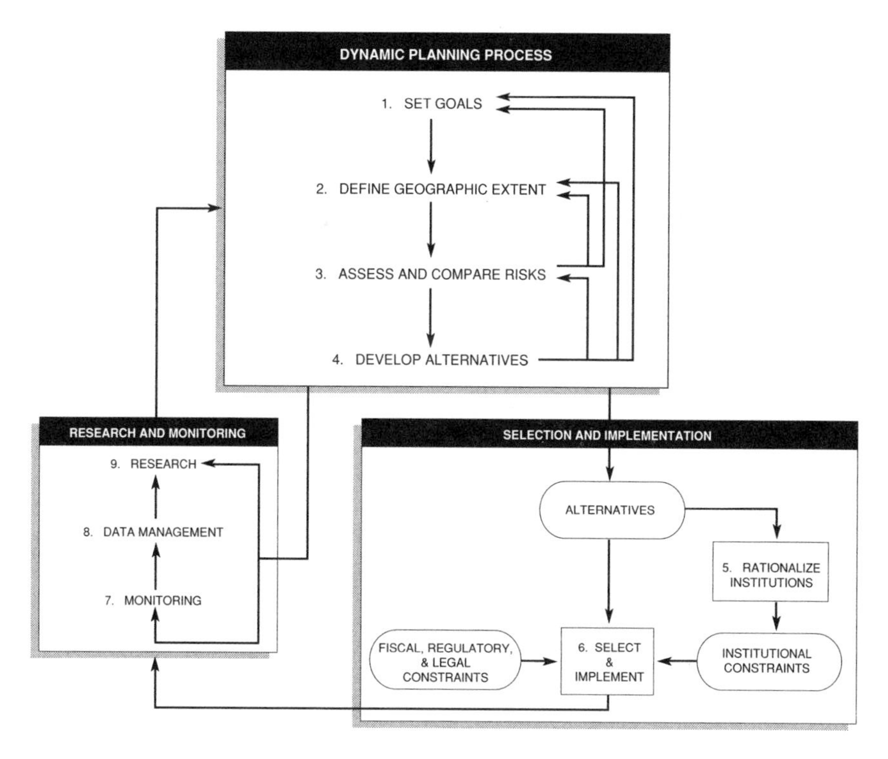

FIGURE ES.1 Process of integrated coastal management.

tives should be evaluated for the region, goals set, risks identified and analyzed, and management alternatives developed and compared. The dynamic planning process should produce two types of results. One is a set of management alternatives to be considered for selection and implementation. The other is an agenda for research and monitoring that is needed to improve understanding and provide feedback on how well the selected management alternatives are working. It is within the selection and implementation process that alternatives should be weighed in regard to objectives, fiscal, regulatory, legal, and institutional constraints, and one should be selected and implemented. Finally, the research and monitoring component should drive the system into the future, bringing new information into the dynamic planning process and developing new methods and techniques for managing coastal resources. Through the continuing ICM process, problems should be tackled in a stepwise, incremental fashion, beginning with those that are of greatest importance as well as those that are easily solved, and then moving on to the next set of concerns.

Benefits, Barriers, and Solutions

Efforts to improve the current system of wastewater management and to implement a system of ICM should afford many benefits as well as encounter many barriers. An ICM system should lead to the development of clearly articulated objectives for the coastal environment. ICM also provides an improved opportunity to meet those objectives by tailoring solutions to specific environmental-quality goals within the region of concern. Flexibility in the combination of management approaches for achieving environmental protection, including increased emphasis on pollution prevention, source control, and economic incentives, provides improved opportunity for achieving environmental objectives. An integrated assessment of the relative risks to the coastal environment and a clear display of management costs and tradeoffs should allow for the implementation of more cost-effective solutions and prevent the diversion of funding from other important activities. Finally, a system of integrated coastal management should be based on expectations for coastal quality within the region. Thus, it provides the opportunity to harness the talent and strengths of individuals within the region and be responsive to local concerns and needs. With ICM, the federal role shifts from that of prescriptive mandate to a partnership with regional authorities in developing a management system that meets coastal quality objectives.

As with any effort to change systems long in place, efforts to develop ICM plans are likely to encounter resistance. Identifying the appropriate geographical region defined by hydrologic and ecologic factors will inevitably cross political jurisdictions and require that there be coordination and cooperation where there may also be conflicting interests. The failure of earlier regional planning efforts mandated by the 1972 act may cause some to dismiss the potential effectiveness of ICM. However, ICM differs significantly from the Area-Wide Planning Studies mandated under section 208 of the Clean Water Act and carried out in the 1970s. The 208 planning process suffered from two fundamental flaws. First, it often was carried out by local agencies having few other water-quality responsibilities. Second, other provisions of the 1972 act, particularly the permitting and facilities funding requirements, forced action so rapidly that they could not be influenced by the planning process.

With ICM, the planning process should be carried out by institutions that are vested with sufficient responsibility, resources, and authority to implement the resulting plan. Building consensus on objectives and goals among interested parties, responsible agencies, and other stakeholders is fundamental to the success of integrated coastal management and will require extensive deliberations and skillful leadership. A regional plan may result in increased public awareness, involvement, and support. In the final analysis, the public is the most important component in making the ideals of an integrated approach a reality.

Implementation

Recommendation: *Improvements in coastal environmental protection in the United States should take place in an incremental manner, building on what has been learned through past efforts and evolving toward a fully integrated and comprehensive approach to coastal protection.*

Moving toward integrated coastal management requires a continuing effort to press forward on scientific, engineering, regulatory, and management frontiers. It will involve risk taking and inevitably experience some setbacks; however, in the long term, ICM is expected to provide the opportunity to apply the most up-to-date information and technologies to coastal problems resulting in efficient and effective coastal protection. In many coastal regions an initial ICM plan could be based primarily on existing information.

Immediate Actions

There are several immediate actions that could be taken to shift the direction of current wastewater, stormwater, and coastal management policies toward one of integrated coastal management. Specifically, existing regional initiatives including those in the National Estuary Program provide an opportunity for implementing the principles of integrated coastal management. The development of Comprehensive Conservation and Management Plans (CCMP) under the National Estuary Program could be done through the ICM process. EPA should encourage states to include ICM concepts in CCMPs by providing supplemental funding to the expanded planning effort.

There are three key areas in which any ongoing activities (including the preparation of a CCMP) directed at the protection and management of coastal waters could be improved. First, public involvement is critical to the success of coastal protection efforts and can be enhanced by increased agency budgets for public-involvement related activities, monitoring programs designed to use citizens, clear communication, and clear lines of authority. Second, scientific and technical information could be applied more effectively to decision-making. This area can be advanced by comparative analyses of risks and management options, good peer review, proper monitoring, directed research, and easily accessible information. Third, improvements could be made in existing institutional arrangements. Consideration should be given to vesting one entity responsible for a coastal region with at least the following functions: responsibility for carrying out the ICM planning process, oversight for budget activities of responsible agencies, responsibility for the design and conduct of monitoring programs, and focal point for public accountability.

Longer-Term Actions

Although some aspects of ICM could be carried out under existing legislation, longer-term strategies are needed that could more fundamentally change the governance of coastal environmental quality, substituting flexibility and local initiative for rigidity and detailed federal control. Several modifications should be made to the two major pieces of federal legislation that address coastal environmental quality.

- Section 320 of the Clean Water Act should be modified to establish, as a supplement to the National Estuary Program, a National Coastal Quality Program that would also apply to those coastal regions that are not estuaries. It would include an integrated planning and management process and supplant the existing CCMP with an Iterative Action Plan that would embody ICM.
- The Clean Water Act and the Coastal Zone Management Act should be amended to provide the flexibility needed to facilitate local ICM initiatives and to better integrate the planning and implementation process between the two statutory systems. An initial effort has been made in this direction with respect to nonpoint sources of pollution.
- In the event that significant new federal funds are authorized to assist states and local governments in complying with the requirements of the Clean Water Act, the availability of these funds should be tied to appropriate use of the ICM process.
- Finally, some of the experience gained in implementing section 301(h) of the Clean Water Act, which provided the opportunity for waivers from secondary treatment for coastal dischargers, might be useful in the development of plans under the proposed National Coastal Quality Program.

Long-Term Implementation

In the next twenty years, it should be the nation's goal to implement a system of integrated coastal management for all of the country's urban shores. Full integration should include all sources of stress to the coastal environment. It should address all environmental media, looking at tradeoffs between disposal of waste to the land, water, and air. It should incorporate the principles of pollution prevention and source control and be a flexible process that facilitates progress and adapts to new information without prescribing the technological means for meeting specified goals. Integrated coastal management should be based in regional objectives and goals for the coastal zone and involve a partnership between federal, state, and local institutions. The lessons learned from the past twenty years of progress clearly point to integrated coastal management as the best direction for the future.

REFERENCES

APWA (American Public Works Association). 1992. A Study of Nationwide Costs for Implementing Municipal Stormwater BMPs. Final Report. Water Resources Committee, Southern California Chapter.

AMSA (Association of Metropolitan Sewerage Agencies). 1990. 1988-1989 AMSA Pretreatment Survey Report.

1

Introduction

Although significant progress has been made in improving the nation's water quality over the past 20 years, many coastal areas continue to suffer from persistent water-quality problems and can expect to encounter new problems in the future. Today's coastal water-quality policies do not provide adequate protection from some types of problems, and in some cases are overprotective for other types of problems. A rethinking of these policies is required if the nation hopes to continue to maintain and improve coastal water quality while keeping pace with coastal population growth.

Around the nation, much of the debate over how to protect and improve coastal waters has focused on urban wastewater and stormwater management. This debate has been fueled by a series of events, some of which have not been demonstrated to be associated with wastewater and stormwater discharges. These events include the mass dolphin deaths that occurred along the Atlantic coast in the summer of 1987, followed by the wash-up of medical wastes along the same coasts in the summer of 1988. Also in 1988, the presidential campaign brought the debate over cleanup strategies for Boston Harbor into the national spotlight. Soon thereafter, the projected $2.4 billion expenditure for upgrades in treatment in San Diego also drew national interest. Brief histories of the San Diego and Boston wastewater management programs and related controversies appear as "case histories" at the end of this chapter. More recently, reports of coral reef die-off in the Florida Keys have fueled interest in wastewater management issues. While each of these situations is unique in the types of problems faced and in the degree of relevance of urban wastewater management, they highlight the need for improved understanding of the impact of human activities and better strategies for preventing and mitigating problems in the coastal environment.

The management of wastewater and stormwater in coastal urban areas is inextricably linked to overall coastal management objectives. While wastewater and stormwater management constitute an immense enterprise, they take place in the context of a multitude of other human activities and natural processes within the coastal zone. In addition, there are many difficult tradeoffs associated with the array of options available for wastewater and stormwater management.

This report, as was requested of the National Research Council by the U.S. Environmental Protection Agency at the direction of Congress, examines issues relevant to wastewater management in coastal urban areas. These issues include environmental objectives, policies, and regulations; technology; management techniques; systems analysis and design; and environmental modeling. The National Research Council was not asked to review past policies or decisions. Instead, it was directed to identify opportunities for improving the current system through which coastal urban wastewater and stormwater are managed. The report addresses marine and estuarine areas in particular and does not consider the Great Lakes.

STRESSES ON THE COASTAL ENVIRONMENT

Most coastal water-quality problems result from human activities associated with populations concentrated along the coasts and from land-use practices throughout coastal watersheds. As the U.S. population grows, it is becoming increasingly urbanized and concentrated along the coasts. In 1990, at least 37 percent of the total U.S. population, or approximately 93 million persons, resided in coastal counties, mostly in urban areas (NOAA 1990a). Population growth in coastal areas is expected to continue more rapidly than in other parts of the nation well into the future. As shown in Figure 1.1, coastal areas are the most densely populated in the United States, rivaled only by the Great Lakes region.

More than 1,400 municipal wastewater treatment plants provide service to these coastal populations and discharge approximately 10 billion gallons of treated effluent per day. Approximately 85 percent of this effluent is discharged into bays and estuaries rather than the open ocean (EPA 1992a). More than 100 municipalities serving approximately 16 million persons have combined sanitary and stormwater sewers that overflow at approximately 1,800 points along the coast. In addition to municipal dischargers, approximately 1,300 industrial facilities are permitted to discharge about 11.3 billion gallons per day of treated industrial wastewater and spent cooling water to marine waters (EPA 1992a).

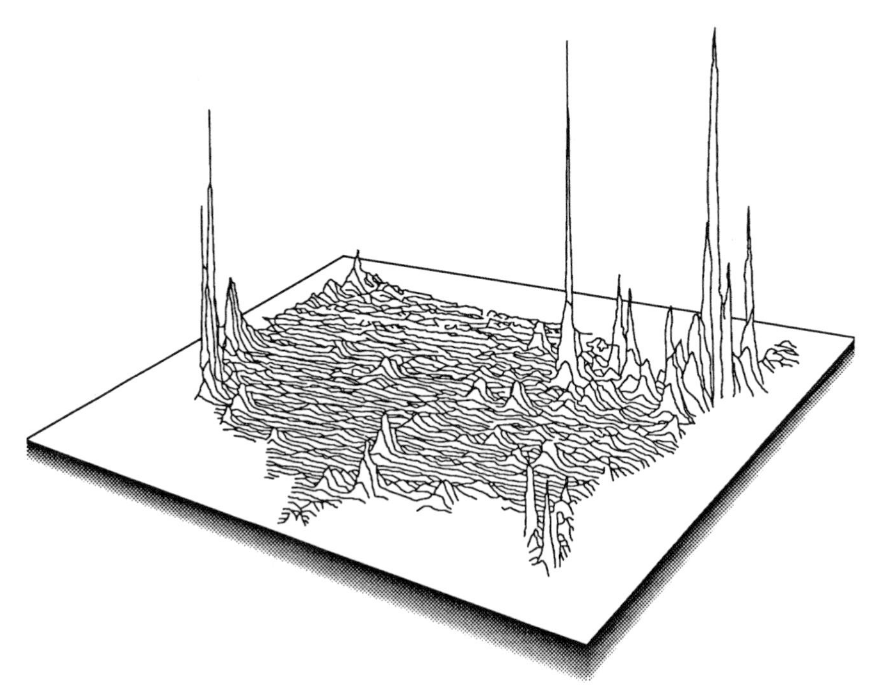

FIGURE 1.1 Distribution of population in the United States by region.

The total quantity of direct discharges to coastal waters (bays, estuaries, and the open ocean) from municipal and industrial facilities does not tell the whole story. Coastal systems receive inputs from a variety of other sources. Urban, industrial, and agricultural runoff, as well as pollutants discharged into rivers upstream of coastal areas, have all been recognized as significant sources of pollutants to marine waters. In general, the volume of runoff and amount of debris and contaminants in runoff increases with increasing urbanization and suburbanization. Scientists are only now realizing the importance of the deposition of pollutants from the atmosphere into water bodies and, in some areas, the potential significance of infiltration of contaminated ground water into coastal waters. Although relatively limited in areal extent, one of the most insidious sources of contamination to the marine environment is that of existing contaminated sediments.

In addition to these direct and indirect inputs of potentially harmful constituents into coastal waters, other human activities can cause stress to marine systems. For example, overfishing, destruction of spawning habitats, and reduction in freshwater flows have all been implicated in the depletion of fishstocks that has taken place around the country. Boating traffic, shipping, dredging and filling, oil and gas development, spills of oil and other hazardous materials, and introductions of nonindigenous species have all been associated with the degradation of coastal environments. Natural events such as hurricanes can cause major perturbations in the coastal zone. It is not only important, but imperative, that the entire range of factors that may have an impact on coastal environmental quality be considered when strategies for protection are developed.

WASTEWATER AND STORMWATER MANAGEMENT

Among the myriad of factors that affect coastal environmental quality, the management of wastewater and stormwater are perhaps the two most critical considerations. Without appropriate control measures, these activities have the potential to wreak serious harm on the coastal environment. As with any other activity that takes place in the environment, the potential for harm depends on the magnitude of the insult, where it occurs, and the characteristics of the stress.

Constituents and Impacts

Municipal wastewater comes from a variety of sources including households, schools, offices, hospitals, and commercial and industrial facilities. Stormwater runoff comes from streets, parking lots, roofs, lawns, commercial and industrial developments, construction sites, farmland, forests, and a number of other settings. While wastewater and stormwater contain a wide variety of constituents, these constituents generally can be described using several characterizations: solids; suspended and dissolved substances that exert a biochemical oxygen demand (BOD) in natural waters; nutrients; pathogens; organic chemicals; metals; oil and grease; and plastics and floatables. Some constituents may fall into more than one of these categories. For example, metals, organics, and pathogens in wastewater are often associated with suspended solids; and organics can be a component of BOD.

In general, each of these categories of constituents can have an adverse effect in the marine (as well as land and air) environment if present in sufficient concentrations. Table 1.1 provides an overview of these categories, examples of the types of constituents, and a summary of the possible impacts associated with the marine environment. The primary concern associated with BOD is, as the name implies, the depletion of oxygen as organic wastewater constituents degrade in the environment. Oxygen depletion associated with BOD can be a serious problem in lakes, rivers, estuaries, and other enclosed water bodies having limited exchange. In most open coastal areas, however, oxygen depletion due to BOD from wastewater is limited and not important. Oxygen depletion in these waters usually results from excess nutrient concentrations, which cause overgrowth of algae. This

TABLE 1.1 Wastewater and Stormwater Constituent Characterizations and the Associated Impacts in the Marine Environment

Characterizations	Examples	Associated Impacts in the Marine Environment
Solids	Particulate matter ranging from large items to fine particles	Most of the larger sized particles will be removed in treatment process. Fine particles remaining in wastewater effluent may be associated with toxic organics, metals, and pathogens. Solids discharged in shallow and nearshore areas, particularly from runoff, may cause excessive turbidity, shading of seagrasses, and sedimentation.
Biochemical Oxygen Demand (BOD)	Oxygen demanded (or required) for the biodegradation of organic matter	In shallow or enclosed aquatic systems, excessive BOD can cause hypoxia and anoxia and suffocate living resources.
Nutrients	Nitrogen Phosphorus Iron Silica	Excessive levels of nutrients increase primary production. At adverse levels, impacts include nuisance algal blooms, dieback of coral and seagrasses, and local- and regional-scale eutrophication. Eutrophication can lead to hypoxia and anoxia, which suffocate living resources.
Pathogens	*Salmonella* *Shigella* *Campylobacter* Enteroviruses Hepatitis E virus and A virus Gastro-intestinal viruses Vibrio species	Exposure to human pathogens via contact with contaminated water or consumption of contaminated shellfish can result in infection and disease.

Toxic Organic Chemicals	Chlorinated pesticides Other halogenated organics Polyaromatic hydrocarbons Surfactants	Toxic organics dissolved or suspended in the water column or accumulated in sediments can result in an array of adverse effects on marine organisms. Many of these compounds are suspected carcinogens and/or reproductive toxicants. They can concentrate in the tissue of fish and shellfish, which may then be consumed by humans. The bioaccumulative effects of these compounds on wildlife are potentially serious (e.g., DDT and the near extinction of the Brown Pelican in the early 1970s).
Metals	Arsenic Cadmium Chromium Copper Lead Mercury Silver Tin	Metals, depending on their form, can be toxic to various marine organisms and humans. Elevated concentrations of metals may be found in shellfish taken from areas where there are highly contaminated sediments.
Plastics and Floatables	Fishing line Condoms Tampon applicators Oil and grease Other floating trash and debris	While the problem most associated with plastics and floatables is aesthetic offense, these materials can also pose severe hazards to marine wildlife. Fish and other marine animals can become entangled in debris or eat it mistaking it for food. Oil and grease can inhibit natural reaeration processes and exacerbate hypoxic and anoxic conditions.

overgrowth leads to hypoxia and sometimes anoxia and other associated adverse effects. In freshwater dominated systems, for example, at the head of an estuary, phosphorus controls primary production, and in saltwater dominated systems in temperate zones, nitrogen is the controlling factor. In the tropics and semi-tropics, phosphorus may be the controlling factor. In mixed systems, such as in the middle of an estuarine mixing zone, both constituents would be of concern. In general, oxygen depletion associated with BOD occurs on a relatively localized scale (it can be controlled by ocean outfalls that achieve high initial dilution), while oxygen depletion and related effects associated with nutrient enrichment are secondary effects that may occur over much larger areas and usually result from a multitude of sources.

Pathogens are microorganisms that can cause disease in humans and are found in wastewater, stormwater, and urban runoff. They include bacteria, viruses and protozoa, and are most often associated with gastrointestinal illnesses and hepatitis. Individuals can be exposed to these organisms through contact with contaminated recreational water and consumption of contaminated shellfish. Disease associated with the consumption of contaminated shellfish result primarily from the ingestion of raw or inadequately cooked clams, mussels, and oysters.

Toxic organics can cause adverse effects in aquatic organisms and humans. Many of these compounds are synthetic chemicals such as pesticides or solvents. Some, such as polychlorinated biphenyls (PCBs), are slow to degrade to innocuous forms, while others degrade relatively rapidly. Metals, like the toxic organics, can also cause an array of adverse effects in aquatic organisms and humans. They may be present in different chemical forms that have widely varying toxicity.

The term solids refers to particles in wastewater that may consist of a mixture of organic and inorganic material. It can refer to everything from gravel to submicroscopic organic colloids. Solids in wastewater are either readily settleable and can be removed in the primary treatment process, or they tend to remain suspended and require more advanced treatment for removal. Toxic organic chemicals, metals, and pathogens often are associated with solids in wastewater and stormwater. Plastics and floatables are materials that cause aesthetic offense and can harm wildlife. Sometimes these items (e.g., condoms) can wend their way through an entire treatment plant; however, almost all trash and debris are deposited in coastal waters by wind, CSOs, runoff, ships, recreational boaters, and other users of the shore.

Anticipated National-Level Priorities for Constituents of Concern

In the collective judgement of the Committee, in general, a wastewater constituent may be considered to be of high concern if it poses significant

TABLE 1.2 Anticipated National-Level Priorities for Constituents of Concern

Priority	Pollutant Groups	Examples
High	Nutrients	Nitrogen
	Pathogens	Enteric viruses
	Toxic organic chemicals	PAHs
Intermediate	Selected trace metals	Lead
	Other hazardous materials	Oil, chlorine
	Plastics and floatables	Beach trash, oil, and grease
Low	Biochemical oxygen demand (BOD)	
	Solids	

NOTE: Within each priority group the order of listing does not indicate further ranking.

risk to human health or ecosystems (e.g., if it contaminates fish, shellfish and wildlife, causes eutrophication, or otherwise damages marine plant and animal communities) well beyond points of discharge and is not under demonstrable control. A wastewater constituent may be generally considered to be of lower concern if it causes only local impact or is under demonstrable control.

In general, it may be anticipated that national-level priorities for wastewater constituents in coastal urban areas over the next several decades will be as summarized in Table 1.2. It is noted, however, that the relative importance of various constituents will likely differ at the local and regional levels depending on site-specific circumstances.

Treatment Technologies and Other Management Techniques

More than 1,400 municipal wastewater treatment plants provide service to the coastal population, discharging 10 billion gallons of treated effluent a day. At an estimated operating cost ranging from $300 to $500 per million gallons of treated effluent, the national expenditure to operate these plants is between $1.1 billion and $1.8 billion per year. Wastewater is collected and conveyed to treatment plants in sanitary sewers. In some older cities, especially in the northeast, most stormwater and wastewater share a common collection system. During storms, the collection system can become overloaded, which may result in the overflow of a mixture of sewage and stormwater into nearby waterways. These events are called combined sewer overflows (CSOs).

Treatment technologies make use of physical, chemical, and biological processes to remove constituents from wastewater. Techniques range from simple screening and settling operations to sophisticated biological, chemical, and mechanical operations that can produce water clean enough to reuse. As a general rule, the quality of treated water increases with more sophisticated technologies, as do costs, land-area requirements, energy requirements, and the amount of sludge produced.

While treatment systems and levels of treatment have dominated the debate over wastewater management in coastal urban areas, a number of other factors in managing wastewater are also of considerable importance. As with any program for managing wastes, the most desirable tactic is to eliminate the production of the waste in the first place. While the complete elimination of waste is obviously not possible in the case of sewage, there are several approaches that can reduce the discharge of some constituents and decrease the volume of water discharged. Phosphate detergent bans in several inland regions of the United States have resulted in significant reductions in phosphorus levels entering treatment plants. Water conservation saves on a scarce natural resource in arid regions and can, in some cases, reduce the volume of water, although not the mass of pollutants, requiring treatment. Major reductions in contaminants from industrial dischargers have been achieved in areas where wastewater treatment agencies have instituted pretreatment requirements.

An array of best management practices is available to reduce the volume and improve the quality of urban and agricultural runoff. Improved management practices (e.g., optimization of pesticide application rates and timing) prevent pollutants from getting into runoff. Public education can play a pivotal role in changing behaviors that can lead to local water-quality improvement, such as appropriate methods for disposal of used automobile oil. Other techniques, such as structural controls, are available to slow runoff, allow more water to percolate into the ground, and filter out contaminants. In addition, weirs, moveable dams, and detention areas can provide storage capacity in storm and combined sewer systems which can reduce the frequency and volume of CSOs.

The location and mechanism of a wastewater discharge plays an important role in determining the extent of impact on marine resources. Contaminant concentrations build up in shallow and/or enclosed systems, whereas deep currents in open systems tend to disperse and flush away discharged material more rapidly. Open-ocean discharges through multiport diffusers are diluted rapidly.

Finally, an important feature of any wastewater and stormwater management system is a monitoring and research program. Monitoring provides information on how well the system is working and where problems may arise; and research can lead to improved methods. It is through monitoring and research efforts that new and improved approaches for managing wastewater and stormwater are developed.

The Role of Government

Numerous agencies at all levels of government have been established to protect the coastal environment from these many stresses and to maintain its desirability as a human habitat; each has its own particular mandate to protect some aspect of the coastal zone and its uses. Human expectations for the coastal environment are diverse and often conflicting. The rich array of uses, needs, and objectives for coastal resources is reflected in the wide variety of institutions established to steward these resources. At the federal level, the Environmental Protection Agency (EPA) holds the greatest responsibility for regulating the quality of the coastal environment. The EPA is responsible for regulating discharges to the coastal environment under the Clean Water Act and the Ocean Dumping Act and for cooperating in the administration of the Coastal Zone Management Act. Under the Clean Water Act, the EPA is responsible for conducting research to ascertain the best and most effective forms of pollution controls.

The National Oceanic and Atmospheric Administration (NOAA) has diverse responsibilities for coastal management. Within NOAA, the Office of Ocean and Coastal Resource Management is charged to assist states in implementing the Coastal Zone Management Act, including implementing new provisions relating to nonpoint source pollution. Since enactment of the federal Coastal Zone Management Act in 1972, 29 states covering more than 90 percent of the nation's shoreline have developed coastal management strategies meeting the requirements of the law. An important element of this system is the requirement that federal actions be consistent with state planning processes. NOAA's National Ocean Service monitors the nation's coastlines for pollution trends, assists the Coast Guard in protecting marine resources during spills of oil and other hazardous materials, and manages National Marine Sanctuaries. NOAA's National Marine Fisheries Service monitors marine animal populations, is responsible for implementing the Endangered Species Act in the marine environment, and holds federal authority over fishstock management. In addition, as trustee for numerous marine resources, NOAA is responsible for conducting research and developing knowledge about coastal areas, which is useful to managers. NOAA recently has taken an active role in litigating recovery of damages and developing and implementing marine resource restoration projects.

The Food and Drug Administration is responsible for seafood safety. The Army Corps of Engineers has authority over engineering projects and dredge and fill operations in the coastal zone. The Coast Guard is responsible for regulating traffic in coastal waters and coordinating response to oil spills from ships as well as other spills of hazardous materials to coastal waters. Several other federal agencies also have interests in coastal urban areas. For example, the National Park Service is the steward for several coastal and marine parks, and the Fish and Wildlife Service oversees coastal and marine wildlife refuges.

Local and regional government agencies plan, construct, own, and operate the public infrastructure to collect, treat, and dispose of municipal wastewater and stormwater. Usually the day-to-day funding and management of these facilities is provided under the authority of a multiple service government such as a city, or a single purpose entity such as a wastewater management agency. Many other agencies and governmental bodies at the state, regional, and local levels also concern themselves with coastal environmental management. These institutions include state legislatures, city councils, county agencies, port authorities, state water and environmental agencies, public health departments, state and regional park services, regional councils of government, and others.

The Role of the Public

A large number of nongovernmental organizations are concerned with the coastal environment. Among these, the most visible are interest groups that represent the views of different sectors of the public. Public interest groups provide a voice for the wide range of expectations that members of the public hold for their coastal environment. They are organized locally, regionally, and nationally. These groups may advocate policy changes to protect the environment; save certain animal species; enhance sport fishing, hunting of waterfowl, and recreational boating opportunities; or protect divers, surfers, and swimmers.

In addition, a multitude of other organizations represent those who have a business or professional interest in the coastal environment. Within the regulated community, commercial fishing, shipping, and other business interests have organized themselves so that their voice is heard in government. Agencies responsible for managing wastewater and stormwater have organized themselves to share information and advocate policies that are protective of the environment while being cost effective for their rate payers. There is, as well, a growing body of environmental professionals, including scientists, engineers, policy analysts, and managers, who may work for any of these organizations, for the government, industry, or academia, and/or as consultants. These professionals have formed organizations that often speak on behalf of the professional community. In addition, they are often sought to provide advice and guidance.

Although the many groups mentioned above do not always speak on behalf of the entire public or community that they represent, their role in defining the issues and establishing policy is extremely important.

THE CURRENT APPROACH TO WASTEWATER AND COASTAL MANAGEMENT

Current wastewater and stormwater management policies are rooted in the 1972 amendments to the Federal Water Pollution Control Act which in 1977 was reauthorized and amended as the Clean Water Act. Prior to 1972, the nation relied on a system of ambient water-quality standards for which states had responsibility. Despite federal incentives, the system proved to be difficult to implement. Many areas lacked adequate information about existing water quality on which to develop standards. Often scientific understanding of the fate and transport of pollutants was not well developed, making it difficult to prove a causal relationship between a particular discharge and poor water quality. Alleged polluters could argue easily that other waste sources were the cause of water quality problems.

Frustration with the slow and ineffective implementation of earlier statutes led Congress to dramatically change the approach to water quality protection with the 1972 act. The 1972 act established the federal objective ". . . to restore and maintain the chemical, physical, and biological integrity of the Nation's waters." Goals and policies to achieve this objective were established, including:

- development of technologies necessary to eliminate the discharge of pollutants into navigable waters, waters of the contiguous zone, and the oceans;
- elimination of discharge of pollutants into navigable waters by 1985;
- protection and propagation of fish, shellfish, and wildlife, and provision of recreation in and on water whenever attainable (This goal is commonly referred to as "fishable and swimmable.");
- prohibition of discharge of toxic pollutants in toxic amounts;
- provision of federal assistance to construct publicly owned treatment works (POTWs); and
- development of areawide wastewater treatment plans.

The 1987 amendments added the goal of

- development of nonpoint source pollutant control programs.

The act established a federal program parallel to state authority over water quality[1] and established a new system of minimum technology-based[2] discharge standards. In addition it required the establishment of more strin-

---

[1]A state can be delegated authority to carry out its own federal program upon meeting criteria that demonstrate equivalency with federal requirements.

[2] Technology-based requirements are performance standards based on the capability of an existing technology, as opposed to performance standards based on receiving water requirements.

gent treatment controls for individual dischargers if technology-based controls proved to be inadequate to meet site-specific, use-based water-quality standards.

In order to meet these goals, the federal law established some significant new provisions. It provided for a national permit system for regulating the discharge of pollutants to the waters of the nation. It also set forth detailed provisions for forcing compliance as a matter of federal law. Perhaps most importantly, the 1972 amendments triggered a new allocation of resources and political will for the achievement of water quality objectives.

The technology-based standard for POTWs was established in 1973. The standard is based on the performance achievable by state-of-the-art secondary treatment technology operating at the time—in this case, well-operated activated sludge treatment plants. Thus, the secondary treatment performance standard was established for POTWs requiring minimum percentage removal and effluent concentration limits for five-day biochemical oxygen demand (85 percent and 30 mg/l respectively) and total suspended solids (TSS) (85 percent and 30 mg/l respectively).

Many of the coastal cities that discharge to the open ocean sought to be exempted from the secondary treatment requirement. They argued that because ocean currents disperse effluent readily and dissolved oxygen depletion is rarely a problem in open coastal marine environments, secondary treatment might not be the most cost-effective method for controlling pollutants from municipal wastewater treatment plants discharging to ocean waters. The 1977 Clean Water Act recognized merit in this argument and established a waiver process by which municipalities could avoid constructing full secondary treatment facilities if, on a case-by-case basis, they could demonstrate compliance with a strict set of pollution control and environmental protection requirements.[3] Coastal dischargers were given a one-time opportunity to enter into the waiver program. Once granted a waiver, coastal dischargers could reapply every five years, with possibly increasingly stringent requirements, to keep the waiver. The opportunity to apply for an initial waiver expired in 1982.

PROGRESS IN MANAGING WATER QUALITY

The success of the policies established in 1972 is evident throughout the nation as a whole. Over the past 20 years more than $75 billion has been expended in capital investments in municipal wastewater treatment.

[3] Section 301(h) of the Clean Water Act requires that applicants demonstrate "the attainment or maintenance of water quality which assures . . . protection and propagation of a balanced, indigenous population of shellfish, fish, and wildlife and allows recreation activities, in and on water," in addition to source control, monitoring, and other requirements.

Of 3,942 major POTWs[4] operating in the United States, 3,116 are now operating in compliance with the Clean Water Act, and an additional 427 have plans to come into compliance over the next five years (EPA 1992a).

There have been notable improvements in the water quality problems targeted by these policies. Where they were once elevated, concentrations of lead, DDT, and PCBs in coastal fish, shellfish, and sediments are decreasing (NOAA 1990b). In Puget Sound, the once toxic and hypoxic waterways of Everett, Seattle, and Tacoma have recovered. Levels of DDT in fish of the Southern California Bight are 1 percent of what they were 20 years ago. The next section describes improvements in New York Harbor and the Delaware River Estuary. Improved treatment processes and phosphate detergent bans have resulted in notably fewer eutrophication problems in inland waters. All of these improvements have taken place in the face of large growth rates in coastal areas. National coastal monitoring programs confirm that, along much of the nonurban coastal zone of the United States, inputs and environmental concentrations of many waste materials and contaminants that were once in excess are now decreasing, or at least have stopped increasing (NOAA 1990b).

With some notable exceptions, many urbanized bays and estuaries are experiencing no such benefits. Blooms of noxious algae periodically plague portions of Long Island Sound, the New York Bight, Puget Sound and some southeastern estuaries. In most urban estuaries, shellfish beds are closed to commercial harvesting due to unacceptable concentrations of bacteria. In many urbanized bays and estuaries, warnings are posted to inhibit public consumption of chemically contaminated fish and shellfish and to deter public bathing at beaches where waters are contaminated.

Eutrophication, shellfish bed closures, and beach closures continue to affect many urban coastal areas. In a 1990 EPA report to Congress, all coastal states except Hawaii reported at least some impairment of designated uses of estuaries (EPA 1992b). Approximately 37 percent of estuarine waters classified for commercial shellfish harvest are closed or under harvest restrictions; sewage treatment plants, septic systems, and urban runoff are the three most frequently cited reasons for shellfish bed closures (NOAA 1991). Improvements in some areas have revealed new problems. For example, the upgrading of New York City's treatment plants has improved water quality in New York Harbor; however, it is now hypothesized that dissolved nitrogen in the cleaner effluent is entering Long Island Sound where it causes eutrophication problems, particularly during the summer (Parker and O'Reilly 1991, Swanson et al. 1991). Off the coast of Los Angeles County, the upgrade to chemically-enhanced primary treatment and

[4]A major publicly owned treatment work is defined as one discharging more than 1 million gallons per day.

partial secondary treatment at one treatment plant has resulted in the disposal of fewer solids into the ocean. There is concern, however, that with reduced deposition of new sediment, the underlying sediments contaminated prior to 1975 may become exposed, releasing DDT, PCBs and other toxic compounds into the water column (Mearns et al. 1991).

The success of specific policies is difficult to determine for coastal regions. There are significant differences in the types of problems wastewater discharges can cause in marine as compared to freshwater environments. For example, phosphorus is of greater concern than nitrogen in freshwater, but in the marine environment, nitrogen is generally more important. The picture is even further complicated by the great differences among marine environments (see Box 1.1). Along the west coast, a narrow continental shelf and strong currents provide a rapid exchange of water with the ocean. Along the east coast, a wide continental shelf results in slower circulation and less exchange with deeper ocean waters. Circulation in enclosed bays and estuaries is more limited than for open coastal waters. Thus, for example, the depletion of oxygen associated with eutrophication from excess nutrient inputs or with the discharge of organic matter (BOD) can be problematic in bays and estuaries but not necessarily in the open ocean or somewhere like the main basin of Puget Sound, where a large amount of oxygen is provided by strong estuarine circulation of coastal waters. Further, while extensive monitoring is required for POTWs operating in the waiver program and discharging at "less than secondary treatment," there is little documentation of the effects of upgrades to full secondary treatment in coastal areas. Nevertheless, secondary treatment and nutrient removal requirements for discharges to estuaries such as south San Francisco Bay and the Chesapeake Bay clearly should help these areas in their efforts to keep pace with the rapid growth and development in their watersheds.

Water Pollution Control Success Stories

New York Harbor

Though one of the most anthropogenically impacted estuaries in the world, there have been significant improvements to the water quality of New York Harbor as a result of a vigorous program planning first initiated in 1907. Today New York City has 14 wastewater treatment plants; eleven of these facilities operate at full secondary treatment, two are being upgraded to secondary treatment, and one is in the upgrade planning phase. A map of the New York Harbor area is shown in Figure 1.2. A total of 1.65 billion gallons per day of wastewater are discharged into New York Harbor by New York City. Today, because sewage that once went directly into

Box 1.1 The Coastal Zone

The 19 coastal states of the coterminous United States include about 8,000 kilometers (5,000 miles) of coastline along the Gulf of Mexico, the Pacific Ocean, and the Atlantic Ocean. Adding Alaska, Hawaii, and island territories, the United States has a total of 20,600 kilometers (12,400 miles) of coastline.

The United States has an Exclusive Economic Zone (EEZ) jurisdiction to a distance of 200 miles (333 kilometers) offshore of the coastline, an area of approximately one million square miles (2.65 million square kilometers). This is about half again as large as the land area of the contiguous United States. The EEZ area contains about 125,000 cubic kilometers of sea water, an amount about 180 times the volume of all inland estuaries and providing a comparable larger dilution capacity for waste.

Despite the dilution capacity of coastal waters, there are large variations in the characteristics of the United States' coastal zone that warrant caution in marine waste disposal. Divisions of the coastal zone that are of concern in wastewater management include: 1) estuaries (i.e., inland extensions of the ocean, progressively diluted by fresh water); 2) open coastal waters overlying the continental shelf; and 3) ocean waters themselves, overlying the deepening slopes, submarine basins, and canyons beyond the coastal continental shelf. Much of the Pacific coastline is straight with deep waters that have open shelf or oceanic characteristics. In contrast, most of the Atlantic and Gulf shoreline of the United States is indented with hundreds of shallow estuaries along which much of the population lives and works. These estuaries connect inland freshwater systems to the ocean. Indeed, there is no dividing line in the estuarine environment between the fluvial and marine world; gradations between sea and river are subtle, occurring sometimes over hundreds of kilometers as a function of river flow, tidal flow, channel gradient, and oceanic conditions. Because estuaries are shallower and more confined than the open coastline, these environments are less able to accept and disperse effluents. In addition, the circulation associated with estuaries often leads to the trapping of particles in the region where the fresh and saline waters meet. This trapping region is a potential site for the accumulation of toxic compounds.

Estuaries provide critical habitat for much of the productivity and diversity of marine fish, shellfish and coastal wildlife of the United States. Most of the major fisheries of the United States (finfish, oysters, clams, shrimp) are based on species that are highly dependent on estuarine habitats for reproduction and growth. Critical habitats include intertidal and subtidal mudflats; thousands of acres of submerged aquatic vegetation and dense salt marshes, which provide refuge for fish, crabs, and shrimp and for complex food chains that lead to large fish; and great populations of shore birds, ducks, geese, and numerous mammals, reptiles, and amphibians. These are the ecosystems in most need of protection from coastal development and pollution.

Continental shelf waters, which often are chemically, physically, and biologically distinct from the adjoining ocean, extend between the coast or

continued

Box 1.1—*continued*

mouths of estuaries to the *shelf break* beyond which ocean depth increases rapidly to the abyss. On the Pacific, this shelf is narrow, between 5 and 20 kilometers, whereas the Gulf is wider, between 20 and 50 kilometers. Along the east coast, north of North Carolina, it is very wide, extending between 80 and 250 kilometers. Along the Pacific coast, the ocean floor is covered by sand near shore and by progressively finer sediments shifting to silt and clay with increasing depth offshore. Much of the Atlantic shelf is covered with relict sands. Beyond the shelf break on both coasts, along most of the edge of the EEZ, are the deep waters of the continental slope, basins, canyons, and border lands where depths greater than 3,000 meters are common.

Marine life is more diverse (has more species) but generally less productive on the coastal shelf than in estuaries. It includes thousands of species of algae and invertebrates; hundreds of species of fishes and sea birds; and dozens of kinds of marine mammals, woven into several major ecosystems: pelagic, near the sea surface; benthic, on the sea floor; and intertidal, along both rocky and sandy shorelines. Although the potential for diluting nutrient and organic wastes is over a hundred-fold greater here than in estuaries, it is still possible to contaminate and harm marine life of the open coastal zone with persistent toxic chemicals. This was aptly demonstrated in the 1960s and 1970s when the pesticide DDT contaminated pelicans and lead to their reproductive failure in Florida, Texas, and California. Although there has been a substantial decrease in this kind of contamination, residues of DDT and other similar chemicals (PCBs) continue to contaminate sea food in several open coastal areas.

waterways is now treated, concentrations of coliform bacteria have decreased an order of magnitude over the last 20 years. This decrease indicates a significant long-term trend of water quality improvement. Figures 1.3a and 1.3b show average summer coliform concentration trends for New York Harbor. For the first time in 40 years, all beaches at Coney Island are approved for swimming. In the summer of 1992, Midland Beach and South Beach on Staten Island were opened for the first time in 20 years. The possibility of opening additional beaches on the Hudson River north of New York City is being evaluated. From 1970 to 1992, there were improvements in dissolved oxygen concentrations in the Hudson River and harborwide, including in such branches as the Kill van Kull, the Harlem River, and the East River. However, on a harborwide basis, the New York State standard for dissolved oxygen is still not being met. Dissolved oxygen trends in New York Harbor are shown in Figure 1.3c (NYCDEP 1991).

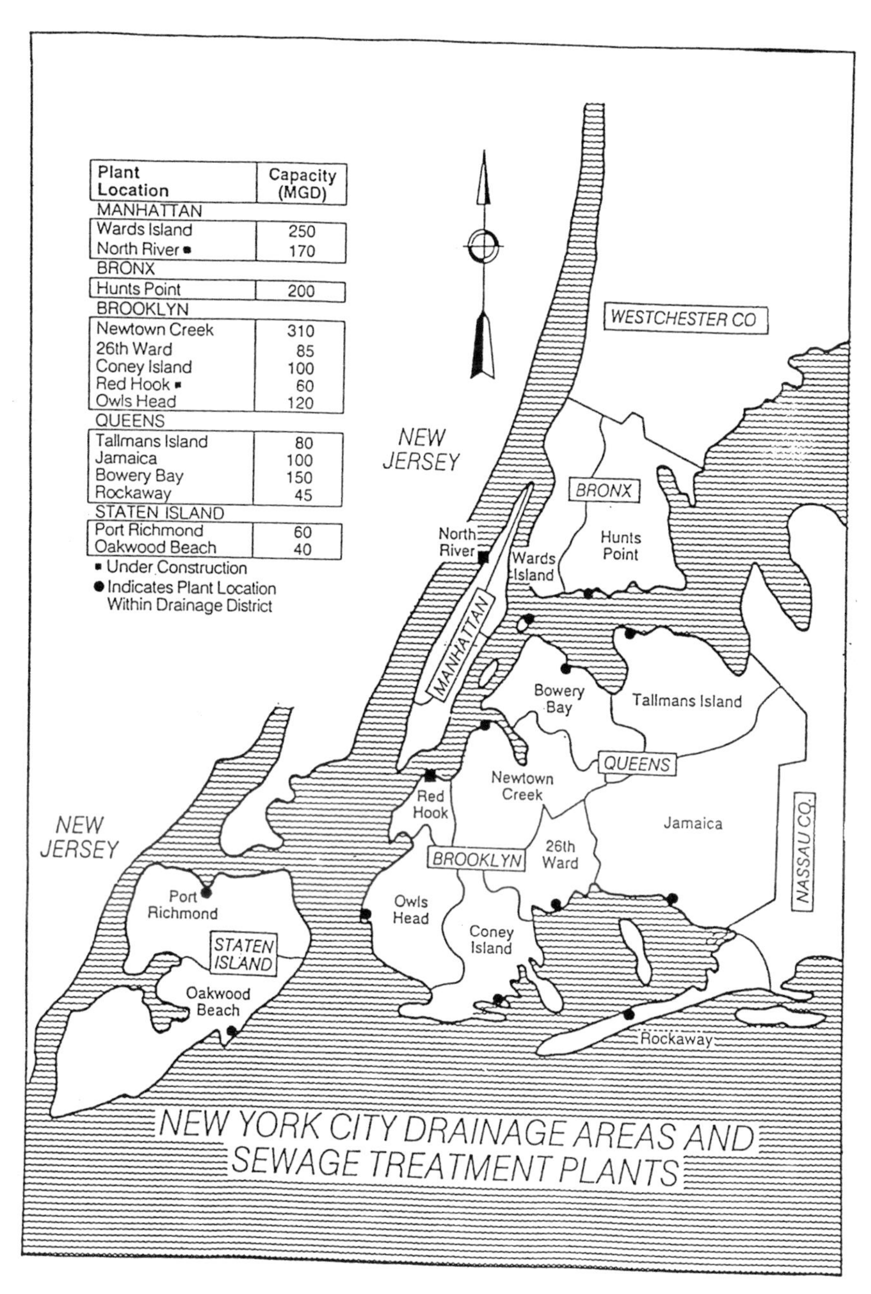

FIGURE 1.2 New York Harbor plant locations and capacities. (Reprinted, by permission, from the New York City Department of Environmental Protection.)

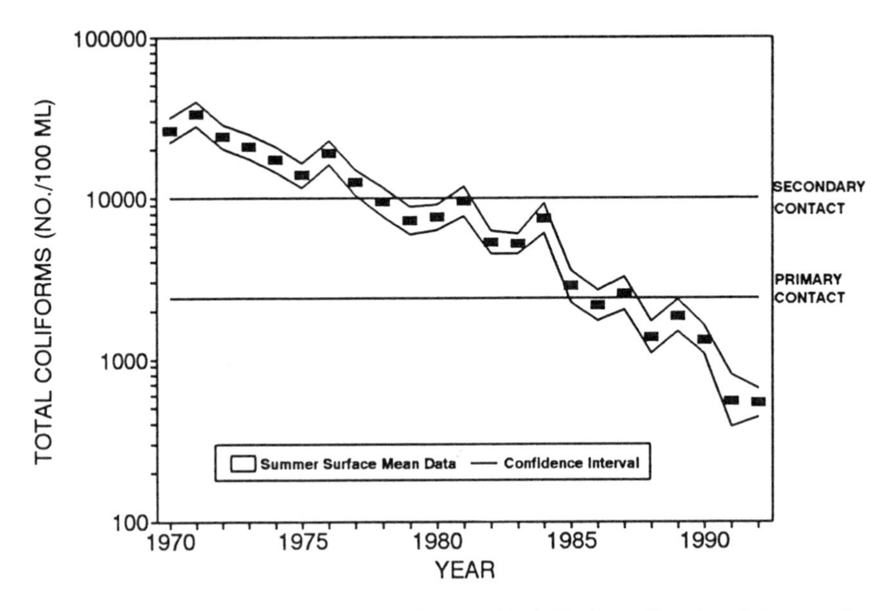

FIGURE 1.3a Total coliforms trends in New York Harbor. (Reprinted, by permission, from the New York City Department of Environmental Protection.)

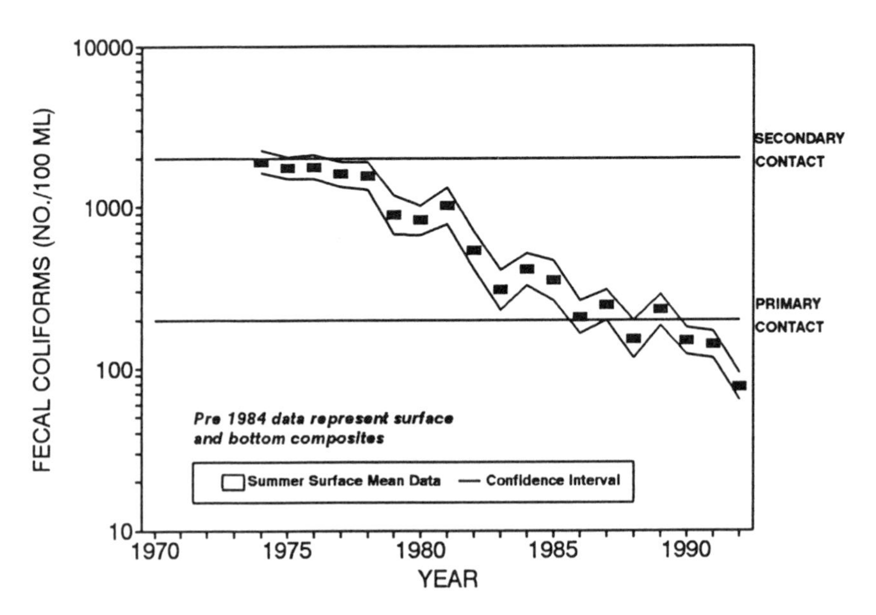

FIGURE 1.3b Fecal coliform trends in New York Harbor. (Reprinted, by permission, from the New York City Department of Environmental Protection.)

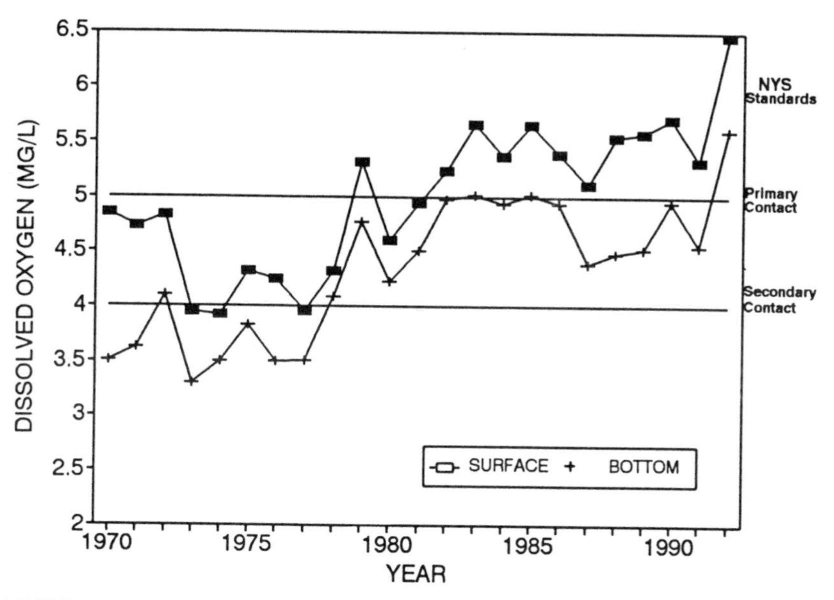

FIGURE 1.3c Dissolved oxygen trends in New York Harbor. (Reprinted, by permission, from the New York City Department of Environmental Protection.)

Delaware River

Intense industrialization and population pressure of the early twentieth century deteriorated the water quality of the Delaware River (Marino et al. 1991). By 1941, a 22-mile portion of the River in the vicinity of Philadelphia and Camden was subject to gross pollution, with average dissolved oxygen concentrations of about 8 percent of saturation during the summer months. In 1950, 282 municipal sewage systems, serving 3.4 million people, discharged into the Delaware River and its tributaries. More than half of the discharges were untreated. By 1960, however, all major cities' sewage treatment plants were operating at primary or secondary treatment levels. In the period from 1977 to 1981, there were significant improvements in Delaware River water quality as a result of treatment plant construction and upgrades. In particular, ambient dissolved oxygen increased, and there were decreases in phosphate, organic nitrogen, and chlorophyll a levels. In 1987, the Camden County plant was upgraded to secondary treatment.

Currently, the status of the Delaware Estuary is significantly improved. Figures 1.4a and 1.4b show trends of improvement over time. Oxygen levels have increased as have populations of some fish (Albert 1987). With these improvements, however, toxics have become a greater concern, and are believed to threaten the living resources in the estuary (DRBC 1989). A

FIGURE 1.4a Historic summer dissolved oxygen profiles for the Delaware Estuary. (Source: Marino et al. 1991. Reprinted, by permission, from Najarian Associates, 1991.).

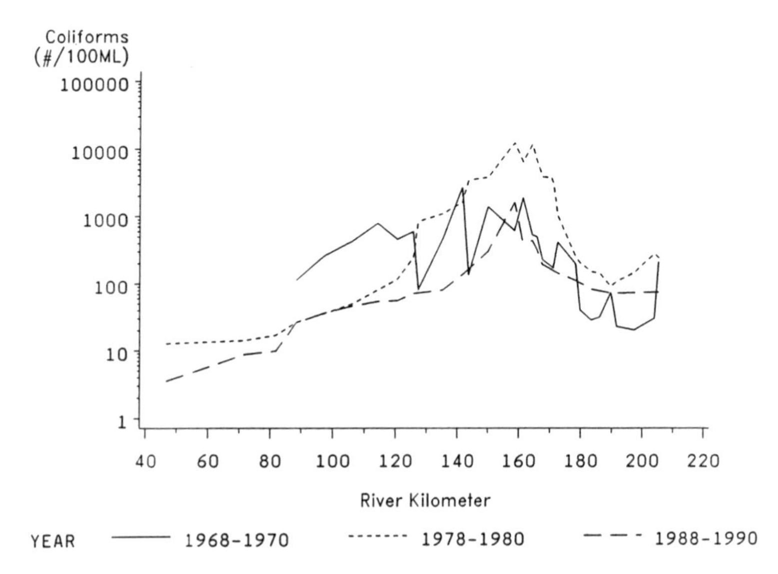

FIGURE 1.4b Historic fecal coliform profiles for the Delaware Estuary. (Source: Marino et al. 1991. Reprinted, by permission, from Najarian Associates, 1991.)

recent, partial assessment of the tidal Delaware River found that finfish were not acceptable for human consumption in 25 of the 84 square miles studied. In the Delaware Bay, shellfish acceptable for human consumption were not supported in 5 percent of the area assessed (DRBC 1992).

CHALLENGES FOR THE FUTURE

Whether because they have failed to modernize their sewage systems, identified new pollution problems, or outgrown the capacity of their current systems, coastal urban centers around the country face the same general need to set priorities for coastal environmental-quality management that are appropriate for long-term conditions. In spite of progress, many coastal areas continue to face serious problems, and virtually all coastal areas can expect to encounter new ones as existing infrastructure ages and population growth continues. The challenges of the future are more complex than those of the past. The coastal problems remaining, such as combined sewer overflows and nonpoint source pollution, are physically and conceptually more complex, and institutionally more elusive, than municipal wastewater treatment issues of the past. In addition, financial resources to address these problems are becoming less available at all levels of government.

Efforts to address coastal water-quality problems over the past 20 years have not solved all our problems, yet progress has been made. While the problems of the future are complex, there is now a greatly improved scientific understanding of physical and ecological processes and improved techniques for managing coastal resources. In addition, advances in computing technologies, such as the modeling of environmental systems and graphical display of information, now afford scientists, engineers, and managers the opportunity to make better use of complex technical information. For example, the use of a three-dimensional eutrophication model in Chesapeake Bay has made it possible to compare the ecological benefits associated with various degrees of point and diffuse source controls (Cerco and Cole 1991, Dortch 1991). Further description of this application is provided in Box 4.1 in Chapter 4.

The Committee believes that faced with the need to solve more complex problems with fewer resources, and armed with a better technical understanding and ability to use complex information, many coastal regions are ready to move toward a system of integrated coastal management. An integrated coastal management system would allow decisionmakers to set priorities amid the complexities of the scientific understanding of, and political objectives for, the coastal resources. Such a system would use regional management structures to address environmental problems on their ecological scales. An integrated system would also recognize that environmental problems require interdisciplinary solutions; a wide variety of skills ranging from aquatic toxicology to resource economics and environmental engineering must be involved.

The experience prior to the 1972 amendments to the Clean Water Act suggests that efforts to manage wastewater on the basis of integrated regional planning were inadequate to meet emerging national objectives for clean water. As a result, the idea of mandatory technology controls became the hallmark of the 1972 act. In retrospect, it appears that the failure of the integrated planning approach can be attributed to a series of extraneous, but important, factors. Since 1972, many of these problems have been overcome. Therefore, it can be supposed that integrated planning is now more likely to be successful. For example, since 1972

- a mandatory permit system that provides a means to apply specific control requirements to particular dischargers has been developed,
- enforcement mechanisms that have the capacity to compel correction of or provide adequate punishment for violations of legal requirements have been established,
- scientific capacity has advanced such that it is now possible to develop relatively accurate predictive models and allocate pollution reduction obligations in a more rational manner, and
- public expectations and support for clean water objectives have grown considerably, allowing for the development of the political will to allocate more adequate resources to cleaning up and protecting the nation's water resources.

Fundamental to an integrated approach are the concepts that control strategies should be driven by an assessment of the priority risks and that risk decisions should place the burden of control on harmful activities. Finally, the implementation of an integrated system should be based in the recognition that coastal management issues will never be completely solved. They require a dynamic approach that first addresses problems that are easily solved or that present the largest opportunity to reduce risk and then moves to address the next highest priority problems. As for any type of management system, feedback is required to help identify successes and failures, identify new issues, make progress on known problems, and indicate where and what improvements can be made. In the case of integrated coastal management, monitoring, research, and public involvement should provide feedback on the management system.

CASE HISTORIES

Boston

Boston Harbor and the major rivers leading into it—the Charles, Mystic, and Neponset—have been used for the disposal of sewage wastes for hundreds of years. The backbone of the present collection and transport system was a combined sewer system completed in 1904 which provided for the discharge of millions of gallons per day of untreated sewage into the harbor. Over the next four decades it became obvious that these untreated discharges presented public health risks for swimmers and shellfish consumers and were causing severe aesthetic problems. To correct these problems, two primary treatment plants that removed about half the total suspended solids (TSS) and about a quarter of the biochemical oxygen demand (BOD) were built—one on Nut Island built in 1952 and having an average daily flow of 110 million gallons per day (MGD), the other on Deer Island built in 1968 and having an average daily flow of 280 MGD. However, the digested sludge produced by these primary treatment plants, approximately 50 tons per day, was also discharged into the harbor until December of 1991. During wet weather, the wastewater system's hydraulic capacity was exceeded causing combined sewage and stormwater to overflow into the harbor through 88 overflow pipes. The discharge from combined sewer overflows (CSOs) would occur approximately 60 times a year, dumping billions of gallons per year of combined wastewater into the harbor and causing frequent closings of nearby shellfish beds and bathing beaches. Figure 1.5 provides a map of the Boston region and includes the location of the Massachusetts Water Resources Authority's wastewater and CSO discharge points.

The Metropolitan District Commission (MDC), the state agency responsible for managing water and wastewater treatment in the Boston metropolitan area until 1984, suffered from insufficient funding to maintain and upgrade the treatment plants. The result was equipment breakdowns, which in turn resulted in the further release of raw and partially treated sewage. The 1972 Clean Water Act mandated an upgrade to secondary treatment (85 percent removal of both TSS and BOD) for all wastewater discharges by 1977. In 1979, the MDC applied for a waiver from secondary treatment as provided by the 1977 Clean Water Act. The MDC proposed a seven-mile outfall to discharge primary effluent into Massachusetts Bay, cessation of sludge discharge into the harbor, and CSO abatement. The MDC studies on water quality determined that the impact of the discharge of primary effluent was acceptable and concluded that secondary treatment would not be cost effective. In June 1983, the EPA denied the waiver, primarily because of concerns about maintaining the dissolved oxygen standard in the bay and protecting the balanced indigenous population of marine life. The EPA also concluded that secondary treatment would result in fewer water-quality exceedences of priority pollutants, one-tenth the loadings of toxics to the sediments around the outfall, and a smaller area of sediment enrichment around the outfall. The MDC modified its waiver request by extending the outfall 9.2 miles into Massachusetts Bay to provide better dilution. The

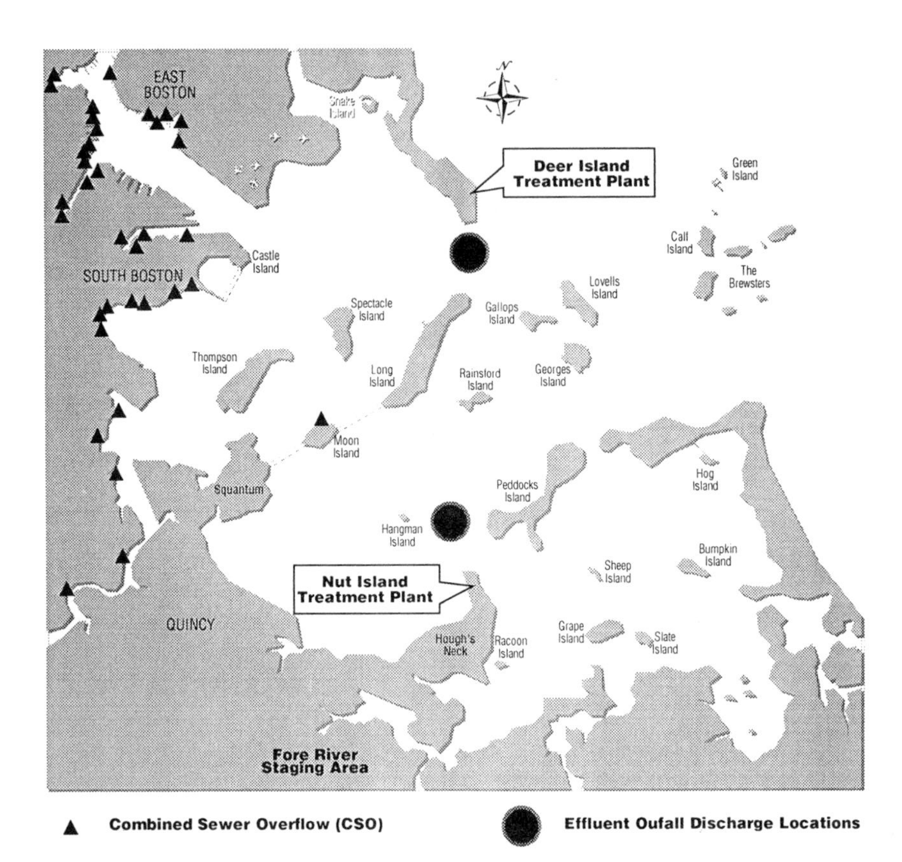

FIGURE 1.5 CSO and effluent outfall discharge locations in Boston Harbor. (Reprinted, by permission, from the Massachusetts Water Resources Authority.)

MDC's application was denied by the EPA again in March 1985. By this time, federal and state grants for construction of sewage treatment facilities were being phased out so that today 90 percent of the costs of the project are borne by the local communities. The MDC took no action to secure federal construction grants for the project prior to 1985. Since 1987, Congress has appropriated $280 million specifically for the Boston Harbor Project. An additional $170 million of State Revolving Fund loans have been earmarked for the project. To date, the combined congressional appropriations and State Revolving Fund loans amount to about 8 percent of the total projected cost of the project.

During the impasse over the need for secondary treatment, two law suits were filed. In 1982, the city of Quincy filed a civil law suit against MDC, and in 1983, the Conservation Law Foundation (CLF), a public interest group, filed a suit against MDC for alleged violations of the Clean Water

Act and against the EPA for failing to bring Boston into compliance with the act. Under pressure from the EPA and the CLF, the state legislature replaced the MDC with an independent authority having control of the regional water and sewer services. The Massachusetts Water Resources Authority (MWRA) was created in December 1984; and in December 1985, the court ordered it to comply with the standards of the Clean Water Act. In 1991, the MWRA, the EPA, the CLF and the court designed a schedule for construction. It included constructing a land-based sludge processing facility to end the discharge of sludge in 1991, constructing a new 1,270 MGD primary treatment plant on Deer Island with a longer outfall by 1995, and constructing a 1,080 MGD secondary treatment plant and solving the problem of CSOs by 1999. The new outfall was selected by MWRA to be 9.5 miles long, including a 6,600-foot-long multiport diffuser—the longest single-leg diffuser in the world—terminating in a depth of water just over 100 feet. Figure 1.6 shows the location of the proposed outfall in Massachusetts Bay.

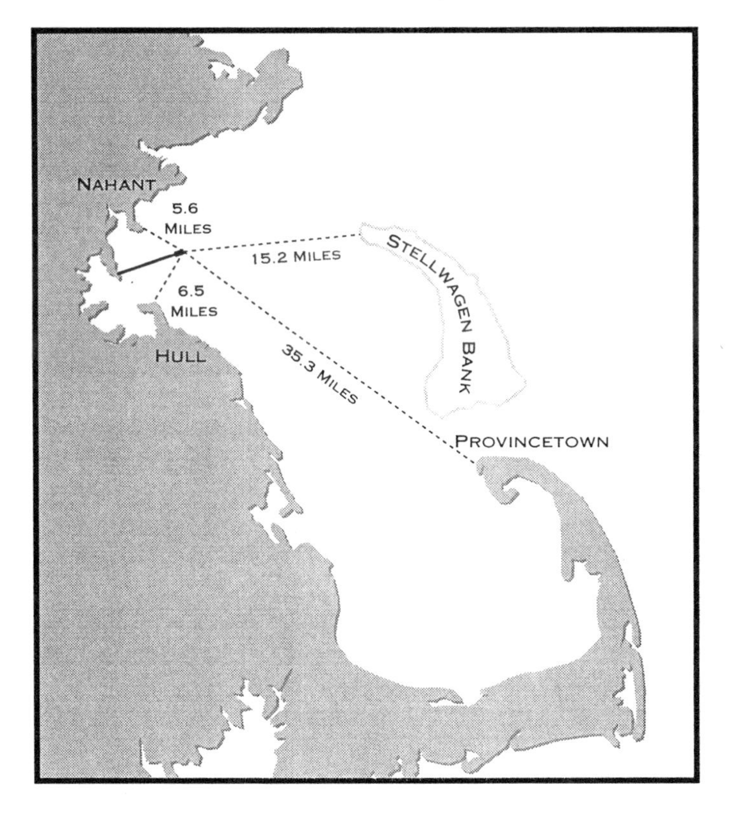

FIGURE 1.6 Location of proposed outfall in Massachusetts Bay. (Reprinted, by permission, from the Massachusetts Water Resources Authority.)

Over the last several years, the MWRA's industrial pretreatment and source reduction programs have significantly reduced the levels of organics and heavy metals in the wastewater. Improvements to the treatment capacity of the existing plants and more efficient use of storage in the system have decreased CSO flows to the harbor, which has reduced bacterial contamination of the harbor to the lowest levels in 50 years. The disposal of sludge into the harbor has ceased, and sludge is now being converted to fertilizer pellets. The outfall and primary treatment plant are under construction, and the first portion of the secondary treatment plant is designed. Plans for CSO treatment are under way. These technical improvements in the wastewater treatment facilities, when complete, are expected to significantly improve the water quality of Boston Harbor.

Recently, the proposed placement of the new outfall, 9.5 miles east of Boston into Massachusetts Bay, has caused substantial controversy in Cape Cod. Concerned about space limitations on Deer Island, integration with other wastewater construction activities, and large increases in sewer rates, in February 1992 the MWRA proposed to pause the construction program after completion of 500 MGD of the secondary plant (sufficient for most dry weather flows) to determine how to best set priorities concerning construction of CSO abatement facilities, the remaining 500 MGD of secondary treatment capacity (needed to treat peak dry weather flows and wet weather flows), and potential additional levels of treatment such as nutrient removal. The MWRA dropped this proposal approximately 5 months later, however it still remains a controversy. In addition, residents of Cape Cod have questioned the sufficiency of secondary treatment to protect Massachusetts Bay. They expressed concern that the discharge would cause nitrogen enrichment of the waters of Cape Cod Bay 35 miles away, resulting in nuisance algal blooms and threatening the endangered North Atlantic right whale and humpback whale. Although the EPA has concluded that such problems will not occur, some Cape Cod groups and others have proposed that the outfall not be built and that the discharge remain in Boston Harbor. There are also concerns within the region that MWRA will retreat from its commitment to secondary treatment following the completion of the outfall and primary treatment plant.

Some civil engineers from the Massachusetts Institute of Technology, including a member of the Committee on Wastewater Management for Coastal Urban Areas, have conducted studies showing that average annual flows in Boston are substantially less than those for which the new treatment plant is designed. They suggest that the new primary treatment plant could be retrofitted for chemically-enhanced primary treatment which would reduce the subsequent BOD loading to the new secondary treatment plant. State and federal requirements could then be met, they propose, by building a biological secondary treatment half the size of the proposed facility. They also argue that with the higher efficiency achieved in the primary treatment plant using chemically enhanced treatment, a portion of the primary treatment plant can be designated to treat wet weather flows, thereby reducing the need for deep tunnel storage capacity to control CSOs. Furthermore, they state that the space saved on Deer Island by reducing the size of the secondary treatment plant would then be available for the construction of nitrogen removal facilities, if needed in the future. They estimate that about $200 million could be saved by building a smaller secondary treatment plant and that several hundred million dollars more could potentially be saved by reducing the capacity of the deep tunnel storage capacity (Harleman et al. 1993). The MWRA is currently funding the MIT civil engineers' studies to further evaluate their claims. MWRA has not, however, presented any such proposals to EPA for consideration. Amid the many controversies, the court-ordered schedule of construction for Boston continues on schedule along with research studies to better understand the environmental processes within Massachusetts Bay.

San Diego

The existing San Diego Metropolitan Sewerage System (SDMSS) was designed and built in the early 1960s to solve water quality problems in San Diego Bay. The creation of the SDMSS involved the consolidation of wastewater from 10 separate communities, previously discharged into the bay or surf zone, into a new primary treatment system at Point Loma with a two and a half mile long, 200-foot deep outfall. The current Point Loma Treatment plant processes between 175 and 185 million gallons of sewage generated each day by more than 1.7 million persons in San Diego and 15 other surrounding cities and sewer districts. During dry weather, an additional 10 to 13 million gallons per day of wastewater from Tijuana is treated at the Point Loma plant. Sludge from the Point Loma treatment plant is dried on Fiesta Island in Mission Bay, a heavily used recreational area, and composted for use in soil conditioners or, as a last resort, disposed of in landfills. Stormwater and wastewater collection systems are completely separate in the San Diego region. Figure 1.7 provides a map of the existing San Diego system.

San Diego applied for a 301(h) waiver from the secondary treatment requirement for its ocean discharge in 1979, for which it obtained tentative approval from the Environmental Protection Agency (EPA) in 1981. In 1983, San Diego submitted a revised waiver application that included revised flow projections as well as plans for treatment of sewage from Tijuana, Mexico. Also in 1983, the State Water Resources Control Board revised the California Ocean Plan to require body-contact bacteriological standards for all kelp beds. Although the Point Loma outfall discharges 6,000 feet be-

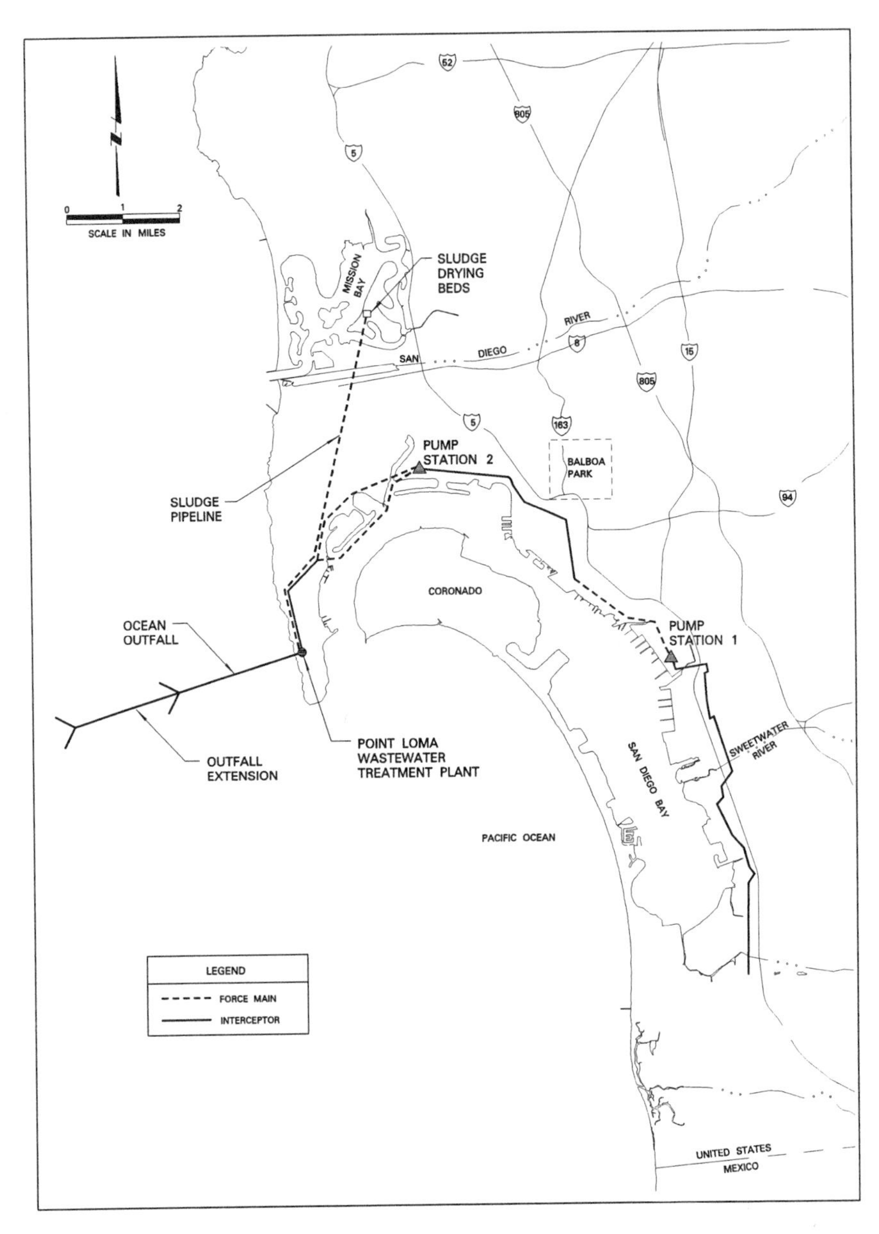

FIGURE 1.7 Locations of San Diego's major wastewater and sludge management facilities. (Reprinted, by permission, from the Clean Water Program for Greater San Diego.)

yond the nearest kelp bed in the area, at times, ocean currents carry the plume into the kelp beds and bacteriologic standards are exceeded. Based on field studies indicating that Point Loma discharges were not adversely affecting divers or the kelp beds, San Diego requested that the Regional Water Quality Control Board exclude or dedesignate the Point Loma kelp beds from the body-contact bacteriological standards requirement.

Between 1979 and 1992, San Diego made several improvements in its treatment system. The initial upgrade from primary to chemically enhanced primary treatment increased the solids removal efficiency from about 55 percent to between 75 and 80 percent in 1988. During this period, San Diego's flow increased from 120 to 185 million gallons per day; thus the total quantity of suspended solids discharged did not change significantly. Currently, tests are being run under court order to determine if the Point Loma treatment plant's removal efficiency can be improved.

In 1986, the EPA tentatively denied both the 1979 and 1983 waiver applications because of the lack of compliance with the California Ocean Plan bacteriological standards in a portion of the kelp beds and interference with the propagation and protection of a balanced indigenous population of bottom-dwelling organisms and fish populations in the vicinity of the Point Loma outfall. In the same year, the Regional Water Quality Control Board indicated that it would recommend against San Diego's prior request to dedesignate the kelp beds as a recreational area that must meet body-contact standards because no alternate standards had been developed to protect divers in the Point Loma kelp beds. Thus in 1987, after two public hearings, the San Diego City Council voted to withdraw the waiver applications and to come into compliance with the Clean Water Act by converting Point Loma to a biological secondary treatment plant. Faced with a water shortage and strong public support, the council also committed to developing an extensive water reclamation and reuse program. The San Diego Clean Water Program was established to carry out these goals.

Following San Diego's withdrawal of its waiver application, the EPA filed suit against the city for more than 20,000 violations of the Clean Water Act and California Ocean Plan. Approximately 3,000 of these alleged violations were raw sewage spills from San Diego's collection system. Between 1987 and 1990, San Diego developed a plan to come into compliance that included upgrading the collection system to prevent spills, complying with bacterial standards in the kelp beds, upgrading the Point Loma plant to secondary treatment, building a second secondary treatment plant, and constructing six water reclamation plants. In 1990, the cost of the plan was estimated to be approximately $2.4 billion which is expected to result in rate payments around $360 per household per year.

By January 1990, the EPA and San Diego entered into a consent decree that put San Diego's plan on a legally enforceable schedule for compliance. However, the federal judge whose approval of the consent decree was also required was troubled by local scientists' contention that secondary treatment would be costly and fail to achieve marked environmental benefits. Following two 1991 hearings on the consent decree and the EPA lawsuit against San Diego, the court found that "the city has been in violation of the Clean Water Act almost continuously since the statute was enacted in 1972" (United States and State of California v. San Diego, No. 88-1101-B[IEG], slip op. at 6 [S.D. Cal. April 18, 1991]). The judge was particularly concerned by the frequency of sewage spills from the collection system, which occurred about once a week. With respect to environmental harm, the court found that the community of organisms in the sediments around the outfall were not adversely affected by the discharge. However, the court also found that the exceedence of bacteriological standards in the Point Loma kelp beds may have adversely affected the marine environment and that viruses in the sewage imperiled divers. (In a later decision, the judge ruled that San Diego could meet bacteriological standards in the kelp beds by extending the outfall rather than by disinfecting the effluent. The extension, which will result in a 4.5-mile-long outfall terminating in water 320-feet-deep, is currently under construction.)

San Diego was fined $3 million for its violations and given the option of paying only $0.5 million and enacting a water conservation ordinance and spending at least $2.5 million on water conservation projects. Finally, the judge deferred his decision on the consent decree until early 1993 which he has since extended until mid-1993. In the interim, he instructed San Diego to conduct tests at the Point Loma plant to determine if it could improve the solids removal efficiency of the chemically enhanced primary treatment process, to complete a master plan for reuse of treated effluent, and to continue all other efforts as if the decree were in effect.

Although the opportunity to apply for a new waiver expired in 1982, in his 1990 and 1991 rulings, the judge has indicated that, since he lacks jurisdiction over the waiver process, his rulings do not affect any rights San Diego might have to reapply for a waiver. In September 1992, the judge modified his interim order, pending the May 1993 hearing on entry into the consent decree, to allow San Diego to proceed with a scaled-down program termed the "consumers' alternative" and granted the city a 19-month extension on all secondary treatment and reclamation projects not included in the consumers' alternative. This $1.3 billion program does not include secondary treatment upgrades, eliminates two of the seven planned water reclamation facilities, delays construction of two, and assumes that two others will be built by other authorities. It contains some capital improvements not included in the original consent decree. The judge suggested San Diego use the additional time to seek administrative and/or legislative relief from the secondary treatment requirement for the Point Loma plant.

San Diego is seeking to reopen its waiver application process although it has not yet applied to the EPA for a reopening. In addition, the city is seeking legislative relief from the secondary treatment requirement in its particular circumstance in the upcoming reauthorization of the Clean Water Act.

REFERENCES

Albert, R.C. 1987. The historical context of water quality management in the Delaware River Estuary. Estuaries 10(3):255-266.

Cerco, C., and T. Cole. 1991. Thirty year simulation of Chesapeake Bay dissolved oxygen. Rotterdam: Balkerna.

Dortch, M.S. 1991. Long-term water quality transport simulations for Chesapeake Bay. Rotterdam: Balkerna.

DRBC (Delaware River Basin Commission). 1989. Attaining Fishable and Swimmable Water Quality in the Delaware Estuary. DEL USA Project, Final Report. West Trenton, New Jersey: Delaware River Basin Commission.

DRBC (Delaware River Basin Commission). 1992. Delaware River and Bay Water Quality Assessment, 1990-1991. 305(b) Report. West Trenton, New Jersey: Delaware River Basin Commission.

EPA (U.S. Environmental Protection Agency). 1992a. Response to Information Request for the National Research Council, Water Science and Technology Board. Washington, D.C.: U.S. Environmental Protection Agency, Office of Water, Office of Wetlands, Oceans and Watersheds.

EPA (U.S. Environmental Protection Agency). 1992b. National Water Quality Inventory, 1990 Report to Congress. EPA 503/9-92/006. Washington, D.C.: U.S. Environmental Protection Agency, Office of Water.

Harleman, D.R.F., L.M.G. Wolman, and D.B. Curll. 1993. Boston Harbor Cleanup Can Be Improved. Boston, Massachusetts: Pioneer Institute for Public Policy Research.

Marino, G.R., J.L. DiLorenzo, H.S. Litwack, T.O. Najarian, and M.L. Thatcher. 1991. General Water Quality Assessment and Trends Analysis of the Delaware Estuary, Part One: Status and Trend Analysis. Eatontown, New Jersey: Najarian Associates.

Mearns, A.J., M. Matta, G. Shigenaka, D. MacDonald, M. Buchman, H. Harris, J. Golas, and G. Laurenstein. 1991. Contaminant Trends in the Southern California Bight: Inventory and Assessment. NOAA Technical Memorandum NOS/ORCA 62. Seattle, Washington: National Oceanic and Atmospheric Administration.

NOAA (National Oceanic and Atmospheric Administration). 1990a. 50 Years of Population Change Along the Nation's Coasts, 1960-2010. Rockville, Maryland: National Oceanic and Atmospheric Administration.

NOAA (National Oceanic and Atmospheric Administration). 1990b. Coastal Environmental Quality in the United States, 1990: Chemical Contamination in Sediment and Tissues. Rockville, Maryland: National Oceanic and Atmospheric Administration.

NOAA (National Oceanic and Atmospheric Administration). 1991. The 1990 National Shellfish Register of Classified Estuarine Waters. Rockville, Maryland: National Oceanic and Atmospheric Administration.

NYCDEP (New York City Department of Environmental Protection). 1991. Harbor Water Quality Survey: 1988-1990. NTIS #PB91-2288i. Wards Island, New York: New York City Department of Environmental Protection.

Parker, C.A., and J.E. O'Reilly. 1991. Oxygen depletion in Long Island Sound: A historical perspective. Estuaries 14(3):248-264.

Swanson, R.L., A. West-Valle, M.L. Bortman, A. Valle-Levinson, and T. Echelman. 1991. The impact of sewage treatment in the East River on western Long Island Sound. In The Second Phase of an Improved Assessment of Alternatives to Biological Nutrient Removal at Sewage Treatment Plants for Alleviating Hypoxia in Western Long Island Sound, J.R. Schubel, ed. Working Paper 56. Stony Brook, New York: Coast Institute, Marine Sciences Research Center.

2

Key Issues Relating to Wastewater and Stormwater Management

In addressing the challenges for the future management of wastewater and stormwater in coastal urban areas identified in Chapter 1, the following eight key issues emerge: regional differences, nutrients in coastal waters, pollution prevention and water conservation, levels of treatment, stormwater and combined sewer overflows, detection of human pathogens, development of management alternatives, and evaluation and feedback. Progress is required in each of these areas if the nation is to continue to advance its clean water goals in coastal areas.

REGIONAL DIFFERENCES

The hydrodynamic and ecological characteristics of coastal zones vary considerably around the country. A description of such variations in the coastal zone is contained in Chapter 1 in Box 1.1. In general, the amount of exchange between coastal waters and the deep ocean is greater in those zones where the continental shelf is narrower. Exchange with deep ocean waters disperses constituents in wastewater and runoff and prevents their concentration in coastal waters.

Even along a particular coastline, one can find dramatically different environments and ecosystems. Estuaries, where fresh water rivers meet marine waters, have unique circulation patterns depending on tides, runoff, and the physical geometry of the system. Estuaries, bays, and sounds can be large or small, shallow or deep. Generally, estuaries, sounds, bays, and other semi-enclosed water bodies have less exchange of marine waters than open coastal areas. Internal recirculation in these systems can lead to the entrapment of pollutants in the sediments and increased concentrations in the water column. Tidal fluxes can also have important influences.

Depending on the circulation patterns and other physical factors, as well as the different ecosystems present, different coastal systems respond differently to wastewater and stormwater inputs. Thus, it is important that wastewater and stormwater management practices be tailored to the characteristics of the particular receiving environment. Because of the wide variations encountered in coastal systems, it is not possible to prescribe a particular technology or approach at the national level that will address all water quality problems relating to wastewater and stormwater management satisfactorily. Any such approach would necessarily fail to protect some coastal regions and place excessive or ineffective requirements on others.

NUTRIENTS IN COASTAL WATERS

Perhaps the most pressing problem in many estuarine and marine systems today is that of nutrient enrichment. While not known to be a problem along most of the open Pacific coast, excess nutrient enrichment, or eutrophication, is a persistent problem in many estuaries, bays, and semi-enclosed water bodies and may be of concern over a large scale in some more open areas along the Atlantic and Gulf coasts. (A complete discussion of coastal nutrient enrichment issues is included in Appendix A; the following is a summary of the key points.)

Nutrients are essential for primary production, the plant growth that forms the base of the food web in all coastal systems. Nitrogen, phosphorus, and a host of other nutrients sustain the production of phytoplankton. In general, in productive freshwaters of the temperate zone, phosphorus is the most important of these elements. It controls the overall rate at which primary production takes place. In nontropical coastal marine waters, however, nitrogen is the most important factor in limiting primary production. At the interface between marine and freshwater, both of these elements are important.

In moderation nutrients can be beneficial, promoting increased production of phytoplankton and, in turn, fish and shellfish. In excess amounts, however, nutrients cause overproduction of phytoplankton, which results in oxygen depletion, which then can reduce the numbers of fish, shellfish, and other living organisms in a water body. Other problems caused by excess nutrient enrichment include nuisance algal blooms, which are aesthetically displeasing and can sometimes carry toxins harmful to fish populations or to humans through consumption of seafood. Nutrient enrichment may also shift the plankton-based food web from one based on diatoms toward one based on flagellates or other phytoplankton that are less desirable as food to organisms at higher trophic levels. The dieback of seagrasses and corals and reduced populations of fish and shellfish have been linked to excess levels of nutrients in coastal waters as well.

Unlike solids and many toxic organic compounds, nutrients are highly soluble in water and therefore highly mobile in coastal systems. Compared with concerns associated with nutrients, problems associated with solids and toxics generally occur more locally. Nutrients can be transported much further, and often there is a time lag between the introduction of nutrients into a water body and the adverse effects associated with eutrophication. Thus, problems related to nutrients can occur on a large scale and may be more difficult to discern.

Nutrients enter coastal waters from every potential point and nonpoint source including: wastewater treatment plants, agricultural runoff, urban runoff, groundwater discharge, and atmospheric deposition. The relative contribution of nutrients from each of these sources varies from area to area depending on local and regional hydrology, land-use patterns, levels of wastewater treatment, and other management practices. Concerns about nutrient enrichment should be addressed in the development of wastewater and stormwater management strategies.

SOURCE CONTROL AND WATER CONSERVATION

No one technology or management control will resolve all wastewater or stormwater management problems. An effective management system must include a suite of technologies and controls tailored to the specific needs of the region. In addition, over the past decade, it has become clear that the best approach to pollution control is pollution prevention wherever possible. It is particularly important that the options not be limited to end-of-pipe treatment strategies.

In the case of wastewater management, municipal systems have found that industrial source control and pretreatment programs can result in significant reductions of metals, toxic organics, and oil and grease (AMSA 1990). Reduction or elimination of these constituents from municipal wastewater treatment plant influent results in better quality sludge, which can then be used for beneficial purposes. It also reduces the discharge of these materials in the effluent to the environment where they may cause harm. Phosphate detergent bans, for example, have resulted in significant reductions of phosphorus from wastewater effluent. It should be noted that an effectively implemented preventive maintenance program in the industrial pretreatment and/or municipal wastewater treatment plants can limit unexpected shutdowns of critical equipment or systems, which could cause bypass of partially treated or untreated waste to receiving waters.

In the case of urban runoff, street sweeping, warnings stenciled on storm drains, and public education efforts are believed to have resulted in improvements in some cities. In new developments, modern stormwater system designs can significantly slow runoff, increasing infiltration into the ground and improving the quality of runoff waters. Retrofitting of parking lots and other drainage areas with treatment and/or control devices can also decrease the amount of pollutants transported from these surfaces to local waters.

Water conservation can achieve benefits in coastal management in several ways. Water conservation reduces the volume of dry weather flows that require treatment, although it does not reduce the mass of pollutants entering the system. Water conservation can, in some cases, delay the need for constructing new treatment capacity for those portions of the plant that are designed on the basis of flow. More significantly, however, water conservation reduces the need for development of new water supplies. Such development can seriously affect the areas where it occurs as well as divert ecologically important freshwater flows from estuaries and other coastal waters.

LEVELS OF TREATMENT

The environmental and human health concerns associated with different wastewater constituents vary from region to region as well as within a region. It is therefore important that the treatment required for wastewater be appropriate for the particular environment to which the wastewater is released. Wastewater treatment and other management controls should be guided by the ecological and human health requirements of the receiving environment, which, in many cases, may be expressed as water and sediment quality criteria, ecosystem indices, or by some other environmentally-based criteria.

In general, suspended solids are of concern because of the metals, toxic organics, and pathogens associated with them. Also, in bays, estuaries, and other shallow nearshore waters sedimentation and shading effects can be a problem. In deep open water, below the depth of light penetration, shading is not a concern; also, outfall diffusers facilitate rapid mixing thereby limiting sedimentation effects to the area in the immediate vicinity of the outfall. Biochemical oxygen demand (BOD) is of concern in bays, estuaries, and semi-enclosed water bodies but is generally not important in the open ocean. Coastal water quality concerns associated with nutrients are discussed above.

There is a wide variety of physical, chemical, and biological treatment processes available for removing each of these types of contaminants. Treatment techniques range from simple screening and settling operations to sophisticated biological, chemical, and mechanical operations that produce water clean enough to reuse. Table 2.1 provides some abbreviated performance

TABLE 2.1 Typical Percent Removal Capabilities for a Range of Wastewater Treatment Processes[1]

	Conventional Primary[2]	Chemically Enhanced Primary[2] (CEPT)		Conventional Biological Secondary Preceded by Conventional Primary[2]	Biological Secondary Preceded by CEPT[2]	Nutrient Removal Preceded by Conventional Biological Secondary and Conventional Primary[3]	Reverse Osmosis[3]
		Low dose	High dose				
Suspended Solids as mg/l TSS	41-69	60-82	86-98	89-97	88-98	94	99
BOD as mg/l BOD_5	19-41	45-65	67-89	86-98	91-99	94	99
Nutrients							
as mg/l TN	2-28	26-48	NA	0-63	NA	80-88	97
as mg/l TP	19-57	44-82	90-96	10-66	83-91	95-99	99

NA = Not available

[1]See Appendix D for more details on the treatment systems and capabilities presented here.
[2]Ranges represent one standard deviation on either side of the mean as determined from two national surveys.
[3]Based on a synthesis of technical literature and engineering models provided by Glen T. Daigger, CH_2M Hill, Denver, Colorado.

information for a series of treatment trains listed in order of increasing sophistication. The materials removed from wastewater end up as sludge and other residuals, which may require additional treatment prior to reuse or disposal. The cost of treatment and volume of sludge produced tend to increase with increasing removal capabilities, as do land area and energy requirements. Generally, the practice is to remove solids, plastics, and floatables, then BOD, and then nutrients. Most of the metals and toxic organics are removed incidentally in these systems, with the metals and some of the toxics ending up in the sludge product and other toxic organics being degraded and/or volatilized to the atmosphere.

The first stage of any treatment plant consists of screening and grit removal to eliminate sand and gravel and other large or heavy items. The next stage is often referred to as primary treatment. Primary treatment simply uses gravity to separate settleable and floatable materials from the wastewater. Other constituents associated with settleable solids are also removed to some degree in the process. The removal capability of a primary treatment system can be improved with the addition of certain chemicals that enhance the tendency of solids to settle. This technique herein is referred to as chemically enhanced primary treatment.

The next level of sophistication in treatment, known as secondary treatment, makes use of both biological and physical processes. Typically, wastewater is first subjected to primary treatment. Activated sludge treatment, the most commonly used biological process, begins by adding oxygen to the system either by vigorous mixing or bubbling. Microorganisms in the mixture feed on organic matter, using the oxygen to convert it to more microorganisms, carbon dioxide, and water. In the second stage, solids, including living and dead microorganisms, are settled out. The solid material produced from both the secondary and primary processes is referred to as sludge or biosolids. Conventional biological secondary treatment is designed specifically to remove BOD and total suspended solids (TSS), but certain amounts of other constituents are also removed incidentally in the process.

The next level of treatment is often referred to as tertiary or advanced treatment and covers a wide variety of physical, biological, and chemical processes aimed at removing nitrogen and phosphorus. Phosphorus removal processes involve either the addition of chemicals to precipitate phosphorus or carefully controlled biological reactions to grow microorganisms having a high phosphorus content and then settle them out of the water. Nitrogen removal involves the carefully controlled biological reactions to convert organic nitrogen and ammonium into nitrate (nitrification) and then into gaseous forms of nitrogen (denitrification).

In addition to nutrient removal, more sophisticated processes such as carbon adsorption and reverse osmosis may be used to remove remaining constituents of concern. Disinfection using chlorine or other chemicals, or exposure to ultraviolet light can be performed at any stage of the process, although it is most effective when the greatest amount of suspended and colloidal solids, which interfere with disinfection, have been removed. The major concern with chlorination is that it is highly toxic to marine organisms. For discharges to some receiving waters, dechlorination may also be required to protect sensitive species.

The solids or sludge removed from these practices must be subjected to another series of treatments. The most commonly used approach for sludge from primary and secondary treatment is anaerobic digestion. This process allows microbes that live in the absence of oxygen to feed on the organic matter, producing more microbes, methane, carbon dioxide, and water. Often, the methane from anaerobic digesters can be captured and used to generate power to operate equipment in treatment plants. Aerobic digestion is used less often because of the high energy costs needed for operation. Composting, a different aerobic process, is used less because of odor problems. Digested sludge contains a large amount of water (typically 95 to 99 percent) and requires dewatering. After dewatering, depending on the content of metals and toxic organics, it may be possible to reuse sludge as a soil amendment with certain crops, on forest land, and for other land-application uses. Otherwise, sludge is generally either landfilled or first incinerated and then the ashes are landfilled. Appendix D contains greater detail on liquid and sludge treatment processes.

Identifying an appropriate series of processes for treatment in a specific situation is complicated because there are often cost and technical tradeoffs when optimizing for the control of a particular class of pollutants. Costs for wastewater treatment increase rapidly with increasing removal efficiencies. Figures 2.1a and 2.1b show performance and cost relationships for removal of TSS and five-day biochemical oxygen demand. A further discussion of these relationships and those for other wastewater constituents is contained in Appendix D.

Technical tradeoffs can be less straightforward. For example, nitrogen discharged from biological secondary treatment is generally in a soluble inorganic form, while nitrogen discharged from a primary treatment process has a high organic content associated with particles. Thus, while biological secondary treatment serves to remove BOD from the discharge to improve local water quality, it also, in effect, mobilizes nitrogen into regional circulation patterns, which may lead to regional scale eutrophication. In the case of primary effluent, the organic nitrogen in the sediments near the outfall also may solubilize over time, but more slowly. In other cases, nitrogen may be of concern while BOD is not. The most practical existing nitrogen removal processes are variations to biological secondary treatment. However, because of the expense associated with nitrogen removal processes, it may be more cost-effective to seek alternative methods for mitigating other

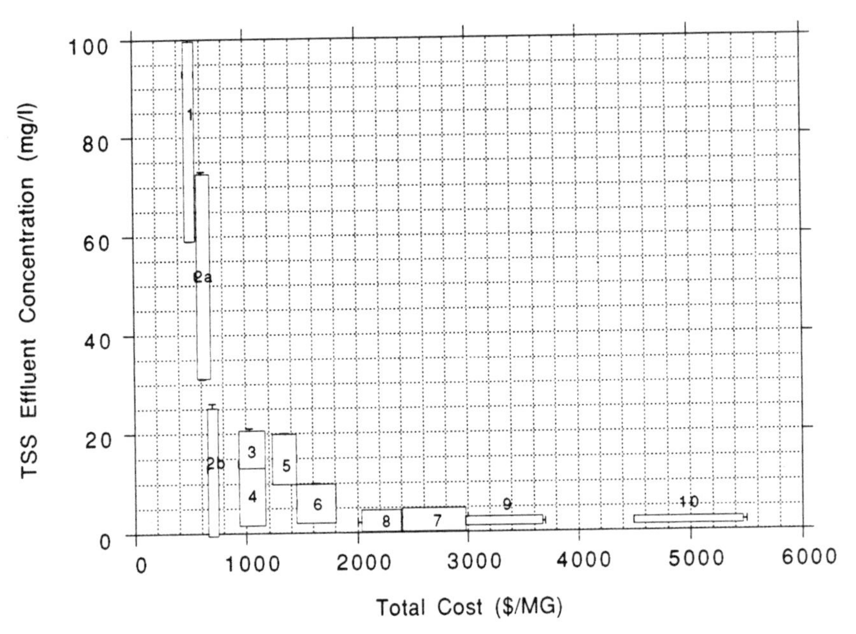

FIGURE 2.1a Total suspended solids performance and cost relationship. NOTE: The 10 different wastewater treatment systems are: (1) primary, (2a) low-dose chemically-enhanced primary, (2b) high-dose chemically-enhanced primary, (3) conventional primary + biological treatment, (4) chemically-enhanced primary + biological treatment, (5) primary or chemically enhanced primary + nutrient removal, (6) system 5 + gravity filtration, (7) system 5 + high lime + filtration, (8) system 5 + granular activated carbon + filtration, (9) system 5 + high lime + filtration + granular activated carbon, (10) system 9 + reverse osmosis. (See Appendix D for further information on the treatment systems and cost and performance ranges presented here.)

sources of nitrogen entering the same receiving waters. Another technical tradeoff associated with nitrogen removal is that while better primary treatment enhances biological secondary treatment and nitrification, it can hinder biological phosphorus removal and denitrification. A fourth example of the tradeoffs inherent in optimizing control of different constituents is that of disinfection. Solids interfere with most disinfection methods. Thus, while the primary concern associated with a particular stormwater outfall or wastewater discharge may be pathogens, suspended solids removal may also be required if pathogens are to be controlled adequately.

In cases of deep ocean discharge where BOD, pathogens, nitrogen, and other nutrients are of little concern, and contributions of toxics and metals associated with solids are low, treatment for removal of these constituents is

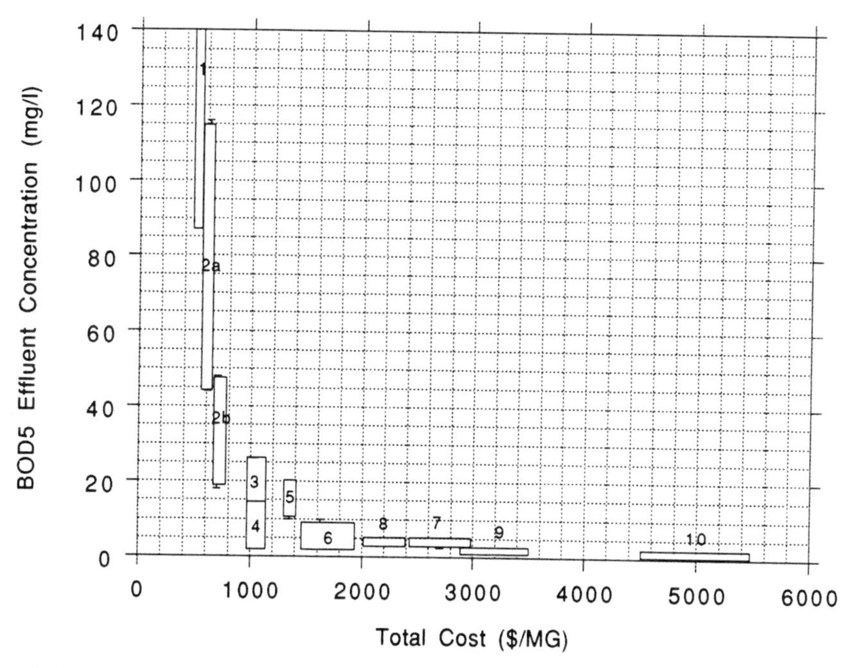

FIGURE 2.1b Five-day biochemical oxygen demand (BOD_5) performance and cost relationship. See note to Figure 2.1a for key to wastewater treatment systems.

unnecessary. At the other end of the spectrum, there may be cases in which wastewater reclamation and reuse are preferable to discharge to coastal waters. For example, as demand for water has increased in some areas or as ecological requirements have led to prohibitions in discharge, wastewater reclamation and reuse have become an important component in water resources planning. Although costly, reclamation and reuse allow water suppliers to supplement short-term needs with reclaimed water while increasing long-term water supply reliability. Reclaimed water is used for landscape irrigation, industrial water supplies, augmentation of ground water supplies, and prevention of salt water intrusion. Depending on the particular needs of a region related to water resources and environmental protection, water reclamation and reuse may play an important role in overall regional strategies for meeting coastal quality objectives. They do not, however, eliminate the need for some ocean discharges.

STORMWATER AND COMBINED SEWER OVERFLOWS

Urban runoff and combined sewer overflows (CSOs) are major contributors to water quality problems in coastal urban areas. Many older cities, primarily in, but not limited to, the northeastern United States, have combined collection systems that carry both stormwater and municipal sewage. Even during small rainstorms, these systems can overflow at designated points, discharging untreated sewage, industrial wastewater, and urban runoff into adjacent waterways.

The way in which urban runoff and CSOs affect receiving waters is much different from continuous, point source loadings. First, rainfall induced loads are not constant, but intermittent, pulsed loads. Pollutant concentrations in these flows vary dramatically during the course of a runoff event, and the total pollution load from any storm is dependent upon the intensity and spatial variability of the rainfall, and the time elapsed since the last rainstorm. In general, the greatest concentration of pollutants is contained in the first flush of stormwater, with concentrations decreasing dramatically for most pollutants as a storm continues. In addition, precipitation in many areas is seasonal so that urban runoff and CSOs may affect coastal waters more during some seasons than others. In central and southern California coastal areas, for example, almost all of the annual rainfall occurs between the first of October and the end of May.

Assessments of the impacts of stormwater and CSO flows on aquatic ecosystems and human health must take into account the variable and intermittent nature of these flows. Fecal coliforms, an indicator of the potential presence of human pathogens (see next section), often can be detected in local receiving waters at levels exceeding health standards for two or three days following a storm event. These levels indicate a potential threat to human health during that period. Heavy metals and toxic organics may be present in toxic concentrations near stormwater and CSO outfalls during a storm but decrease in concentration rapidly as stormwater mixes with receiving waters. They may, however, accumulate in the sediments near stormwater and CSO outlets. Nutrients, on the other hand, pose no immediate threat during a wet weather event but may contribute to the overall loading of nutrients to the region, an issue which is discussed above.

Reducing pollutant loads from urban runoff and CSOs is significantly more challenging and potentially more costly than removing pollutants from municipal and industrial wastewaters. Wastewater treatment processes are designed to treat relatively constant and continuous flows, and they perform poorly when subjected to the extreme variations characteristic of stormwater flows. Thus, different types of treatment and control methods are needed. However, relatively little effort has gone into the development of stormwater and CSO treatment and control technologies. The federal government sponsored research on treatment of CSOs in the late 1960s and 1970s, but the results of those efforts were disappointing. Funding for CSO research was greatly reduced by 1974 and had disappeared by 1981. Some research on structural controls for improving stormwater runoff was conducted in conjunction with the Area Wide Planning Studies mandated under Section 208 of the Clean Water Act and conducted in the early and mid-1970s; since then only a limited amount has been done. While, intuitively, runoff source control techniques, such as covering chemical storage areas, spill response and containment programs, elimination of illegal dumping, removal of floor drain connections to storm drains, street sweeping, household hazardous waste collection programs, and public education, are expected to result in reduced pollutant loadings to coastal waters, little quantitative research has been conducted to determine their overall effectiveness in practice.

Stormwater and CSO impacts are largely a function of regional and local hydrology and existing system capacity. Thus, while the capture of a one-year, one-hour storm in one city may limit CSO events to four times a year, in another city, the overflow frequency will be more or less. Figure 2.2a shows the percent capture of annual runoff achieved by various-sized detention basins. In addition to showing how variations in regional hydrology affect runoff capture capacities, this figure shows that the most cost-effective facility (taken as the knee of the curve) varies for different regions of the country. Figure 2.2b shows the annual overflow frequency obtained with various sized detention basins. Stormwater and CSO abatement requirements should be based to the greatest extent possible on an understanding of regional and local hydrology. They should also be designed in conjunction with other regional environmental protection programs to achieve the most cost-effective combination of structural and nonstructural controls.

Currently, pollutant removal efficiencies of treatment facilities for CSOs and urban runoff cannot be stated with sufficient confidence to design a facility plan that will limit pollutant loads to a prescribed level. The difficulty in making such predictions stems from the high variability in hydraulic and pollutant loadings that a facility will experience from storm to storm and within a particular storm. Given the cost of constructing these facilities on a large scale in urban areas ($20 to $60 million per square mile for combined sewer areas and $0.6 to $3.8 million per square mile for stormwater facilities [APWA 1992]), a serious, well-funded research program of runoff characterization and new technology demonstrations is needed. Descriptions of various CSO and stormwater controls are given in Appendix D.

In the absence of the ability to predict pollutant discharge concentrations accurately, there have been proposals to legislate national technology-based requirements mandating the capture and treatment of runoff from all storms up to a certain size and frequency. The difficulty with such requirements is the same as that for technology-based wastewater treatment requirements: they are likely in some cases to result in costly overcontrol, in others undercontrol with continued adverse environmental effects, and in relatively few cases will they likely meet the environmental protection requirements of a particular region in a cost-effective way.

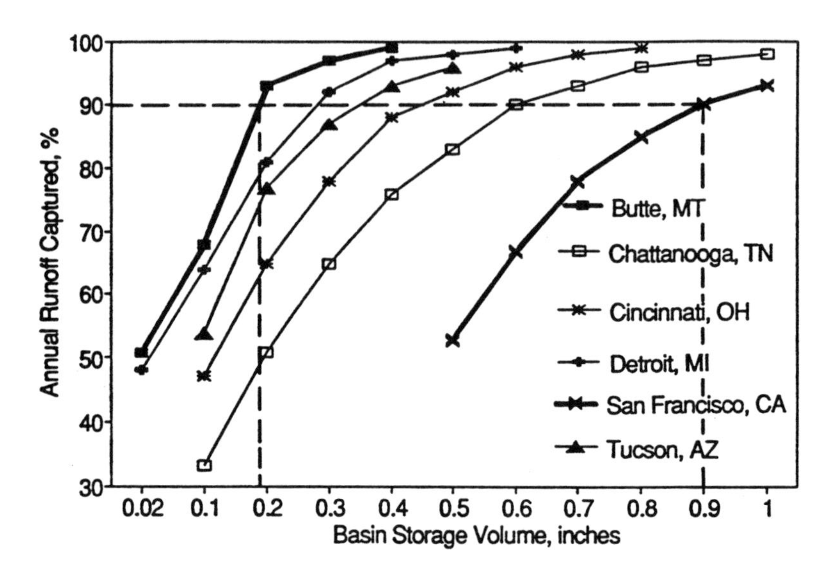

FIGURE 2.2a Runoff capture efficiency versus unit storage volume. (Source: Roesner et al. 1991. Reprinted, by permission, from American Society of Civil Engineers, 1991.) NOTE: Basin storage volume is in watershed inches, which is equivalent to acre-in/acre of drainage area. Dividing basin storage volume numbers by 12 converts the abscissa to acre-ft of storage required per acre of tributary area.

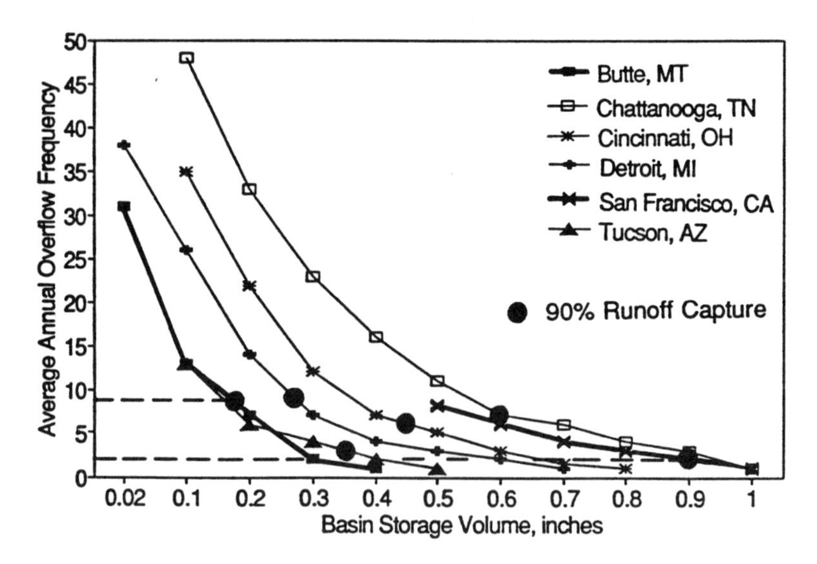

FIGURE 2.2b Basin overflow frequency versus unit storage volume. (Source: Roesner et al. 1991. Reprinted, by permission, from American Society of Civil Engineers, 1991.) See note to Figure 2.2a.

It is important that regulatory programs for various wet weather sources be coordinated and that management programs for urban runoff, CSOs, and wet weather overflows in sanitary systems be integrated. Where integration is lacking, coastal quality objectives may not be met by the sum of the activities of the individual programs. If they are, the cost is likely to be much greater than would be for an integrated program. For example, the study of CSOs in Boston showed that even if overflow events were reduced to one per year, Boston Harbor would still have periods of high coliform levels due to stormwater discharge (and likely wet weather overflows from sanitary systems) from communities having separated stormwater and sanitary sewers (CH_2M Hill 1992). Milwaukee, on the other hand, having jurisdiction over both the combined sewer area and the separate sewered areas, was able to use a single, integrated facility plan to solve both CSO problems and sanitary sewer overflow problems at considerable savings in comparison to possible independent solutions (MMSD 1980).

The current permitting structure at the federal and state levels provides the opportunity to address CSOs and sanitary system overflows simultaneously, provided both systems are owned by the same permittee. Stormwater discharges, however, are regulated under a different permit and, at least at the federal level, by different entities within the Environmental Protection Agency. The permitting process should be improved to provide flexibility that would allow for and encourage the development of integrated regional solutions for all components of wet-weather pollution sources.

DETECTING HUMAN PATHOGENS

Over 100 different human enteric pathogens, including viruses, parasites, and bacteria, may be found in treated municipal wastewater and urban stormwater runoff. Many of these pathogens can survive for up to several days in water, and even longer in fish and shellfish. The transmission of disease to humans from consumption of contaminated fish and shellfish has been well documented. In addition, several epidemiological investigations have demonstrated recreational contact with contaminated waters through activities such as windsurfing, surfing, swimming, and diving in polluted waters is associated with increased diarrhea, skin, ear, and respiratory infections (Cabelli et al. 1983, Richards 1985, DeLeon and Gerba 1990, Balarajan et al. 1991, Alexander et al. 1992, Fewtrell et al. 1992). More detail on the risks associated with microbial pathogens in coastal waters is contained in Appendix B.

While effective disinfection methods can greatly reduce the number of pathogens in wastewater treatment plant effluent, they may not inactivate them completely. The efficiency of disinfection technologies depends on the concentration of the disinfection agent, contact time, and the characteristics of the water being disinfected. In addition, pathogen inputs to coastal waters from uncontrolled nonpoint sources may result in pathogen concentrations more than sufficient to impair human health.

Intestinal bacteria have been used for more than 100 years as indicators of fecal contamination in water and of overall microbial water quality. These bacteria normally live in the intestinal tract of humans and other warm-blooded animals without causing disease. If found in significant concentrations in water, they are considered to indicate the potential presence of human and/or animal fecal waste. The most commonly measured indicators are total coliforms and a subset of this group, the fecal coliforms which are considered to be more predictive of fecal contamination. Generally greater than 90 percent of the coliforms found in feces of warm blooded animals are a specific fecal coliform *Escherichia coli* (*E. coli*). In addition to the coliform bacteria, fecal streptococci and enterococci have been used to monitor water quality and are also natural flora of the intestines of animals, including humans.

In the United States, bacteriological standards for shellfish harvesting waters and recreational waters are set by each state. Standards for shellfish growing waters are generally more consistent nationally and reflect the Food and Drug Administration requirement for interstate transport of harvested shellfish with less than 14 fecal coliforms per 100 milliliters and with no more than 10 percent greater than 43 per 100 milliliters. Recreational water-quality standards vary from state to state. About 50 percent of the states use a standard of less than 200 fecal coliforms per 100 milliliters; some use a total coliform standard in lieu of or in addition to the fecal coliform standard (Kassalow and Cameron 1991). The EPA's ambient water-quality criteria document for bacteria recommends a using an enterococcus standard based on epidemiological evidence of the relationship between enterococci and gastrointestinal illness (EPA 1986). However, only about 30 percent of the states now use enterococcus standards. The EPA recommended criterion is 35 enterococci per 100 milliliters; state standards range from 3 to 52 enterococci per 100 milliliters.

Although bacterial indicators have been used extremely successfully in the development of strategies for controlling bacterial diseases such as cholera and typhoid, they have limited applicability in the control of many nonbacterial pathogens and some waterborne bacterial agents in both seafood and recreational waters. In the United States, for example, outbreaks of illnesses continue to occur due to enteric pathogens (e.g., hepatitis A) as a result of consumption of shellfish taken from waters near wastewater discharges but meeting coliform standards. The state of Florida recently issued a statement to the public advising that the state could not guarantee the safety of raw shellfish. No epidemiological study to date has concurrently evaluated the association between disease incidence and the presence of both bacterial indicators and nonbacterial pathogens, such as viruses and protozoa.

The use of a single bacterial standard for determining either marine water quality or seafood safety is inadequate for several reasons (Gerba et al. 1979, Cheung et al. 1990, NRC 1991). Indicator bacteria cannot be used to distinguish between human and animal fecal contamination. When used in marine, particularly tropical, waters, current methods may enumerate primarily non-*E. coli* and nonsewage-related bacteria.

In addition, low levels of bacteria may be less of a health concern than low levels of viruses and protozoa because of the higher infectious dose needed to initiate an infection. Current indicator bacteria enumeration methods use relatively small sample volumes (i.e., 100 milliliters). Thus, viruses and protozoa may be present at levels of concern although indicators are absent from the sample.

Finally, indicator bacteria have patterns of survival in the environment that differ from those of nonbacterial pathogens. The survival of pathogens in the marine environment is influenced by several factors, including temperature, sediments, nutrients, light, dissolved oxygen, and the type of microorganism. Die-off rates for enteric microorganisms are higher in saline waters than in freshwaters. In general, the survival of coliforms in marine waters is shorter than that of other enteric microorganisms. For this reason, the absence of coliforms does not guarantee the absence of pathogenic microorganisms in marine waters, sediments, shellfish, or fish. Studies have found limited correlation between indicator bacteria and the presence (or absence) of enteric viruses (Gerba et al. 1979). A study of virus levels in shellfish from shellfish beds meeting bacteriological standards on Long Island found virus counts ranging from 10 to 200 virus plaque units per 100 grams shellfish. (Levels that might be associated with risks of acute illness range from 0.001 to 0.1 virus plaque units per 100 grams.) A more detailed discussion of the survival of enteric microorganisms in marine waters can be found in Appendix B.

The need for a new surrogate to determine the presence of pathogenic organisms in water is apparent. Efforts to overcome the deficiencies of the current system for virological water-quality protection, have been directed toward the investigation of the coliphage as an indicator of recreational and marine water quality (O'Keefe and Green 1989, Borrego et al. 1990, Palmateer et al. 1991). Coliphage are bacterial viruses that infect *E. coli* as their host. The coliphage are easily assayed and results are obtained more rapidly than for bacteria assays. They are similar in size and structure to the human enteric viruses and thus mimic their fate in the environment. Methods are being developed to examine large volumes of water for coliphage to eliminate the disadvantage of the 100-milliliter sample size used in bacterial assays.

New methods are also being developed for the rapid examination of waters for individual pathogens. Immunological approaches as well as molecular techniques using gene probes and polymerase chain reaction are being investigated and their application to environmental samples is progressing (Richardson et al. 1988, Bej et al. 1990, Abbaszadegan et al. 1991). These methods will be useful in the performance of sanitary surveys and the development of a risk based approach for determining acceptable levels of specific pathogens in shellfish, recreational waters, and effluents from wastewater treatment plants.

Methods for directly measuring the presence of bacteria, viruses, parasites and other pathogens associated with sewage have been developed and are available (Bendinelli and Ruschi 1969; Van Donsel and Geldreich 1971; Goyal et al. 1979; Ellender et al. 1980; Vaughn et al. 1980; Schaiberger et al. 1982; Wait et al. 1983; Rose et al. 1985, 1988, 1991; Volterra et al. 1985; Richardson et al. 1988; Perales and Audicana 1989; Bej et al. 1990; Colburn et al. 1990; DePaola et al. 1990; Knight et al. 1990; Abbaszadegan et al. 1991; and Desenclos et al. 1991). Depending on the type of organism and detection method used, costs may range from $50 to $1,000 per sample. Public health agencies, wastewater management agencies, environmental engineers, and others responsible for monitoring wastewater treatment impacts have been slow to use them because they are more expensive and more complex than traditional methods and because coliform monitoring has such a long tradition. As described in detail in Appendix B, however, the risk of disease transmission related to wastewater management practices is potentially large, and needs to be better understood. The Environmental Protection Agency, public health agencies, and wastewater treatment agencies should vigorously pursue the development and implementation of improved techniques to measure more directly the presence of pathogens, particularly in marine and estuarine waters.

DEVELOPING MANAGEMENT ALTERNATIVES

Wastewater and stormwater management strategies should be developed in the context of each other and of other important sources of perturbation in the coastal zone. Environmental regulation in the coastal zone should be flexible, encouraging the development and implementation of innovative alternative strategies which promise greater overall efficiency and efficacy than existing approaches. In most areas it is a combination of sources and human activities that leads to degradation of coastal waters. Successful management strategies must take into account the full range of sources. In the process of identifying approaches for controlling inputs, it also will be possible to identify and explore the full range of potential alternatives for controlling those sources. A management plan should then consist of a combination of alternatives that can achieve desired results in the most cost-effective manner. For example, there may be cases where it is more effective and/or efficient to control agricultural and other nonpoint source runoff rather than upgrading wastewater treatment systems. In some cases, the cooperative efforts of stormwater agencies, wastewater agencies, water supply agencies, and other agencies concerned with the region's resources can result in the development of efficient and mutually beneficial solutions.

In order to assure continued public support for environmental protection and coastal resource management programs, it is important to develop strategies that make effective use of public funds. If the costs of proposed programs are perceived to be too high or unfairly allocated in relation to the benefits gained, public support will erode and may be lost, particularly for future programs. Strategies that take into consideration cost-effectiveness are more likely to be supported by the public and foster support for future protection programs.

EVALUATION AND FEEDBACK

The effectiveness of management approaches can only be determined, and corrected if necessary, if there is adequate monitoring, evaluation, and feedback. Monitoring of water and sediment quality and other ecosystem parameters is important not just for determining compliance with regulatory requirements but also for developing a better understanding of the coastal system that management efforts aim to use and protect. Monitoring data, research results, and other information about management systems requires careful evaluation to determine if predictions are accurate, whether unanticipated problems have developed, and where improvements may need to be made. Evaluation efforts should take into account the effectiveness of specific efforts as well as that of the whole integrated management effort. They should focus on lessons that can be learned from implementation experiences in order to improve the potential for success of future efforts. For example, the ability to use mathematical models to predict the behavior of sewage effluent in coastal systems has advanced dramatically over the past 20 years. Comparatively little effort has been put toward the verification of these model predictions, however. To make good use of these tools and identify where they need improvement, it is important that follow-up monitoring studies be conducted.

Feedback of information from monitoring, research, and evaluation indicates how management strategies should be revised to meet the goals and objectives of the region. It is important that management approaches be sufficiently flexible to be able to make changes and improvements in response to new information. Evaluation of the effectiveness of management efforts should be an integral part of the overall approach and should take place on an ongoing basis.

Monitoring studies that are performed often limit evaluations of trends on a year-to-year basis and do not look at long-term (10 to 20 year) trends. Some preliminary evaluations of data for southern California indicate that long-term improvements are occurring despite dramatic increases in wastewater flows due to population growth (SCCWRP 1989, 1990; LADPW 1991; Mearns et al. 1991). Source control programs have been very effective in reducing toxicant loads. Predictive models used for developing source control programs need to be linked to environmental monitoring data so that the loads and indicators can be correlated and success or failure measured quantitatively.

SUMMARY

The eight issues identified above form a snapshot of current needs in wastewater and stormwater management in coastal urban areas. Coastal systems around the United States are complex and diverse; environmental processes and human activities within these systems are dynamic and interconnected. Addressing these issues in a comprehensive manner will require some innovation and some change. Ideally, coastal resources and the human activities that affect them should be managed on a thoroughly integrated basis in the context of the complex functions and interrelationships that form the coastal environment. Ultimately, this integration should be done across the spectrum of pollutants and other sources of stress, between the various environmental media, and over the time and spatial scales of impacts.

Many coastal areas, particularly those in the National Estuary Program (see listing in Table 2.2), have developed or are in the process of developing integrated programs focused on water quality protection. In addition, efforts throughout the country to develop watershed management plans and waste-load allocation programs are structured to take into account the diversity of pollution sources.

Because of their relative importance and the degree of experience and extent of related physical and institutional structures already in place, the management of wastewater and stormwater is often a good place to begin the development of an integrated coastal management program. The following chapter discusses the principles and methodology on which such a program should operate. Chapter 4 provides a technical description and examples of how the methodology for such a management system should be applied.

TABLE 2.2 Estuaries Participating in the National Estuary Program

Convened 1985-1987
Puget Sound, Washington
Buzzards Bay, Massachusetts
Narragansett Bay, Rhode Island
Long Island Sound, New York and Connecticut
Albemarle-Pamlico Sound, North Carolina
San Francisco Bay, California
Convened 1988
New York-New Jersey Harbor, New York and New Jersey
Delaware Inland Bays, Delaware
Santa Monica Bay, California
Galveston Bay, Texas
Delaware Bay, Delaware
Convened 1990
Casco Bay, Maine
Mass Bays, Massachusetts
Indian River Lagoon, Florida
Tampa Bay, Florida
Barataria-Terrebonne Bays, Louisiana
Convened 1992
Peconick Bay, New York
Corpus Christi, Texas
San Juan Harbor, Puerto Rico
Tillamook Bay, Oregon

REFERENCES

Abbaszadegan, M., C.P. Gerba, and J.B. Rose. 1991. Detection of *Giardia* with a cDNA probe and applications to water samples. Appl. Environ. Microbiol. 57:927-931.

Alexander, L.M., A. Heaven, A. Tennant, and R. Morris. 1992. Symptomatology of children in contact with seawater contaminated with sewage. J. Epid. Commun. Hlth. 46:340-344.

AMSA (Association of Metropolitan Sewerage Agencies). 1990. 1988-89 AMSA Pretreatment Survey Final Report. Washington, D.C.: Association of Metropolitan Sewerage Agencies.

APWA (American Public Works Association). 1992. A Study of Nationwide Costs for Implementing Municipal Stormwater BMPs. Final Report. Water Resources Committee, Southern California Chapter. Chicago, Illinois: American Public Works Association.

Balarajan, R., V.S. Raleigh, P. Yuen, D. Wheller, D. Machin, and R. Cartwright. 1991. Health risks associated with bathing in sea water. Brit. Med. J. 303:1444-1445.

Bej, A.K., R.J. Steffan, J. Dicesare, L. Haff, and R.M. Atlas. 1990. Detection of coliform bacteria in water by polymerase chain reaction and gene probs. Appl. Environ. Microbiol. 56:307-314.

Bendinelli, M., and A. Ruschi. 1969. Isolation of human enterovirus from mussels. Appl. Microbiol. 18:531-532.

Borrego, J.J., R. Cornax, M.A. Moringo, E. Martinez-Manzanares, and P. Romero. 1990. Coliphages as an indicator of faecal pollution in water: Their survival and productive infectivity in natural aquatic environments. Water Res. 24:111-116.

Cabelli, V.J., A.P. Dufour, L.J. McCabe, and M.A. Levin. 1983. A marine recreational water quality criterion consistent with indicator concepts and risk analysis. Journal of the Water Pollution Control Federation 55:1306-1314.

Cheung, W.H.S., K.C.K. Chang, and R.P.S. Hung. 1990. Health effects of beach water pollution in Hong Kong. Epidemiol. Infect. 105:139-162.

CH_2M-Hill. 1992. Final Combined Sewer Overflow Facilities Plan and Final Environmental Impact Report. Volume II, Recommended Plan. Prepared for the Massachusetts Water Resources Authority, Boston, MA, September 20, 1992.

Colburn, K.G., C.A. Kaysner, C. Abeyta, Jr., and M.M. Wekell. 1990. *Listeria* species in a California coast estuarine environment. Appl. Environ. Microbiol. 56:2007-2011.

DeLeon, R., and C.P. Gerba. 1990. Viral disease transmission by seafood. Pp. 639-662 in Food Contamination from Environmental Sources, J.O. Hriagu, and M.S. Simmons, eds. Somerset, New Jersey: John Wiley & Sons, Inc.

DePaola, A., L.H. Hopkins, J.T. Peeler, B. Wentz, and R.M. McPhearson. 1990. Incidence of *Vibro parahaemolyticus* in U.S. coastal waters and oysters. Appl. Environ. Microbiol. 56:2299-2302.

Desenclos, J.C.A., K.C. Klontz, M.H. Wilder, O.V. Nainan, H.S. Margolis, and R.A. Gunn. 1991. A multistate outbreak of hepatitis A caused by the consumption of raw oysters. Am. J. Public Health. 81(10):1268-1272.

Ellender, R.D., J.B. Map, B.L. Middlebrooks, D.W. Cook, and E.W. Cake. 1980. Natural enterovirus and fecal coliform contamination of Gulf coast oysters. J. Food Protec. 42(2):105-110.

EPA (U.S. Environmental Protection Agency). 1986. Ambient Water Quality Criteria Document for Bacteria. EPA A440/5-84-002. Washington, D.C.: U.S. Environmental Protection Agency.

Fewtrell, L., A.F. Godfree, F. Jones, D. Kay, R.L. Salmon, and M.D. Wyer. 1992. Health effects of white-water canoeing. Lancet 339:1587-1589.

Gerba, C.P., S.N. Singh, and J.B. Rose. 1979. Waterborne viral gastroenteritis and hepatitis. CRC Crit. Rev. Environ. Control 15:213-236.

Goyal, S.M., C.P. Gerba, and J.L. Melnick. 1979. Human enteroviruses in oysters and their overlying waters. Appl. Environ. Microbiol. 37:572-581.

Kassalow, J., and D. Cameron. 1991. Testing the Waters: A Study of Beach Closings in Ten Coastal States. New York, New York: Natural Resources Defense Council.

Knight, I.T., S. Shults, C.W. Kaspar, and R.R. Colwell. 1990. Direct detection of *Salmonella* spp. in estuaries by using DNA probe. Appl. Environ. Microbiol. 56:1059-1066.

LADPW (City of Los Angeles, Department of Public Works). 1991. Marine Monitoring in Santa Monica Bay, Annual Assessment Report for the period July 1989 through June 1990. Los Angeles, California: Environmental Monitoring Division, Bureau of Sanitation, City of Los Angeles, Department of Public Works.

Mearns, A.J., M. Matta, G. Shigenaka, D. MacDonald, M. Buchman, H. Harris, J. Golas, and G. Lauenstein. 1991. Contaminant Trends in the Southern California Bight: Inventory and Assessment. Administration Technical Memorandum NOS/ORCA 62. Seattle, Washington: National Oceanic and Atmospheric Administration.

MMSD (Milwaukee Metropolitan Sewerage District). 1980. MMSD Wastewater System Plan Planning Report. Milwaukee, WI: Program Office, Milwaukee Water Pollution Abatement Program.

NRC (National Research Council). 1991. Seafood Safety. Washington, D.C.: National Academy Press.

O'Keefe, B., and J. Green. 1989. Coliphages as indicators of faecal pollution at three recreational beaches on the Firth of Forth. Water Res. 23:1027-1030.

Palmateer, G.A., B.J. Dutka, E.M. Janson, S.M. Meissner, and M.G. Sakellaris. 1991. Coliphage and bacteriophage as indicators of recreational water quality. Water Res. 25:355-357.

Perales, I., and A. Audicana. 1989. Semisolid media for isolation of *Salmonella* spp. from coastal waters. Appl. Environ. Microbiol. 55:3032-3033.

Richards, G.P. 1985. Outbreaks of shellfish-associated enteric virus illness in the United States: Requisite for development of viral guidelines. J. Food Protec. 48:815-823.

Richardson, K.J., A.B. Margolin, and C.P. Gerba. 1988. A novel method for liberating viral nucleic acid for assay of water samples with cDNA probes. J. Virol. Methods 22:13-21.

Roesner, L.A., E.H. Burgess, and J.A. Aldrich. 1991. The hydrology of urban runoff quality management. In Proceedings of the 18th National Conference on Water Resources Planning and Management/Symposium on Urban Water Resources, May 20-22, 1991. New Orleans, LA. New York, New York: American Society of Civil Engineers.

Rose, J.B., H. Darbin, and C.P. Gerba. 1988. Correlations of the protozoa, *Cryptosporidium* and *Giardia* with water quality variables in a watershed. Water Sci. Tech. 20:271-276.

Rose, J.B., C.N. Haas, and S. Regli. 1991. Risk assessment and control of waterborne Giardiasis. Am. J. Public Health. 81:709-713.

Rose, J.B., C.P. Gerba, S.N. Singh, G.A. Toranzos, and B. Keswick. 1985. Isolating viruses from finished water. Journal of the American Water Works Association 78(1):56-61.

SCCWRP (Southern California Coastal Water Research Project). 1989. Annual Report 1988-89, P.M. Konrad, ed. Long Beach, California: SCCWRP.

SCCWRP (Southern California Coastal Water Research Project). 1990. Annual Report 1989-90, J.N. Cross, ed. Long Beach, California: SCCWRP.

Schaiberger, G.E., T.D. Edmond, and C.P. Gerba. 1982. Distribution of enteroviruses in sediments contiguous with a deep marine sewage outfall. Water Research 16:1425-1428.

Van Donsel, D.J., and E.E. Geldreich. 1971. Relationships of *Salmonellae* to fecal coliforms in bottom sediments. Water Research 5:1079-1087.

Vaughn, J.M., E.F. Landry, T.J. Vicale, and M.C. Dahl. 1980. Isolation of naturally occurring enteroviruses from a variety of shellfish species residing in Long Island and New Jersey marine embayments. J. Food Protec. 43(2):95-98.

Volterra, L., E. Tosti, A. Vero, and G. Izzo. 1985. Microbiological pollution of marine sediments in the southern stretch of the Gulf of Naples. Water, Air and Soil Pollution 26:175-184.

Wait, D.A., C.R. Hackney, R.J. Carrick, G. Lovelace, and M.D. Sobsey. 1983. Enteric bacterial and viral pathogens and indicator bacteria in hard shell clams. J. Food Protec. 46(6):493-496.

3

Integrated Coastal Management

The Committee believes that whether because of any of the issues discussed in the preceding chapter or because of other concerns, most coastal cities now, or in the near future will, face the need to address complex coastal quality management issues. With increasing population pressures, increasing recognition of the importance of nonpoint sources in coastal waters, and decreasing availability of public funding at the federal, state, and local levels, coastal cities face the need to establish objectives and set priorities for protecting coastal resources. The Committee therefore proposes here a framework toward which it believes coastal environmental quality management should evolve.

DEVELOPING A SUSTAINABLE VISION

Most indicators suggest that, throughout the country, human impact upon urban coastal areas continues at a level of severity which threatens the biological integrity of many marine systems and seriously impairs their capacity to produce a full range of goods and services valued by people.

Given the importance of coastal areas to society, managing their social and economic uses in a sustainable fashion should be a central tenet of government policy. In the absence of controls, coastal resources are unpriced and widely accessible, and the market fails to reveal social values or restrain use. There is a need for a complete system of resource valuation capable of identifying the consequences and opportunity costs associated with various patterns of resource use. Unfortunately, such a comprehensive system does not exist. Some work has been done, particularly in the area of human uses of water resources, but it is significantly incomplete, especially with respect to the full range of human values, the consequences of health effects, and ecological values. Thus, there is no comprehensive set of economic tools capable of developing and implementing an optimal coastal management strategy that meets society's goals.

In the absence of such a capability, reversing trends of degradation in an effective and efficient manner is more difficult. It requires, at the least, that society formulate a clear vision of the coast's future and identify the tasks necessary to achieve that vision. The concept of integrated coastal management (ICM) is a starting point for that vision. This chapter addresses the conceptual principles and methodology for ICM. Chapter 4 addresses specific applications of the principles and methodology outlined here. ICM is associated here with two general objectives: 1) to restore and maintain the ecological integrity of coastal ecosystems and 2) to maintain important human values and uses associated with those resources.

Recently, in *Reducing Risks: Setting Priorities and Strategies for Environmental Protection*, the Environmental Protection Agency (EPA), through its Science Advisory Board, explored the problem of making its management programs more relevant to ecological imperatives. It concluded that the "EPA should target its environmental protection efforts on the basis of opportunities for the greatest reduction of risk" and the "EPA should attach as much importance to reducing ecological risk as it does to reducing human health risk" (EPA 1990). In a substantial sense the methodology for integrated coastal management set forth in this report is a further step in the general direction proposed in the EPA report.

The principles and methodology set forth are a further elaboration of work done by others (see bibliography at the end of this chapter). There appear to be few examples of where these ideas have been systematically applied in the conscious effort to protect coastal resources. Nonetheless, there are several examples where the concepts have, to some extent, been applied over time. Notable in this regard are the examples of the Great Lakes and the Chesapeake Bay. Some detail about one aspect of the Chesapeake Bay Program—the experience in controlling nutrients—is presented in Chapter 4. Chapter 5 provides a discussion of the benefits and barriers associated with the implementation of ICM and provides some recommendations for implementation. The evolution of the Great Lakes Water Quality Agreement from 1978 through 1987 consisted of an iterative planning and implementation process. As a result, toxics joined nutrients as pollutants of concern and the management area was expanded and now includes the entire watershed. A more complex understanding of pollution sources and prevention strategies in the Great Lakes has evolved and the mechanisms for governance have become more sophisticated.

OBJECTIVES OF INTEGRATED COASTAL MANAGEMENT

- To restore and maintain the ecological integrity of coastal ecosystems, and
- To maintain important human values and uses associated with those resources.

An integrated approach to coastal management is an increasingly important component of the international agenda for wise environmental management. At the United Nations Conference on Environment and Development held in Rio de Janeiro in June 1992, integrated coastal management was a significant element of the policy document dealing with the problems of the marine environment which was agreed to by governments. Similarly, the World Bank is currently in the process of developing guidelines for integrated coastal management to be utilized by countries receiving assistance.

PRINCIPLES AND METHODOLOGY FOR A SYSTEM OF INTEGRATED COASTAL MANAGEMENT

Important changes in the two decades since the passage of the Clean Water Act suggest that integrated coastal management is achievable. Two developments, one social and the other technical, are worth highlighting:

- Today, a far greater proportion of the public cares about the environment in general and coastal areas in particular than 20 years ago. Although public *concern* over specific problems was high then (e.g., the first Earth Day was in 1970, the same year EPA was established), the general public is now more *aware* of the complexity of environmental issues. This awareness is a potentially powerful political force that can be effective in driving admittedly complex processes to useful conclusions.
- Additionally, important technical progress has been made over the past 20 years in source reduction, treatment systems, outfalls, and modeling of systems. Although scientific knowledge about coastal processes is far from complete, it is far advanced over that of a generation ago. Beyond mere data, scientists and engineers also understand in a more sophisticated way the nature of how coastal processes operate. Finally, modern computing and other data management tools have given scientists and others, for the first time, the capacity to organize, analyze, and display this complex of information in a way that is accessible to technical analysts and lay persons alike.

Integrated coastal management is an ecologically-based approach to environmental management and is therefore in many ways a departure from the technology-driven strategy that has characterized the national effort since the Federal Water Pollution Control Act of 1972. This is especially so in urban areas, where a major emphasis has been appropriately on the construction of sewage treatment plants. In fact, to a certain extent ICM is analogous to the water-quality based efforts that preceded the 1972 amendments. Although that approach was supplanted by the technology emphasis of the 1972 amendments, two decades have passed and it is now time to reconsider the validity of an ecologically-based strategy, especially in highly complex situations. As noted above, the public attention to environmental issues and the scientific understanding is vastly more pronounced than was the case in the 1960s. Therefore the basic likelihood of environmental action is enhanced. In addition, there is now a well-developed regulatory system with permits, inspection, and enforcement procedures, capable of assuring that identified obligations are, in fact, met. This system is being complemented by the development of an economically-based incentive system, which may add even greater force to the compliance imperative. Perhaps most critically, it is now necessary to ensure that scarce public economic resources are spent on those management options that will have the maximum positive impact. Society can no longer afford to spend large amounts of money solving unimportant problems.

The following discussion of integrated coastal management sets forth a set of interlocking management principles and a related process. The fundamental objective of this system is to allow for improved identification of important priorities and better allocation of resources to the identified problems. An integrated system is interlocking and iterative, that is, each of the principles builds on and influences others. The following discussion identifies these six principles and suggests how they could be applied in a coherent and logical fashion.

Principles

The following six principles should underlie the development of an integrated coastal management system and the institutions necessary to implement it:

Coastal management's overall objective is to maintain ecological processes and meet human needs for goods and services. Accordingly, management actions need to be developed on the basis of the best science available about ecological functions as well as a comprehensive understanding of human needs and expectations, which are both tangible and intangible.

Management objectives should be expressed as water- and sediment-quality based, and other environmentally and health based goals. Using environmentally and health based goals allows flexibility in the methods used to achieve those goals, while assuring that ecosystems and human health are protected at the desired level.

Comparative assessment of both risk scenarios and available management options should drive the selection of management strategies. This approach will ensure that a rational basis is used for focusing action, and human and financial resources on the most important problems. A dynamic and iterative planning process should integrate risk-based analysis with evolving scientific understanding and human expectations. The process is a continuing one and allows for incremental decision making when that is the prudent course. On the other hand, it also provides a context for making the high cost or risk decision most likely to be correct in the face of scientific uncertainty. This integration requires a stable institutional base for the planning process.

A transdisciplinary perspective is critical to coastal problem solving. While specific disciplines such as chemistry, biology, toxicology, microbiology, and hydrology and their integration through environmental engineering principles will always be central to this transdisciplinary perspective, increasingly, fields such as economics, law, and sociology are important components of marine management. Maintaining systems for routine exchange and analysis among professions is essential to help managers gain a comprehensive understanding of coastal problems. With a linked analytical system, disciplines could assist more effectively in formulating a transdisciplinary management response. Such a system should integrate scientific, engineering, and other information into management practice in a timely and flexible fashion.

This system should function in a context that is responsive to scientific uncertainty about the functions of coastal ecosystems. Accordingly, it is based on the premise that how these natural systems respond to stresses from pollution events, overfishing, sedimentation, or encroachment on habitat is not often well understood or subject to human intervention. Management actions instead must recognize that it is the human activities at the source of these stresses that must be reduced or changed.

Integrated coastal management is driven by science and engineering together with public expectations. Especially important, therefore, is the requirement that there be significant opportunities for public participation and involvement throughout the process. Appropriate public values are not only economic or recreational but also those arising out of ethical and aesthetic concerns for protection of the environment.

PRINCIPLES OF INTEGRATED COASTAL MANAGEMENT

- Management actions are based on the best available scientific information about ecological functions, as well as a comprehensive understanding of human needs and expectations, which are both tangible and intangible.
- Management objectives are expressed as water- and sediment-quality based, and other environmentally and health based goals.
- Comparative assessment of both risk scenarios, and available management options drive the selection of management strategies.
- A transdisciplinary perspective is critical to coastal problem solving.
- The ICM process functions in a context that is responsive to scientific uncertainty about the functions of coastal systems.
- ICM is driven by science and engineering together with public expectations.

Process

The process of integrated coastal management is composed of the following elements, which are applied in the sequence they are discussed. Elements at the bottom of the list are not less important than earlier ones and should feed back into them. The flow chart in Figure 3.1 gives a simple schematic view of the relationship among the elements.

Set Goals

At the outset of the development of a management plan, it is important to develop a well-informed understanding of the goals and expectations held for a coastal region and the range of environmental problem areas that require further attention. Management has two primary objectives: to protect the fundamental functions and biological richness of the ecosystem and to maintain important human uses. The starting point of integrated management is to identify the problems threatening these goals. Two important tasks must be completed to meet these objectives: 1) define critical (important) environmental processes in time and space using existing information and data; and 2) define and rationalize the variety of human expectations about uses and benefits to be derived from the coast.

Environmental Processes. All coastal systems are not the same. There will be variations in the functions of natural processes. In a particular system, the important elements of these processes must be identified with sufficient precision that management can protect them. The definition of processes and issues inevitably leads to a system for setting priorities, forcing decisions about issues that are more important to address than others in order to maintain ecological and human health.

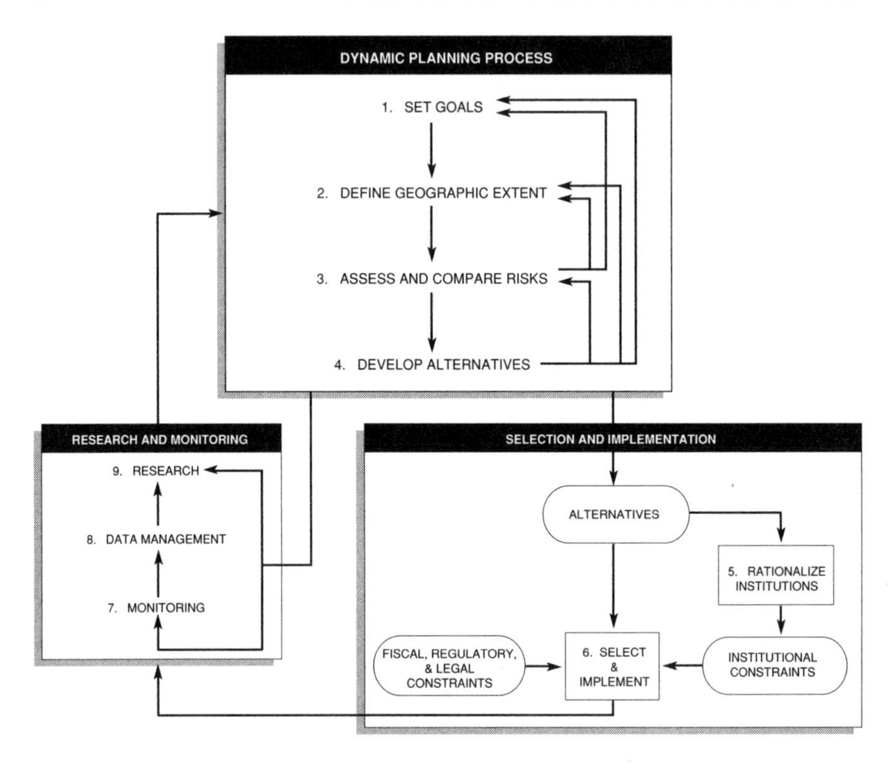

FIGURE 3.1 Process of integrated coastal management.

Human Expectations. Societal values and needs with respect to a particular coastal system also vary from region to region. Often there are conflicts among various interests, and changes in values and needs will occur over time. Initial management steps must seek to understand the existing range of expectations. This understanding will further contribute to an appreciation of those elements of the natural systems that are important for protection.

Once important environmental processes and human expectations have been identified, the anthropogenic conditions that threaten their maintenance can be determined.

Define Geographic Extent of Concerns

It is important to address environmental problems at the scale on which they occur. Thus, integrated coastal management must be based on adoption of a relevant environmental domain with appropriate aquatic, terrestrial, and atmospheric components. The starting point for defining the geographic extent of an issue often will be the coastal area of concern, some or all of its watershed, and other areas as dictated by important related marine and terrestrial processes. For example, concerns about pelagic fisheries likely would suggest concern over a large area. The process of defining the geographic extent of an issue should also take into account sources of problems and other demands on the system, such as for the products produced in the coastal area, which create stress.

Only rarely will there be a perfect coincidence between the variety of ecological domains, the various sources of degradation, and the boundaries of administrative jurisdictions. However, it is possible to delineate the most important environmental processes and to identify the most immediate or significant sources of degradation at a scale that is consistent with appropriate management actions. An overall goal in adopting relevant environmental domains is to minimize the number of significant causes and effects taking place outside the domain and maximize the effectiveness of management measures that can be taken within the domain.

In many situations, the result of this analysis may produce both a variety of environmental domains and parallel sets of variable jurisdictional boundaries. As the problem analysis progresses, decisions can be made about the most logical way in which to draw areas for specific management and concern as a function of knowing what problems are important and what management options can be used to address them.

Assess and Compare Risks

An assessment and comparison of risks to ecological systems and human health across the full spectrum suggested above should be completed before management options are selected. The risk comparison should guide decisions for setting priorities. Ideally, risk management decisions should place the burden of control on activities that may significantly harm humans or the environment. While risk assessment methodologies are not fully developed and comparisons of human risks with ecological ones are difficult to make, even a qualitative examination driven by the goals set for coastal management will substantially improve the priority setting process. The continued refinement of risk assessment and comparison methods using the basic concepts of dose, exposure, and hazard is crucial to integrated management.

When an integrated understanding of environmental degradation and deterioration in human use has been developed, choices for priority attention can be made on the basis of the relative importance of issues within the total complex of problems. In addition to deciding which problems are important to solve, this analysis should include a component that attempts to define an understanding of what level of protection or management is required in order to meet established goals.

An integrated understanding of the sources of degradation provides the basis for choosing those control options that will yield the greatest net benefits. A comprehensive display of the relative risks and their causes is the essential ingredient to a strategic approach to management. It is also critical to decisions about allocating societal resources to those problems that are of most immediate importance.

The science of natural systems often will be understood imperfectly. A comparative risk analysis can contribute to maximizing the probability that more important problems will be addressed first. However, uncertainty may occasionally be so significant as to not allow for a clearly rational choice; and social values may appropriately demand that preventive action be taken. Where such value laden choices are made, they are likely to be better informed if made within the context of strategic assessments of comparative risks and with input from the public, who ultimately provide the necessary resources.

Develop and Compare Alternatives for Risk Management

Coastal problems cannot be managed successfully as separate issues, such as pollution or wetland loss or fisheries depletion. There are at least two reasons for the need to integrate such pressing issues. In the first instance, apparently separate issues are interrelated. Thus, fisheries declines could be due to overfishing, pollution, or failed reproduction due to loss of tidal wetlands. Secondly, resources are scarce. In the foregoing case of fisheries loss, it would be ideal to define the most likely cause of the loss and then design management options addressing that specific problem. Thus, risk management strategies must be devised to address the most important elements in the complex structure of problems in an integrated fashion—one that assesses important sources of risk and achievable management alternatives.

Risk management decisions should be made through the consideration of priority problems in light of available management options, human and financial resources, administrative and legal structures, and other factors relevant to solutions. Consideration must also be given to the tradeoff between expenditures and benefits both with respect to different solutions to the same problem and solving different problems.

Not all problems need to be addressed using the same kinds of strategies. While a standards-based regulatory system may be essential to providing a basic framework of governance, other techniques, such as economic incentives, streamlined management, land-use policy, and education will be appropriate for particular problems.

Too often society perceives that the problem of a coastal environment is singular or one-dimensional. Thus, enormous efforts are made to improve a sewage treatment plant or to eliminate the presence of a particular chemical, yet problems remain. As suggested above, more frequently, degradation of resources is due to the combined effects of a number of human actions or to a condition that is not the most obvious. While correcting one problem might mitigate environmental harm for a brief period of time, long-term protection and restoration may not be accomplished. This situation does not mean that all problems need to be addressed with equal vigor simultaneously. It does mean that choices must be made about which problems to solve in a context that considers the multiplicity of causes and effects and maximizes the expenditure of resources on those issues that are important.

These four steps described together constitute the Dynamic Planning Process. This process whereby values, ecological processes, comparative risks, and strategies are developed and assessed must be considered as a dynamic and continuing planning process. Such a process is needed to capture interactions where one action may lead to another, to recognize new problems, to respond to new knowledge, and to recognize and correct mistakes. This process is an iterative exercise in which choices are made about how to anticipate and resolve conflicts and set priorities among multiple uses before environmental harm is done. In order to give reality to the planning process, it must be tied to some system of allocating uses in the coastal environment. These systems can include a broad spectrum ranging from land-use controls to regulatory systems with permits to protected-area programs such as marine sanctuaries.

The dynamic nature of the planning process is the core concept that allows for feedback between the various elements of the methodology. At one scale there are a series of interactions among the planning process, the conduct of scientific activity, and the establishment of implementation programs. Within the planning process itself there are numerous iterative steps between the various functions. Perhaps the strongest connections exist between the comparative risk assessment and comparative risk management activities.

Institutional Arrangements

Institutions and mechanisms of governance must be arranged in a fashion that is capable of meeting the demands that an ecosystem-based management initiative will dictate. Ideally, one might strive to create a single management institution responsible for all aspects of management within an entire coastal ecosystem. In a world with existing institutions, government boundaries, and variable ecological boundaries, however, the development of an ideal institution will rarely be possible. But, communication among institutions and an improved system of coordination can go far to remedy the current, highly fragmented governance structure.

It may also be necessary to create new institutions and structures to meet certain management needs. For example, a significant land-use management objective may only be able to be implemented with a change in the relation between local and state governments. While theoretically achievable, if such change is politically impossible, it imposes a constraint on management options.

Development, Selection, and Implementation

Once alternative management options have been developed, choices must be made about the alternative to be implemented. Inevitably the choices result from a political decision-making process that may involve executive and legislative authorities within several jurisdictions and entities having different missions and levels of authority. This political process is an integral part of coastal management. First, fiscal, regulatory, or institutional realities may pose constraints that make certain options impossible to achieve. Inflexibility in environmental laws and regulations make it difficult to implement an ICM approach effectively. The existence of such constraints will force reconsideration of management options. In addition, failures to use scientific and related information may result in insufficient attention to the coastal problems so that hard choices are avoided. Finally, if public expectations and values are ignored it may be difficult to implement recommended alternatives.

Monitoring

Integrated coastal management must include a comprehensive monitoring effort that focuses on factors of significant ecological, human health, and resource use importance, or the processes that are crucial to them, and the control measures that have been put in place. In general, such a monitoring system will not only measure the status of water quality in a chemical and physical sense but will also take the pulse of the biologic regime. The results of monitoring function as a feedback mechanism to modify management actions, direct new research, and provide information for public accountability. The products of monitoring are the essential glue that allows integrated coastal management to take advantage of the new factors of public concern about, and the technical capacity to understand, environmental problems.

Information Management

Monitoring and research develop the data that drives an integrated data management system. Integrated coastal management can only be accomplished if monitoring and other data for environmental systems are managed in a way that allows managers and other interested parties to appreciate and make decisions about the whole. As noted, in the last decade, for the first time, significant capacity has been developed that permits a much greater degree of comprehensive understanding. This new capacity consists of three essential elements: significant increases in raw data; enhanced models for analyzing this data; and powerful hardware and software to manage the data. While data and models will never be complete or perfect, the increased capacity to manage them does provide an ability to integrate available data in a fashion that substantially enhances the prospect of integrated management. It is this information that helps to continually refine the understanding of the nature of particular coastal problems, expands the capacity to carry out comparative risk assessment and management, and ultimately allows for the development of new options for implementation when needed. Finally, it provides the picture as to whether efforts have succeeded or failed.

Research

An ongoing peer-reviewed research program to continually refine the capacity to carry out the various elements of the dynamic planning process is essential. Attempting to carry out each step of the process will suggest questions to which there are only incomplete answers. Furthermore, research coupled with monitoring will develop new information that might suggest additional questions. To provide for appropriate research efforts, the system must foster the formulation of these issues and the making of decisions. The need for more research need not become an excuse for doing nothing. On the contrary, display of problems, comparative risks, and management options with an understanding of uncertainty can allow decisions to proceed where problems are severe.

CONCLUSION

Integrated coastal management is a rigorous and difficult process. It is needed for situations that are scientifically or governmentally complex, costly, risky or otherwise fraught with a degree of uncertainty. Accordingly, it need not be used for those problems that, upon initial examination, present a relatively simple solution.

While integrated coastal management may be useful most often in complex ecological systems that extend far beyond the limits of an urban area, it is also a useful analytical and management methodology when decisionmakers are faced with problems having a predominantly urban theme. In urban areas, sources of human perturbation of the marine environment and their effects are often highly complex. Urban areas also rarely affect adjacent marine resources in splendid isolation from events in freshwater watersheds or the distant ocean. Finally, public resources are always scarce and must be allocated to correcting those problems having the highest likelihood of important environmental benefits. Integrated coastal management provides a context for considering all of these complexities and then deciding what is important to be done in the urban setting.

Integrated coastal management is an iterative process. As discussed, there is continual feedback among the various components of the methodology. Equally importantly, the entire process can be used for a particular situation with an increasing level of precision over time. For example, a quick analysis might be carried out in a matter of days using existing information. The result of such a rapid exercise can set up the dimensions of a more protracted process by identifying gaps and problems that require focus.

For the elected official this apparently complicated process should have utility because it will produce clearer choices. These choices allow, and may even force, the political process to allocate resources to the most important problems. In essence, the process allows the political decisionmaker to strike a balance between the expectations of various publics with respect to the facts as presented by technical professionals and to reach a conclusion about implementing achievable management options. While those responsible for political choice and implementation need not necessarily understand, or even participate in, every aspect of the science or planning of integrated coastal management, it is a process that allows for a rational display of choices and provides for their refinement and modification over time.

REFERENCES

EPA (U.S. Environmental Protection Agency). 1990. Reducing Risk: Setting Priorities and Strategies for Environmental Protection. SAB-EC-90-021. Washington, D.C.: Science Advisory Board (A-101).

Background Reading

Bailey, R.G., and H.C. Hogg. 1986. A world ecoregions map for resource reporting. Environmental Conservation 13:195-202.

Born, S.M., and A.H. Miller. 1988. Assessing networked coastal zone management programs. Coastal Management 16:229-243.

Bower, B.T., C.N. Ehler, and D.J. Basta. 1982. Coastal and Ocean Resource Management: A Framework for Analysis, Analyzing Biospheric Change Programme. Delft, The Netherlands: Delft Hydraulics Laboratory.

Caldwell, L.K. 1988. Perspectives on Ecosystem Management for the Great Lakes. New York: State University of New York Press.

Christie, W.J., M. Becker, J.W. Cowden, and J.R. Vallentyne. 1986. Managing the Great Lakes basin as a home. Journal of Great Lakes Research 12(1):2-17.

Cicin-Sain, B. 1990. California and ocean management: Problems and opportunity. Coastal Management 18:311-335.

Colborn, T.E., A. Davidson, S.N. Green, R.A. Hodge, C.I. Jackson, and R. Liroff. 1990. Great Lakes, Great Legacy? Washington, D.C.: The Conservation Foundation and the Institute of Research on Public Policy.

Derthick, M. 1974. Between State and Nation. Washington, D.C.: The Brookings Institution.

Donahue, M.J. 1986. Institutional Arrangements for Great Lakes Management: Past Practices and Future Alternatives. Lansing, Michigan: Michigan Sea Grant.

Donahue, M.J. 1988. The institutional ecosystem for Great Lakes management: Elements and interrelationships. The Environmental Professional 10:98-102.

Eichbaum, W.M. 1984. Cleaning up the Chesapeake Bay. Environmental Law Reporter, News & Analysis 14(6):10237-10245.

Eichbaum, W.M., and B.B. Bernstein. 1990. Current issues in environmental management: A case study of southern California's marine monitoring system. Coastal Management 18:433-445.

Godscholk, D.R. 1992. Implementing coastal management 1972-1990. Coastal Management 20:93-116.

Haigh, N., and F. Irwin. 1990. Integrated Pollution Control in Europe and North America. Washington, D.C.: The Conservation Foundation and The Institute for European Environmental Policy.

Hansen, N.R., H.M. Babcock, and E.H. Clark II. 1988. Controlling Nonpoint-Source Water Pollution—A Citizen's Handbook. Washington, D.C.: The Conservation Foundation and The National Audubon Society.

Howarth, R.W., J.R. Fruci, and D. Sherman. 1991. Inputs of sediment and carbon to an estuarine ecosystem: Influence of land use. Ecological Applications 1:27-39.

Juhasz, F. 1991. An international comparison of sustainable coastal management policies. Marine Pollution Bulletin 23:595-602.

King, L.R., and S.G. Olson. 1988. Coastal state capacity for marine resources management. Coastal Management 16:305-318.

Kumamoto, N. 1991. Management and administration of enclosed coastal seas. Marine Pollution Bulletin 23:477-478.

Moriarity, F. 1983. Ecotoxicology: The Study of Pollutants in Ecosystems. London: Academic Press Inc.

Sorensen, J.C., S.T. McCreary, and M.J. Hershman. 1984. Institutional Arrangements for Management of Coastal Resources. Columbia, South Carolina: Research Planning Institute.

4

The Process

As discussed in Chapter 3, integrated coastal management (ICM) is a dynamic and continuing process for managing coastal systems in a manner that is responsive to scientific information and human expectations. With a focus on wastewater considerations, this chapter describes the application of the various steps in the process and the tools and methods needed to implement the process for managing coastal environments. While the Committee on Wastewater Management for Coastal Urban Areas is not aware of any particular situation in which integrated coastal management is being implemented at the fullest possible extent, it has identified several examples where elements of ICM are being developed and used. These examples are described throughout this chapter.

DYNAMIC PLANNING

The bulk of problem analysis and assessment takes place within the dynamic planning process (see items 1-4 in Figure 3.1). The power of dynamic planning lies in the bringing together of all relevant data and points of view to identify issues, and the use of a comparative risk assessment approach. Dynamic planning maximizes the use of information in the decision-making process. Most important, it ensures that the major risk management decisions are informed by a complete risk assessment.

Set Goals

In a large coastal area with multiple problems and inputs, the setting of goals is a complex and iterative process involving the balancing of expectations from different sectors of the community. It is seemingly simple but sometimes difficult to identify the important issues relative to wastewater in our coastal environment. This difficulty is due in part to our ignorance of all the goods, services, and other values the coastal environment provides and in part to our individual goals, biases, and perceptions. Coastal resources are, for the most part, a public commons. It is therefore very important that the dynamic planning process be an open and public one that involves all sectors of the communities that may be affected.

Identify Resources

The first step in setting goals for coastal resources in a region is to identify and inventory those resources. This inventory should take a broad interpretation of what may be considered resources in order to arrive at a truly comprehensive starting point for integrated coastal management. It should encompass both the natural and the built environment.

The most obvious resources of a region may be recreational areas (e.g., areas for boating, swimming, scuba diving, surfing) and fisheries. Also of importance would be ecological habitat, birds, wildlife, areas for aesthetic enjoyment, and other environmental attributes. Ports, shipping channels, and other features of the built environment should be included in the inventory as well.

Review Existing Scientific Knowledge

It is important that the goal-setting process be informed by the best available scientific information for a region. The point of this step of the process is to understand what is known about a region as well as to identify what is not known. This review should also serve to bring all participants in the goal-setting process to some common understanding of what is known about a region's resources and environmental characteristics and processes. However, incomplete and imperfect scientific knowledge is not an excuse for delaying action until more research is done. The ICM process should be used to determine if reasonable management decisions can be made, based on existing knowledge.

Assess Human Expectations

A key to the success of dynamic planning is the development of an adequate understanding of human expectations for coastal resources. Expectations may differ considerably from person to person. Often these different perspectives will identify issues that are quite different. Although there may be conflicting objectives or goals behind the issues, frequently the underlying desires will be similar. For example, long-term viability of commercial fisheries and protection of rare species often appear to be in conflict in the short-term, but in the long-term both rely on protection of the ecosystem. Increasingly there is a variety of sophisticated social science techniques for assessing public expectations and values. Use of these techniques can be valuable and informative in addition to the traditional techniques of public hearings and comment which may often elicit only a relatively narrow, albeit important, perspective.

Public Expectations. The public communicates its expectations in the form of societal values (e.g., ecosystem preservation, protection of endangered species, and pristine beach fronts) and human needs (including recreational uses, fisheries, coastal development, transportation, manufacturing, agriculture, and waste management). Often values and needs will conflict with each other so it is important to understand them well. Out of such understanding those interests that may not have been immediately obvious can become more apparent. Additionally, principles for accommodating apparently conflicting uses and values can be developed.

Public expectations also will change over time. Identification of new health hazards, results from risk assessments, data from monitoring programs, and results of research into ecosystem impacts lead to changes in how issues are defined over time and the identification of new problems. Issues formerly of concern are usually dropped from consideration when they no longer need as much attention. New scientific information, depending on how it is communicated to the public, can change public expectations and drastically shift public attitudes toward single issues in exclusion of others. Public expectations also differ over time and among various subgroups within the population. Recently, there has been concern about *environmental inequity* expressed by primarily poor and minority populations who have become increasingly alarmed that their adverse environmental exposures may be greater than for more affluent populations (EPA 1992a, b).

To identify public expectations, it is necessary to involve the public in the planning process from the outset and continuously. To ensure that all issues are on the table at the outset, efforts need to be made to reach diverse groups and individuals who are concerned (NRC 1989a).

While public expectations are quite diverse, a common theme often can be identified. That common interest is appreciation or use of resources. Various parts of the public tend to identify issues relative to wastewater management in terms of whether the coastal resources with which they are concerned are protected. Consumers want to be assured that seafood is plentiful and safe to consume. Surfers, divers, and swimmers want to be certain that it is safe and pleasant to be in the water and walk on the beach. Commercial and recreational fishermen expect that the productive quality of the coastal waters is protected from pollution. Residents in the region may be concerned about the effect of water quality on property values and the local economy. Some want the coastal environment protected for wildlife such as marine mammals and shore birds or simply want to know that the environment is viable, healthy, and sustainable for future generations.

Professional Perspectives. Just as public objectives may vary depending on the particular resource use that a segment of the population values in the marine environment, so will professional objectives vary depending on the particular expertise and interests of the professional in question. As illustrated by several examples, the range of views is vast. A public health practitioner will want to maximize the degree to which human health is protected. Traditionally this philosophy has been articulated through practices that erect the maximum number of barriers between humans and those stressors that could adversely affect human health. An environmentalist may expect maximum protection of the environment and that it remain unaltered. At the other extreme, one might find private entrepreneurs who will strive to minimize the cost of resource utilization in favor of its exploitation. In the middle might be the scientist who favors management objectives that are clearly related to well-understood scientific cause and effect relationships or an economist intent on developing marine-related resources and finding a balance between economic benefits and protection of the environment. One might also find the consulting engineer or government official who must define a wastewater control strategy that is practicably achievable, economically acceptable, and approaches the environmental objectives of the most interests.

Political Decision Making. The objectives of political decisionmakers often will be unstated because the political environment is one in which the process of decision making tends to dominate the need to articulate the goals of the outcome of the process. Political leaders are often freed from the need to articulate their ultimate objectives for wastewater management. There are, however, at least two circumstances in which their objectives become clear. One happens when there is a public outcry to protect a particular resource, such as "Save the Bay!" The second is when there is a dramatic need to exploit the marine environment for the sake of human welfare, resulting in a cry to "Save Our Jobs!" Although in the short-term they may appear to be in conflict in the political process, in the long-term (usually longer than the term of office of the relevant political leaders) these two objectives usually complement each other.

Defining Issues and Setting Goals

The last step in the goal-setting process is the synthesis of the information and expectations assessed in the preceding three steps into a set of issues. General goals then should be set around each of these issues. While these goals define a starting point for addressing a region's problems, they can and probably should be revised as new information is discovered and public expectations shift. As the dynamic planning process proceeds, some perspectives may change and the established goals may need to be revisited. Multiway dialogues must be established to bring together the various points of view.

As the foregoing discussion suggests, there is not a simply stated set of goals for wastewater management. Therefore, it follows that the selection of issues will depend somewhat on the viewpoint of the particular participants involved. These viewpoints will generally fall into one of the two general objectives for coastal protection stated in Chapter 2: 1) to restore and maintain the ecological integrity of coastal areas and 2) to maintain important human uses associated with those areas. Both views are valid when analyzed from the stance of the societal values each seeks to protect. The range of viewpoints held will determine how tradeoffs among competing interests will be established.

The development of a rational set of goals, and thereby selection of issues, depends on the skilled blending and balancing of several quite different values, including:

- *economic interests*, such as those of coastal developers or commercial fishers;
- *personalized expectations*, such as those of scuba divers, swimmers, or sport fishers;
- *rigorous scientific demands*, such as those of the basic scientists;
- *conservative analyses*, such as those of the ecological and public health sciences;
- *preservation interests*, as posed by environmentalists; and
- *fiscal considerations*, as posed by public agencies, ratepayers, and taxpayers.

At this stage, if a large number of issues has been identified, it may be necessary to do a *risk screening* in order to reduce the universe of concerns to the most major ones. For example, for a Pacific coastal area such as Santa Monica Bay, there has been no concern about dissolved oxygen in the water column, but there are significant public concerns about maintaining safety of bathing beaches, particularly near storm drain outlets. In Long Island Sound, the reopening of the extensive contaminated shellfish beds is not a high priority for most people, although it may become important in the future. On the other hand, eutrophication and associated hypoxia present a clear and growing danger to the fish and shellfish stocks of the sound, a danger about which the public is far less aware.

Define the Geographic Extent of Concerns

Once issues have been identified and goals have been set, it is time to define the geographic extent of the associated problems. The importance of this step in the dynamic planning process is that coastal problems occur on different scales. No problem can be addressed adequately and effectively if it is not tackled on the scale at which it occurs. Wastewater and stormwater associated effects occur across the spectrum of scales from very localized changes in benthic populations around the end of an outfall to large-scale nutrient enrichment due to point and nonpoint source inputs occurring over hundreds of square kilometers. Problem domains should encompass the resources affected by the issue of concern and the probable contributory sources. With the environmentally-based identification of the geographic extent of an issue, there also needs to be an involvement of the administrative authorities responsible for the relevant activities in these regions. If these authorities were not a part of the original goal-setting process, goals should be revisited with their involvement.

Resources

For each issue identified in the goal-setting process, there will be a relevant geographic extent of concern. These domains may relate to marine phenomena, such as current transport and upwelling; geographic boundaries, such as drainage area or ridge line; hydrologic phenomena, such as river transport; atmospheric fallout; animal behaviors, such as migration and breeding patterns; and regions for human expectations, such as the demand for products, housing, or other goods from the coastal area.

Sources

Known or presumed sources of contaminants must also be taken into account in defining environmental domains. Where are the outfalls and CSOs? From which portions of the watershed are nutrients being discharged? What are the significant diffuse or nonpoint sources of contaminants and nutrients? Are there septic tanks or other sources of pathogens? Are there aerial inputs of nutrients or contaminants of concern and, if so, from where? Changes in human activities may alter the contributions of various sources over time. Research and monitoring can improve understanding of the relative importance and regions of impact for various sources.

Administrative Authorities

While the inclusion of all important environmental processes and sources of stress to the coastal resources of concern inevitably will lead to the definition of large areas of geographic extent for certain problems, the need to define areas over which management strategies can be effectively coordinated and implemented may require that areas be narrowed somewhat. In the initial analysis, however, these areas should be defined as large as necessary to include the important processes and sources of concern. Later, based on an understanding of these functions, areas of geographic extent can be narrowed in a well-informed manner.

Assess and Compare Risks

A central principle of ICM is that the setting of priorities for action and allocation of effort toward addressing problems should be guided by an understanding of the relative magnitudes of risks to ecologic and human health. Thus, the third major step in the dynamic planning process is to assess and compare risks.

Assessing Risk

Risk assessment is a tool to distill large amounts of scientific and technical information into a form that indicates where the greatest threats to human and ecosystem health are likely to occur. It is an analytic tool that can be used to estimate potential adverse impacts of urban wastewater and stormwater on the various organisms, populations, communities, and ecosystems inhabiting coastal waters, as well as on the various uses we make of the coastal environment. Risk assessments have been used extensively to determine human cancer risk (NRC 1983). More recently, risk assessments have been used to address other human health outcomes such as reproductive toxicity and developmental impairment. Of late, the risk assessment paradigm has been extended beyond human health to broader environmental and ecosystem impacts (EPA 1990, NRC 1993). The results of such assessments can inform risk managers of the probability and extent of environmental impacts resulting from exposure to different levels of stress. This process allows the maximum amount of available scientific information to be used in the decision-making process.

The risk assessment process consists of four steps: hazard identification, exposure assessment, dose-response assessment, and risk characterization. Hazard identification involves defining the inherent ability of some stress to cause harm. Exposure assessment involves quantifying the likely dose of the agent that may be expected to reach the target organs or the magnitude of the stress on the system (e.g., a sediment or water column concentration). The dose-response assessment involves estimating the adverse effect or response due to an exposure. The next step, risk characterization, involves the calculation or estimation of potential impacts based on hazard and exposure, i.e., risk is a function of exposure times hazard,

$$\text{Risk} = f\,[(\text{exposure})(\text{hazard})]$$

The process of determining risk to the environment from anthropogenic stresses involves a greater multiplicity of effects or endpoints, more complexity, and often more uncertainty than assessing human health risk. Also, ecological risk assessments involve various levels of biological organization and there is great regional variability among populations, communities, and ecosystems. For these and other reasons, a universally accepted methodology for ecological risk assessments has not been constructed yet.

Identify Hazards to Ecosystems and Human Health. The identification of hazards to ecosystems and human health should, in effect, take place within the goal-setting and domain definition processes. It is the identification of issues of concern and affected resources that point to the hazards of concern in the region.

Screen for Priority Issues. At this point in the process, the number of hazards identified may be too large to manage effectively. If so, two techniques may be used to narrow down the list of identified issues to one that contains the most significant hazards. It may be possible to screen the issues based on what is already known about their relative importance in the region. Some issues may be agreed upon as being less important than others. Initial efforts could then be focused on the ones of greatest concern with the understanding that those of less concern will be addressed at a later date.

A review of the issues may reveal that many of them have a common root cause. For example, regional-scale eutrophication, seagrass dieback, and nuisance algal blooms all result from excess nutrient enrichment. Thus, it may be appropriate to group these issues together in conducting a risk analysis on nutrient loadings.

Determine Dose-Response Relationships. The dose-response relationship is the one relation between the dose of an agent administered or received and the incidence of an adverse effect in the exposed population (NRC 1983). This step is perhaps one of the most important in the dynamic planning process because the results produced are useful in many ways. For example, once the dose-response relationship is determined, it is possible to establish exposure levels which will produce a particular level of response. This approach was taken in the setting of a goal of 40 percent reduction of nutrient loadings to the Chesapeake Bay (see Box 4.1). A general approach for assessing the dose-response relationship for nutrients and eutrophication is presented in Appendix A.

Box 4.1 SETTING GOALS AND DEFINING DOMAINS FOR NUTRIENT CONTROL IN THE CHESAPEAKE BAY

The Chesapeake Bay Program provides an example of dynamic planning at the regional level that addresses problems occurring across multiple jurisdictions. With specific regard to nutrients, the program has now gone through three iterations of the goal-setting process.

The Chesapeake Bay Program is the cooperative effort of the District of Columbia, Virginia, Pennsylvania, Maryland, the Chesapeake Bay Commission, the U.S. Environmental Protection Agency (EPA), and other federal agencies to restore the Chesapeake Bay. The original Chesapeake Bay Program, begun in 1978, targeted three specific issues of concern: nutrient enrichment, toxic substances, and the decline in submerged aquatic vegetation. These issues were identified as the major concerns facing the region based on existing scientific information.

In 1983, with the signing of the Chesapeake Bay Agreement, participants agreed to a major action program addressing a wide range of issues, including nutrient reduction. While many specific actions were undertaken, no overall goal for nutrient reduction was established at that time. From 1983 to 1987, program participants developed a state-of-the-art three-dimensional hydrodynamic water quality model of the watershed and conducted research to develop a better understanding of nutrient sources and their impact on the bay. As discussed further in the Assessing Risks section of this chapter and in Appendix A, nutrient enrichment can cause anoxia and hypoxia, dieback of seagrasses, and nuisance algal blooms. While the bay program was not following a formalized framework for integrated coastal management, the approach taken in regard to nutrients clearly illustrates the application of the ICM concepts presented in this report. From the mid-1980s on, the program has evolved to embody important elements of ICM, including reevaluation and feedback.

The Chesapeake Bay is the largest estuary in the contiguous United States. Nutrients enter the bay from both point and nonpoint sources throughout the watershed. Point sources include municipal and industrial wastewater discharges. Nonpoint sources include runoff from cropland and farm wastes, urban and suburban runoff, ground water discharges, and atmospheric deposition. Because the sources of nutrients to the bay occur throughout the watershed, the Chesapeake Bay Program defined its domain of analysis as the watershed that is shown in Figure 4.1. This domain includes the entire drainage area of the bay, which extends beyond the jurisdictional domains of the program participants into the states of West Virginia, New York, and Delaware. Thus, although those states chose not to be involved in the program, the analysis was designed to develop an understanding of nutrient inputs that derive from those states as well.

The Chesapeake Bay Model is a computer simulation of processes in the watershed and the bay itself. This model was developed and then used to determine the level of nutrient loadings at which deleterious oxygen depletion in the mainstem of the bay would be stopped. Using loading estimates for 1985 as the base year, it was predicted that a 40 percent reduction in nutrient loadings would mitigate the hypoxia and anoxia in the mainstem sufficiently to encourage recovery of the bay's living resources. It is important to note, how-

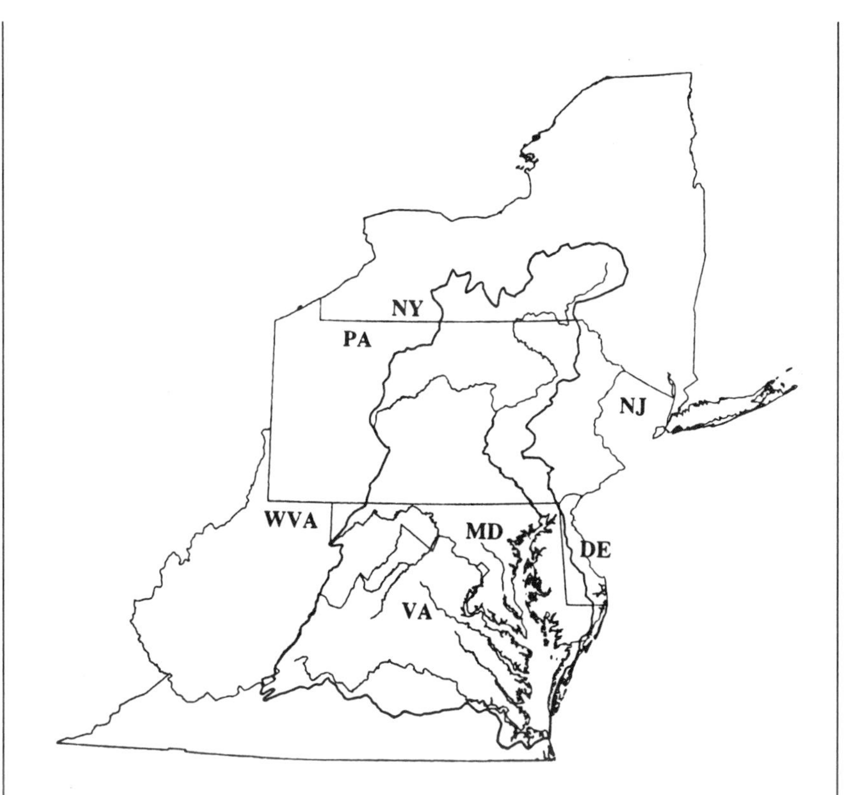

FIGURE 4.1 Chesapeake Bay watershed (Source: CBP 1992).

ever, that nutrient inputs from the atmosphere were not accounted for in the model. The use of the bay model in this way was, in effect, a risk assessment on nutrients to determine the dose-response curve for loadings and oxygen depletion. No comparison of risks was done between nutrients and other stressors that affect the bay.

Based on the information gained through research and monitoring and risk assessment and modeling, specific goals for nutrient reduction were set in the 1987 Chesapeake Bay Agreement:

> By July 1988, to develop, adopt, and begin implementation of a basinwide strategy to equitably achieve by the year 2000 at least a 40 percent reduction of nitrogen and phosphorus entering the mainstem of the Chesapeake Bay. The strategy should be based on agreed upon 1985 point source loads and on nonpoint loads in an average rainfall year.

Because of considerable uncertainty in the 1985 model loading estimates, the agreement built in another iteration to this goal-setting process. It required that an evaluation of the 40 percent reduction target be undertaken in 1991. This reevaluation, completed in 1992, concluded that the 40 percent target reduction is appropriate.

Using existing levels of exposure, such as concentrations of a constituent of concern in the water column, sediments, or shellfish, one can determine the likelihood that an adverse effect will occur.

Characterize Exposure. Exposure characterization is the step in which the degree to which the critical elements of the ecosystem or humans are exposed to various sources of concern is determined. Exposure characterization can be very complex in the context of the coastal zone. The key factor to take into account when characterizing sources and exposure in the coastal zone is that environmental concentrations of a constituent of concern will vary considerably depending on where the source enters the system and how many different sources a particular constituent is associated with. For example, seepage from septic systems adjacent to a shallow and enclosed bay is likely to result in locally increased concentration of nutrients and, if sited inappropriately, pathogens. If the bay also receives stormwater runoff that contains significant concentrations of these contaminants, the problem would be compounded. It may also be difficult to determine the relative contributions of the two sources.

Characterizing exposures to humans can also be confounding because of the multitude of behavioral factors associated with human exposures. These are discussed further in the section below on human health risks.

Assessing Human Health Risks. The World Health Organization states that "health is a state of complete physical, mental and social well-being and is not merely the absence of disease or infirmity" (WHO 1948). Rene Dubos defined health as "expressions of the success or failure experienced by the organism in its efforts to respond adaptively to environmental challenges" (Dubos 1965). In the coastal urban environment, human health issues of concern include not only acute and chronic toxicity but also other contributors to human well-being, such as nutritional value of fish and shellfish stocks, recreational opportunities, and contributions of the coastal ecosystem to mental well-being. As an example of the latter type of effect, algal blooms or fish kills that diminish the recreational opportunities in the coastal area would create stress as well as economic consequences for those whose livelihood depends on recreation. While recognizing the full breadth of human health affected by damage to the coastal environment, the approach used here will focus on assessing risks for acute and chronic illnesses caused by exposure to hazardous chemicals and microbiological stressors. Within integrated coastal management, other stressors will be considered as part of other human expectations (such as economic value of a recreational resource) even though there may be direct or indirect health consequences.

Adverse human health effects can range from minor to severe to fatal, and are usually classified as either acute or chronic. Acute effects or illnesses occur with short-term exposures, are of short latency, and usually recovery occurs. Examples are acute toxicity from exposure to a toxicant and acute gastroenteritis. Chronic effects or diseases usually result from long-term or repeated exposures, may have longer latency periods, and have longer duration. Examples include cancer, neurotoxicity, and infections associated with chronic diseases such as hepatitis A and liver disease, and coxsackie viruses and diabetes. Developmental effects and reproductive toxicity, while conditions of a long-term nature, may result from short-term exposure to harmful agents.

Two principles guide the evaluation of human stressors: 1) the dose makes the poison and 2) there is specificity between agent and effects. Other issues that must be considered are latency (time between exposure and effect); possibility of secondary spread (i.e., from person to person); and the possibility of additivity, synergism, or antagonism between multiple exposures. All of these factors are used in developing risk assessment models to extrapolate from high to low doses in humans and to extrapolate between animal species.

In coastal urban wastewater and urban runoff, the two major classes of contaminants that are of potential concern to human health are hazardous chemicals and infectious agents. These include metals and organic chemicals that may pose varying risks depending on the method of disposal and ultimate environmental fate (i.e., disposal in the ocean, land disposal, or incineration). The toxicity of metals may vary by route of exposure, and by physical and chemical form such as valence state, whether in organic or inorganic state, whether sorbed or dissolved, and whether hydrated or complexed. For example, inorganic, but not organic, arsenic is a carcinogen (Gibb and Chen 1989). Cadmium is considered to be carcinogenic by the inhalation but not the oral route (Life Systems, Inc. 1989; IRIS 1993). Hazardous organic chemicals have entered coastal waters from a number of sources, most of which are due to industrial and agricultural activities. Many of these, like DDT and PCBs, have since been banned but continue to be present in sediments and the tissues of aquatic organisms and water fowl.

Infectious agents of concern include bacteria (e.g., campylobacter, salmonellae, v. cholerae), viruses (e.g., poliovirus, coxsackie, echovirus, adenovirus, and hepatitis A), and parasites (e.g., cryptosporidium, giardia, and entamoeba). Further information on infectious agents is contained in Appendix B. Exposure takes place while swimming in contaminated waters or eating contaminated shellfish. Several diseases can result. These range from subclinical infection to acute, self-limited respiratory, gastrointestinal, skin, or ear infections to extreme gastrointestinal and liver disease (e.g., cholera and viral hepatitis) and other potentially terminal diseases.

Most of these pathogens are derived from human feces; their presence in the environment is often associated with a source of domestic wastewater (e.g., septic tanks, combined sewer overflows, or sewage treatment plant discharges with an inadequate reduction of pathogen levels before disposal). For illness to occur, there must be ingestion, aspiration, or inhalation of a sufficient number of viable organisms. There are, as well, some zoonotic (animal derived) pathogens that may be present in urban runoff water including the protozoa giardia and cryptosporidium and the bacteria salmonellae and campylobacter. It is unknown to what extent these zoonotic organisms pose a threat in coastal areas.

There are also poisonings from shellfish toxins elaborated by microorganisms. Such poisonings include neurotoxic shellfish poisoning, paralytic shellfish poisoning, ciguatera poisoning, scromboid poisoning, and domoic acid poisoning. It is unknown to what extent wastewater disposal impacts shellfish poisoning problems.

Risk assessors must not only look at risks to the general population ("average" exposures) but also at special populations that may be at greatest risk. For exposures in the coastal ecosystem, there are important developmental, immunological, and behavioral differences between individuals that can affect exposure and risk. For example, infants and toddlers who play in the ocean are more likely to swallow or aspirate seawater contaminated with pathogens when their heads are submerged than are adults. Methyl mercury in fish is most toxic to humans during gestation because of greater vulnerability of the developing brain. Newborn infants and the elderly are more susceptible to infectious diseases because of being relatively immunodeficient.

Human behavior is tied to the expectations people hold for the coastal environment. Behaviors that may lead to greater risk of exposure include swimming, scuba diving, consuming fish caught from local piers, eating raw fish and shellfish, and eating organ meats of fish and shellfish (such as the crab hepatopancreas, which is an ingredient in some Chinese recipes). These factors should be considered in the construction of exposure scenarios for human health risk assessments.

Recently, methods have been developed to assess risks from exposure to infectious agents in seawater and in shellfish (Cabelli et al. 1983, Fleisher 1991, Rose and Gerba 1991). These studies use human epidemiological data and water quality or shellfish monitoring for either indicators of infectious agents or for the agents themselves in order to establish the relationship between disease and water contamination. However, the limitations of epidemiological studies and monitoring for indicators are a source of some uncertainty in these risk assessments. Extrapolation from animal studies to humans has not been used for infectious agents.

Excess lifetime risks of disease incidence can be identified via clinical observation alone for very large risks (between 1/10 and 1/100). Most epidemiological studies can detect risks down to 1/1,000 and very large studies can examine risks in the 1/10,000 range. Risks lower than 1/10,000 cannot be studied with much certainty using epidemiological tools. For many diseases such as cancer, risk assessors often must extrapolate from experimental animal studies to humans because of inherent limitations in studying rare endpoints in human epidemiological studies. For example, Figure 4.2 shows the relationship between event levels of risk and the ability to identify cancer risk in the human population. Since regulatory policy generally strives to limit risks below 1/100,000 for life-threatening diseases like cancer, these lower risks are estimated by making inferences about the shape of the dose-response curve and extrapolations from effects to humans at higher doses or from animal testing. Imperfect though this system is, it has the advantage of incorporating all of the available information and creating usable estimates of risk that can be helpful for decision making.

Assessing Ecological Risks. For problems related to wastewater inputs to the coastal environment, ecological risk assessment can best be broken into two parts: effects due to excessive nutrient inputs and effects from toxic substances. The effects of nutrient inputs to coastal ecosystems have received a tremendous amount of scientific study over the past two decades. The role of nutrients in coastal waters and the risks associated with excess

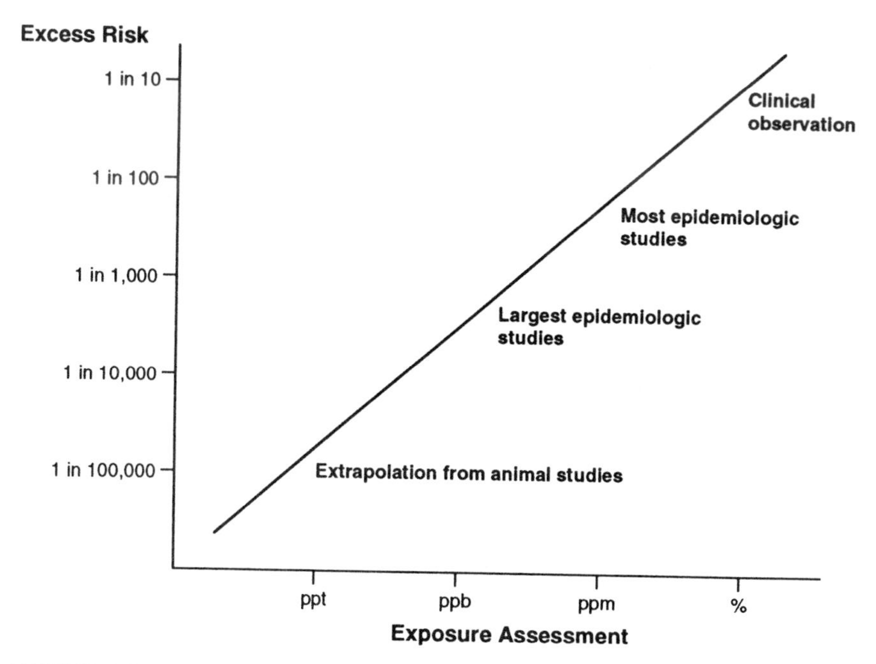

FIGURE 4.2 Sensitivity of epidemiology in detecting risks of regulatory concern.

inputs are well known and described in Appendix A. In moderation, nutrient inputs to estuaries and coastal seas can be considered beneficial. They result in increased production of phytoplankton (the microscopic algae floating in water), which in turn can lead to increased production of fish and shellfish. However, excess nutrients can be highly damaging, leading to effects such as anoxia and hypoxia from eutrophication, nuisance algal blooms, dieback of seagrasses and corals, and reduced populations of fish and shellfish. Eutrophication also may change the plankton-based food web from one based on diatoms toward one based on flagellates or other phytoplankton which are less desirable as food to organisms at higher trophic levels (Doering et al. 1989). Coastal waters receive large inputs of nutrients from both point and nonpoint sources. This is particularly true for estuaries, many of which receive nutrient inputs at rates up to 10,000 times higher per unit area than heavily fertilized agricultural fields (Nixon et al. 1986). As a result of these inputs, many estuaries and coastal seas throughout the world are increasingly experiencing such problems as anoxia and nuisance algal blooms (see Appendix A).

The degree of risk posed by nutrients varies among regions and among different types of ecosystems. In general, the more enclosed the water body and the less water available for dilution, the greater the threat. However, even large areas on the continental shelf can sometimes become anoxic from excess algal production as was demonstrated in the New York Bight in 1976 (Mearns and Word 1982). The degree of density stratification and mixing in the water body are also critical in assessing its sensitivity to low-oxygen events. The degree of risk from nutrients is further affected by the dominant organisms present; for instance, coral reefs and sea grass beds are particularly sensitive to nutrient inputs.

In most estuaries and coastal seas of the temperate zone, nitrogen is the primary element of concern which controls eutrophication. This phenomenon is in sharp contrast to eutrophication in lakes where phosphorus is often the limiting element. Phosphorus is also limiting in tropical lagoons and may be limiting in some temperate estuaries during at least some times during the year. Exactly which element is more critical is a result of differences in the ratio of nitrogen to phosphorus in external inputs, to differences in rates and controls of nitrogen fixation, and to differences in recycling of elements from bottom sediments (Howarth 1988). The relative abundances of other elements can also be important in controlling eutrophication. For instance, some evidence suggests that toxic algal blooms frequently become prevalent only after the ratio of silicon to nitrogen or phosphorus becomes low, that is when silicon is in relatively short supply (Smayda 1989, Officer and Ryther 1980). Silicon is required by diatoms but not by other phytoplankton species; thus as long as silicon is available, diatoms can outcompete other species are suppress blooms of toxic algae. High levels of iron may also be involved in the formation of toxic algal blooms (Graneli et al. 1986; Cosper et al. 1990). These issues are discussed in detail in Appendix A.

The effects of toxic substances on ecological systems have proven more difficult to study than the effects of nutrients or than human health effects. In general, the science of ecological risk assessment for toxic substances is not as well developed as that for human health risk assessment. The two techniques, however, involve the same fundamental principles. In the case of ecological risk assessment, the causative agent is generally referred to as a stressor, and adverse effects are identified as stresses on an ecosystem. Sometimes, ecological risk assessment can be easier than human health risk assessment. For example, controlled experiments can be performed directly on the systems of concern eliminating the need for extrapolation from high doses to low doses or sensitive subpopulations. These advantages are, for the most part, however, outweighed by the greater complexity in organization and response of ecosystems to stresses.

Several indicators of the health of an ecosystem can be used to assess the hazard of a particular stress. These indicators include rates of primary production or other processes; trophic structure; survival of sensitive species; species diversity; and population of fisheries and shellfish stocks, as well as endangered species of birds and mammals. Specific measures include population counts, growth, survival, reproduction, and recruitment.

Responses to toxic chemical stresses can take place at four levels of biological organization: 1) biochemical and cellular, 2) organismal, 3) population, and 4) community and ecosystem. Within each of these levels, there are multiple potential endpoints that could be considered. Not all responses are disruptive in nature, and they do not necessarily result in degeneration at the next level of organization. Only when the compensatory or adaptive mechanisms at one level begin to fail do deleterious effects become apparent at the next level (Capuzzo 1981). However, failures at various levels are often exceedingly difficult to discern, and so changes in populations or in ecosystems may occur without any change at the organismal level ever being detected. In general, ecological risk assessments should be performed using the most sensitive measure of stress. For aquatic ecosystems, change in community structure is an important ecological concern and appears to be sensitive to toxic chemical response (Schindler 1987, Howarth 1991). A variety of toxic agents predictably cause changes, with loss of sensitive species (e.g., amphipods) which will result in domination by weedy or opportunistic species (e.g., capitellid worms). Species diversity usually decreases although this is generally a very insensitive measure of change compared to loss of sensitive species (Howarth 1991). Table 4.1 shows the response levels of marine organisms to chemical contaminants at the four levels of ecological organization.

TABLE 4.1 Responses of Marine Organisms to Chemical Contaminants at the Four Levels of Biological Organization (Adapted from Capuzzo 1981. Reprinted, by permission, from Oceanus, 1981, Volume 24:1.)

Level	Types of Responses	Effects at Next Level
Biochemical-Cellular	Toxication Metabolic impairment Cellular damage	Toxic metabolites Disruption in energetics and cellular processes
	Detoxication	Adaptation
Organismal	Physiological changes Behavior changes Susceptibility to disease Reproductive effort Larval viability	Reduction in population performance
	Adjustment in rate functions Immune responses	Regulation and adaptation of populations
Population	Age/Size structure Recruitment Mortality Biomass	Effects on species productivity and coexisting species and community
	Adjustment of reproductive output and other demographic characteristics	Adaptation of population
Community and Ecosystem	Species abundance Species distribution Biomass Trophic interactions	Replacement by more adaptive competitors Reduction of secondary production
	Ecosystem adaptation	No change in structure and function

Ecosystems vary in their sensitivity to stress by type and region, and therefore ecological health risk assessments should be specific to the setting of concern. Sensitivity is determined by both physical and ecological parameters. Due to dilution, coastal systems that are more open to hydrologic flows (e.g., most of the Pacific coast with the exception of enclosed bays) may be less sensitive to effects of toxic substances although, of course, many toxic substances can be bioaccumulated. Regional differences in community or ecosystem structure will also make a difference. For example, marine ecosystems already subject to natural stresses such as large river plumes may be more resistant to the effects of toxic chemical stress. Areas receiving only small nutrient and sediment inputs and having few major storms that affect the bottom are the most sensitive to the effects of toxic chemical stress (Howarth 1991). Some types of ecosystems, such as coral reefs, are also notoriously sensitive to the effects of toxic substances (Jackson et al. 1989).

The types of stress exhibited in ecosystems are not easily recognizable as resulting from one specific stressor or another. Ecological effects such as population shifts or declines can result from a variety of stressors acting independently or synergistically. It can be difficult to tease apart the effects of a variety of co-occurring stressors. Often concentrations of toxic substances and nutrients tend to covary. Areas subjected to overharvesting of resources and habitat alteration also tend to be those receiving excess nutrients and toxic substances.

The assessment of ecological systems must take into account the spatial and temporal scales at which the effect of concern occurs. The spatial scale of a given effect should correlate with the geographic extent defined in the second step of the dynamic planning process. The temporal scale includes the expected timing and duration of a particular stress, such as pulse loadings from a stormwater discharge, as well as the time required for an ecosystem to recover once the stress has been removed.

The ecological risk assessment process is guided by many questions: What level of biological organization and which potential endpoints should be considered? One of the difficulties in determining risk at the population, community, or ecosystem level is that the myriad of physical, chemical, and biological interactions among individuals and populations is not known or well understood. Since the number of interactions increases with the complexity of the biological system, the uncertainties in risk characterizations may increase accordingly. The basic elements for consideration in ecological health risk assessment are shown in Table 4.2.

The fact that ecological risk characterizations are difficult and the results relatively uncertain at higher levels of biological organization does not imply that they cannot be conducted. For instance, both commercial fisheries quotas and migratory waterfowl hunting bag limits are the results of ecological risk assessments that determined the numbers of individuals that could be lost (caught or killed) without having an unacceptable effect at the population or higher level. For many such assessments, a long history of trial and error and professional judgements are important factors in limiting uncertainty. Applying pesticides that are awaiting registration to microcosms, mesocosms, and field plots helps determine risks at the community level and higher. In these experiments, effects of the chemical on various

TABLE 4.2 Basic Elements for Consideration in Ecological Health Risk Assessment (Source: EPA 1991)

1. STRESS - The type, properties, temporal and spatial patterns, and interactions of the stresses are of fundamental importance in defining the temporal and spatial dimensions and the potential types of ecological effects.
2. ECOLOGICAL ORGANIZATION - Ecological organization represents the level of biological complexity (for both ecological endpoints and indicators) at which the ecological risk assessment is conducted. In theory, the scale of ecological organization chosen for the ecological risk assessment is dependent upon both the spatial and temporal scales of the stress and the co-occurring ecosystem component affected by the stress.
3. ECOSYSTEM TYPE - Ecological assessments are currently ecosystem specific, that is, assessments describe the risk of ecological effects for aquatic, terrestrial, or wetlands categories of ecosystems and/or their respective sub-categories.
4. SPATIAL SCALE - Spatial scale delineates the area over which the stress is operative and within which the ecological effects may occur. Indirect ecological effects may greatly expand the spatial scale required for the assessment.
5. TEMPORAL SCALE - Temporal scale defines the expected duration for the stress, the time-scale for expression of direct and indirect ecological effects, and the time for the ecosystem to recover once the stress is removed.

interactions between and among individuals and populations may be observed without having a complete knowledge of the interactions beforehand.

The EPA uses a relatively simple method for assessing risk to aquatic organisms from single chemicals. It is called the Quotient Method, whereby the concentration of a chemical in a water body is compared to a previously determined safe or acceptable concentration for that substance, e.g., a water quality criterion or a water quality standard. If the quotient, Q,

$$Q = \frac{[\text{Environmental Concentration}]}{[\text{Acceptable Concentration}]}$$

for a particular situation approaches or is greater than one, that body of water is considered to be at risk from that chemical; the smaller the quotient the lower the risk. Even with the inherent problems of extrapolating toxicities from one species to another and using chronic to acute ratios to determine the acceptable concentrations, the Quotient Method has some merit because it is simple and easy to understand (Table 4.3). It is not, however, capable of determining risk from nonchemical stresses. Also, it is not often easy to decide what is an "acceptable concentration." The Water Quality Criteria are based on a variety of different types of studies (Table 4.3), but virtually all of these are laboratory based and generally of short duration. Studies of effects of toxic substances in natural ecosystems over longer periods of time have frequently found major effects that were not predicted by such short-term laboratory studies (Schindler 1987, Jackson et al. 1989, Howarth 1991).

Many of the chemicals in urban wastewater that have the potential to adversely affect aquatic organisms are hydrophobic and sorb, or attach themselves, to sediments. The resulting contaminated sediments can affect organisms living in direct contact with the solids as well as those residing in the overlying waters since the sediments themselves can act as a source of the toxic substances. The Quotient Method, as now used, does not assess hazardous substances in sediments directly. This limitation and the fact that there are numerous coastal areas that have contaminated sediments (NRC 1989b) have led to the development of methods to establish acceptable chemical levels in bottom materials. These methods, sometimes called Sediment Quality Criteria or Sediment Quality Values, can use models to predict whether a given mass loading of chemicals from an effluent will likely result in toxic sediments. These models also can be used to gauge the existing or potential adverse biological impacts of existing contaminated sediments.

TABLE 4.3 Water Quality Assessment Methods (Adapted from Rand and Petrocelli 1985)

Method	Concept
Median Lethal Concentration (LC_{50})	The concentration of a substance in water that results in death of 50 percent of the test organisms when exposed for a specified time, e.g., 48 hours.
Median Effective Concentration (EC_{50})	The concentration of a substance in water that results in some sublethal effect on 50 percent of the test organisms when exposed for a specified time.
No Observed Effect Level (NOEL)	The highest concentration of a substance which results in no adverse effect on the exposed test organisms relative to controls.
Water Quality Criteria	The concentrations of a substance in water that correspond to various effects levels.
Water Quality Standard	The concentration of a substance or the degree or intensity of some potentially adverse condition that is permitted in a water body. An effluent standard refers to a concentration or intensity of impact permitted in an effluent.

The EPA has compiled ten methods that have the potential to assess sediment quality relative to chemical contaminants (EPA 1989). Some of the methods involve chemical analyses that allow for the establishment of chemical specific criteria (e.g., an acceptable level of phenanthrene in sediments). These methods should allow risk to be assessed using the Quotient Method with sediment concentrations rather than water concentrations. Other methods involve only biological observations that limit the results to assessment of whether a sediment is toxic. Still others combine chemical and biological measurements. Brief descriptions of the ten methods are given in Table 4.4.

Aesthetics. Adverse aesthetic impacts include unpleasant sights, noxious odors, and unpleasant tactile sensations (such as from contact with the algae *Pillayella litoralis*). Adverse aesthetic impacts discourage recreational uses and thus can have significant economic impacts. Multiple sources of materials, including combined sewer overflows (CSOs) and urban runoff, as well as commercial ships, recreational boaters, and beachgoers can cause aesthetic offense. Aesthetic impacts can be quantified, albeit through indirect methods. For example, many jurisdictions survey beaches for plastics and other floatable solids and report numbers of objectionable items per unit length per time period.

The range and volume of plastic wastes that end up in the world's oceans are enormous. Typical are a variety of bottles, ropes, and fishnets. There are no reliable estimates of the total volume of such wastes nor the contribution from urban areas to the marine environment. Beach surveys finding condoms and plastic tampon inserters do not identify reliably the source of debris as wastewater, stormwater, recreational boaters, or beachgoers. Plastic debris is not only of aesthetic concern but also can carry pathogens, be mistaken for food and harm marine animals that ingest it, or entangle organisms and strangle them. The Center for Marine Conservation's Coastal Cleanup program cleared 4,347 miles of beaches and waterways of almost 3 million pounds of trash in 1991. Approximately 60 percent of the debris was plastic (Younger and Hodge 1992).

Some of the materials that cause the most aesthetic offense in the coastal marine environment are those that both mobilize public concern and cause significant environmental threats. Floatables, oil and grease, and materials that wash up on shorelines are visible signs of patterns of waste disposal and general human conduct that may also have other impacts on coastal water. As such, they are powerful symbols of more widespread problems. Garbage and syringes washing up on the New Jersey beach in the summer of 1987 did more to alarm the public than did numerous scientific studies. Similarly, when a lawyer jogging along Quincy Bay realized he was treading in human feces, the lawsuit that led to the beginning of the effort to

TABLE 4.4 Sediment Quality Assessment Methods (Source: EPA 1989)

Method	Concept
Bulk Sediment Toxicity	Test organisms are exposed to sediment which may contain unknown quantities of potentially toxic chemicals. At the end of a specified time period, the response of the test organism is examined in relation to a specified biological endpoint.
Spiked-Sediment Toxicity	Dose-response relationship are established by exposing test organisms to sediments that have been spiked with known amounts of chemicals or mixtures of chemicals.
Interstitial Water Toxicity	Toxicity of interstitial water is quantified and identification evaluation procedures are applied to identify and quantify chemical component responsible for sediment toxicity. The procedures are implemented in three phases to characterize interstitial water toxicity, identify the suspected toxicant, and confirm toxicant identification.
Equilibrium Partitioning	A sediment quality value for a given contaminant is determined by calculating the sediment concentration of the contaminant that would correspond to an interstitial water concentration equivalent to the U.S. EPA water quality criterion for the contaminant.
Tissue Residue	Safe sediment concentrations of specific chemicals are established by determining the sediment chemical concentration that will result in acceptable tissue residues. Methods to derive unacceptable tissue residues are based on chronic water quality criteria and bioconcentration factors, chronic dose-response experiments or field correlations, and human health risk level from the consumption of freshwater fish or seafood.
Freshwater Benthic Community	Environmental degradation is measured by evaluating alterations in freshwater benthic community structure.
Marine Benthic Community Structure	Environmental degradation is measured by evaluating alterations in marine benthic community structure.
Sediment Quality Triad	Sediment chemical contamination, sediment toxicity, and benthic infauna community structure are measured on the same sediment. Correspondence between sediment chemistry, toxicity, and biological effects is used to determine sediment concentrations that discriminate conditions of minimal, uncertain, and major biological effects.

continued

TABLE 4.4 *Continued*

Method	Concept
Apparent Effects Threshold	An AET is the sediment concentration of a contaminant above which statistically significant biological effect (e.g., amphipod mortality in bioassays, depressions in the abundance of benthic infauna) would always be expected. AET values are empirically derived from paired field data for sediment chemistry and a range of biological effects indicators.
International Joint Commission	Contaminated sediments are assessed in two stages 1) an initial assessment that is based on macrozoobenthic community structure and concentration of contaminants in sediments and biological tissues, and 2) a detailed assessment that is based on a phased sampling of the physical, chemical, and biological aspects of the sediment, including laboratory toxicity bioassays.

clean up Boston Harbor was filed. In many places, the aesthetic quality of the shoreline has deteriorated, and most of the causes represent serious threats to the ecological integrity of the marine coastal environment.

Compare Risks

A comparison of risks can be helpful in guiding the setting of priorities. Those risks determined to be more important or most unacceptable can then be the focus of risk management options. A clearly articulated comparison or ranking can help to focus efforts on the more important issues.

A determination of highly accurate and precise risk estimates may not be possible but the available data may be adequate to allow the various risks to be compared and ranked on a relative basis as highest or lowest or high, medium, or low. Risk comparisons could be applied to answer questions such as, does biochemical oxygen demand (BOD) from urban stormwater pose as much risk to a receiving water body as nutrients from publicly owned treatment works (POTWs) and agricultural runoff? In this case, a knowledge of the relationship between nutrients and algal blooms can lead to an estimate of the oxygen consumed by respiration and by decomposition of the excess algae, which can then be compared with the relatively more certain estimate of the oxygen consumed by BOD. Another comparison might be, does coastal habitat alteration (e.g., increased sedimentation due to urbanization, wetlands filling, and shoreline alteration) pose as much risk as hazardous chemicals from wastewater?

The risk comparison process should be carried out in the context of the goals established for the region and with the participation of the interested parties in the region. Since the comparison of risks is not a precise quantitative ranking exercise, it might be helpful to adopt a set of guidelines for comparing risks. These guidelines would help insure that each risk receives adequate scrutiny and that judgements about whether one is more important than another are consistent within a region and based in science. While comparison guidelines will likely vary from region to region due to differences in expectations among the residents of a region and due to differences in ecosystems and the types of problems faced, there are at least four generic criteria applicable to both human health and ecological risks that should be used in comparing risks (EPA 1990):

1. the area or numbers of individuals affected,
2. the level of biological organization affected and the importance within the area,
3. the temporal dimension of the effects and potential for recovery, and
4. the risk estimate itself.

As an example of how these criteria could be used for human health risks, data from Santa Monica Bay were used to compare the risk to humans from swimming with the risk from eating contaminated fish. Using the data, Table 4.5 was constructed. Both of these risks would merit public health concerns. The risk to swimmers is for a generally milder disease but

TABLE 4.5 Comparing Human Health Risks from Swimming Versus Eating Contaminated Fish from Santa Monica Bay

Activity	Exposure	Hazard	Range of Risk Estimate	Number Exposed
Swimming	Swallow 100 ml water	Enteric virus infection	1×10^{-2} to 2×10^{-4}	44×10^{6} person-days/yr
Eating Fish:				
caught from boats	23 g fish/day for 70 years	Cancer[1]	3×10^{-4} to 3×10^{-6}	320,000 persons/year[2]
caught from piers	23 g fish/day for 70 years	Cancer	4×10^{-4} to 7×10^{-4}	unknown

[1]Based on consumption of five different species of fish in a hypothetical "average creel" (Pollock et al. 1991).

[2]Based on 80,000 anglers surveyed in 1987 by the California Department of Fish and Game. Fish caught were shared by an unknown number of persons, here assumed to be four per angler (MBC-AES 1988).

TABLE 4.6 Comparative Risk Criteria (Source: MBC-AES 1988)

Activity	Area or Number Affected	Level of Biological Organization	Temporal Dimension	Risk Estimate
Swimming	44×10^6 person-days/year[1]	Human	Short (infection)	1×10^{-2} to 2×10^{-4}
Eating Fish	320,000 persons/year	Human	Long (cancer)	10^{-4} to 10^{-5}

[1]For 1987.

the probability of contracting the illness is high. Among the large number of persons using the beaches, it can be expected that many infants and elderly individuals would be likely to develop severe manifestations or sequelae. The cancer risk, on the other hand, is below the level of detection for the population at risk (approximately 96 excess cases are expected even assuming all 320,000 exposed persons in the Santa Monica region consume 4.6 grams per day for 70 years of Santa Monica Bay fish—an unlikely scenario). A comparison of the risks relative to the comparative risk criteria is given in Table 4.6. Although there are several important information gaps, it appears that in this analysis the swimming-associated risks would be of higher priority. The analysis not only provides relative risk information but also points to weaknesses in the database—the paucity of information about numbers of anglers and lack of data on microbial and chemical shellfish contamination. This information can be fed back to the planning of research and monitoring plans, with risks to be assessed as new data are collected. In fact, for Santa Monica Bay, a survey of anglers is under way to develop estimates on numbers who catch fish and the quantities caught to gain information about vulnerable populations at risk (e.g., infants, the elderly, the immunosuppressed). This information should help refine the comparative assessment. An ideal comparative risk assessment would include health, ecologic, and aesthetic impacts. Table 4.7 shows an array that was developed as part of the Santa Monica Bay Restoration Project. A description of the project and some of the issues to be addressed appears in the Santa Monica Bay case example beginning on page 114. Table 4.7 provides a semiquantitative assessment of the relative importance of a number of stressors for a range of ecosystem components. This model illustrates the numbers of stressors that may need to be considered for a comparative risk assessment and assists in screening for the stressors of concern, including urban wastewater. A more detailed analysis of individual risks would be needed for the next level of priority setting.

Risks, once assessed, must be communicated to all those who are con-

TABLE 4.7 Comparison of the Relative Importance and Understanding of Stressors to Critical Components of Santa Monica Bay

SOURCES OF PERTURBATION \ VALUED ECOSYSTEM COMPONENTS	Intertidal	Phytoplankton	Zooplankton	Soft Bottom Benthos	Hard Bottom Benthos	Kelp Beds	Wetlands & Estuaries	Commercial Shellfish	Pelagic Fish	Demersal Fish	Fish Eggs & Larvae	Marine Mammals	Marine Birds	Human Health	All
Storms	Major (High)			Major (Moderate)	Some (Moderate)	Controlling (High)	Some (High)							Some (Low)	
El Ninos	Moderate (Moderate)	Controlling (Moderate)	Controlling (Moderate)	Major (Low)	Some (Low)	Controlling (High)	Some (Low)		Controlling (Moderate)	Major (Moderate)		Some (Low)	Major (Moderate)		
California Current		Controlling (Moderate)	Controlling (Moderate)						Controlling (Moderate)	Controlling (Moderate)	Controlling (Low)				
Upwelling		Controlling (Moderate)	Controlling (Moderate)			Controlling (High)			Controlling (Moderate)		Controlling (Moderate)		Some (Moderate)		
Basin Flushing				Controlling (Low)											
Mass Sediment Flows				Controlling (Low)						Some (Low)					Net effect of each source on all components
Blooms/Invasions				Major (High)		Some (High)			Some (High)	Some (High)					
Diseases						Some (Low)		Some (High)		Some (Low)					
Ecological Interactions	Controlling (Moderate)	Controlling (Moderate)	Controlling (Moderate)	Some (Moderate)		Controlling (Moderate)	Some (Low)	Major (Moderate)	Moderate (Low)		Major (Low)		Moderate (Low)		
Power Plants							Some (Moderate)		Some (Moderate)		Some (Low)				
Wastewater Outfalls		Some (Low)		Controlling (High)	Some (Moderate)	Some (Low)		Some (Low)		Moderate (Moderate)		Some (Low)	Moderate (High)	Some (Low)	
Dredging				Some (High)		Some (Low)	Moderate (High)								
Rivers/Storm Water Runoff	Major (Moderate)					Moderate (Moderate)	Moderate (High)			Some (Low)				Moderate (Moderate)	
Commercial Fishing						Major (High)		Controlling (Moderate)	Major (Low)	Moderate (Moderate)		Some (High)	Some (High)	Some (Low)	
Sport Fishing								Moderate (Moderate)	Moderate (Moderate)	Moderate (Moderate)		Some (Low)		Moderate (Moderate)	
Marine Commerce/Boating							Controlling (High)					Some (Low)	Some (High)		
Habitat Loss/Modification	Controlling (High)					Moderate (High)	Controlling (High)	Some (Low)		Some (High)	Some (Moderate)	Some (Moderate)	Moderate (High)		
Oil Spills	Controlling (High)						Major (Moderate)						Moderate (High)	Some (Low)	
Oil Seeps	Moderate (Moderate)			Some (Moderate)											
Atmospheric Input	Some (Low)	Some (Low)	Some (Low)								Some (Low)			Some (Low)	
All				Net effect of all sources on each component											

KEY

Potential Importance	Understanding
Controlling	High
Moderate	Moderate
Major	Low
Some	

cerned about the process, especially members of the public and stakeholders. Risk communication in clear terms can help give a sense of the relative ranking of various risks.

Santa Monic Bay: A Case Example

Santa Monica Bay in Southern California is currently being evaluated for management options under authority of the Santa Monica Bay Restoration Project, an activity supported by EPA under the National Estuary Program. Santa Monica Bay is a 690 square kilometer indentation along the Southern California Bight. The project is an ambitious and difficult undertaking. It is bounded on the northwest by Point Dume and on the southeast by Point Fermin as shown in Figure 4.3. The bay extends from the shoreline of Los Angeles and the adjacent cities of Santa Monica, El Segundo, and Redondo Beach westward to depths greater than 500 meters. Offshore, the bay drops off into the 750-meter-deep Santa Monica Basin. Two princi-

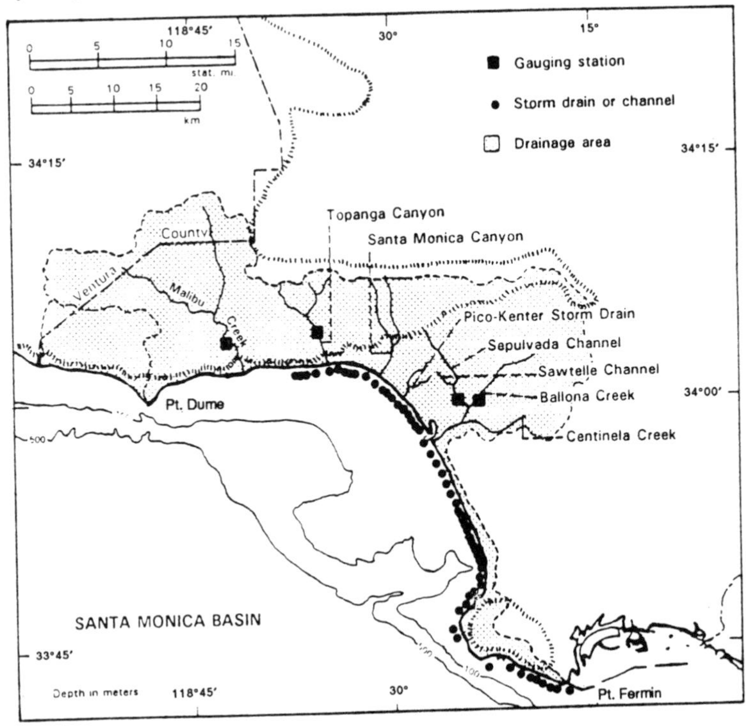

FIGURE 4.3 Natural drainage into Santa Monica Bay. Each dot represents one storm drain or channel (modified from Los Angeles County Department of Public Works maps). (Source: SMBRP 1993. Reprinted, by permission, from Santa Monica Bay Restoration Project, 1993.)

pal inlets of Santa Monica Bay are Marina del Rey, the largest marina in the nation, and the adjoining wetlands of Ballona Creek and King Harbor in Redondo Beach. The shoreline is entirely within the county of Los Angeles.

The Santa Monica Bay Restoration Project defines "the Bay" to include part of the San Pedro Bay shelf, namely the western end along the Palos Verdes Peninsula. Unlike the bay itself, the shelf along the Palos Verdes Peninsula is steep and narrow; within 5 kilometers of shore it plunges into the 850-meter-deep San Pedro Basin.

The present land drainage area of the bay is extremely narrow, extending no more than 24 kilometers inland. The actual drainage area is 850 square kilometers (328 square miles) and includes communities such as Malibu, Santa Monica, west Los Angeles, Beverly Hills, and part of Hollywood in the northeast corner of the drainage triangle; to Westchester, El Segundo, Manhattan Beach, Hermosa Beach, Redondo Beach, and part of Torrance along the central shoreline; and Palos Verdes Estates, Rancho Palos Verdes, and San Pedro on the southern shoreline.

The Bay's Ecosystem

Santa Monica Bay encompasses all the major habitats and ecosystems of the midlatitude Pacific coastline of the United States. The pelagic ecosystem is the largest and supports epipelagic and mesopelagic populations of fish, invertebrates, and algae. Inshore waters of the bay support one of the major areas for eggs and juveniles of keystone species such as northern anchovy and Pacific sardine. Inshore, shallow-water rocky outcrops support related species that compete for space with a large variety of green, red, and brown algae; seaweeds; and kelp. Sandy beaches support dense colonies of mole crabs and other small crustaceans, polychaetes, and clams while rocky intertidal areas are rich in additional seaweeds and attached and free-living invertebrates. Kelp forests are critical habitat for many species of fish, invertebrates, birds, and mammals and have been an economic (harvested) resource as well. The Ballona wetlands have been reduced by the construction of Marina del Rey from 8.5 square kilometers to 0.65 square kilometers. A smaller wetland occurs at the mouth of Malibu creek.

Major Uses of the Bay

The principal uses of Santa Monica Bay include recreation (swimming, boating, diving, fishing), sport and commercial fishing, aesthetic enjoyment, coastal developments, shipping, industrial cooling water, and waste disposal (domestic and industrial). The full market value of residential, commercial, and industrial development along the bay exceeds $30 billion.

The 22 beaches of Santa Monica Bay attract 44 million person days of visitors each year, principally during holidays, weekends, hot weather periods, and the summer. Since 1985, however, attendance has dropped due, possibly, to increasing news coverage of environmental conditions (MBC-AES 1988). Surfing occurs primarily along the Malibu coast, where waves usually impact the coastline at a better angle than in the center of the bay. Diving is a popular year-round sport principally along the Palos Verdes and Malibu coasts. Boating is also a year-round activity. In 1986, tourism was estimated to contribute $232 million, more than 3,000 jobs, and $4.2 million in tax revenues to Santa Monica's economy (MBC-AES 1988).

A major bait purse seine fishery operates within the bay. Sport anglers contribute nearly $4 million to the local economy. In 1987, nearly 80,000 sport anglers took almost half a million fish, mostly Pacific bonito, chub mackerel, and barred sandbass.

The three major industrial and municipal disposal uses of the bay area are power generation, oil refining, and wastewater disposal. The Hyperion Treatment Plant of the City of Los Angeles discharges 370 million gallons per day (MGD) of treated wastewater 8.5 kilometers (5 miles) offshore and 60 meters deep, below the seasonal pycnocline. The Whites Point outfalls of the County Sanitation Districts of Los Angeles County discharge 360 MGD of treated wastewater 60 meters deep, 3 kilometers offshore, near the east end of the Palos Verdes Shelf. Other sources of contaminants to Santa Monica Bay include return cooling water from three power generating stations, one oil refinery, 68 storm drainage channels, over 7,200 private vessels at two marinas, 92 tanker off loadings per year, oil spills, 2,200 metric tons per year of litter, oil seeps, stormwater runoff, and aerial fallout from smog, brushfires, and other sources.

A special situation of considerable concern exists off Whites Point. During the 1950s and 1960s, the pesticide DDT was discharged to the ocean through the Los Angeles County outfalls. The emissions were discovered in 1969 and promptly controlled (Carry and Redner 1970). However, by that time, marine biota in the Los Angeles area and beyond was contaminated and hundreds of metric tons of the pesticide had accumulated in surface sediments. Since then, deposition from natural and sewage solids have buried this field under 20 to 25 centimeters of sediment. These contaminated sediments, however, apparently continue to be a chronic source of DDT to fish and wildlife in the bay (Mearns et al. 1991).

Identification of Issues of Concern

Issues of concern regarding the environmental quality of Santa Monica Bay deal with human health and marine resources as well as the scope and costs of additional pollution control, clean-up, and abatement activities.

During 1990 and 1991, the Santa Monica Bay Restoration Project conducted workshops with the stakeholders and the public and identified five classes and numerous subclasses of issues or resource use conflicts. The five basic concerns identified were:

1. Swimming and water contact: Is it safe to swim in the bay?
2. Seafood contamination: Is it safe to eat seafood from the bay?
3. Wetlands: How can wetlands be restored and protected?
4. Marine ecosystems and habitats: Are marine ecosystems protected?
5. Fish and shellfish stocks: Are fisheries protected?

Swimming and Water Contact Issues

The area of concern for water contact issues is defined by where people come in contact with potentially contaminated water. This domain includes the entire bay shoreline out to where people can conceivably swim or dive. The State of California defines the offshore swimmable boundary as the 30-foot depth line and 1,000 feet from shore. The California Ocean Plan (CSWRCB 1990) requires that water quality within kelp beds, which are popular diving spots, must meet bathing water standards. However, boaters may come in contact with water not only within marinas, boat harbors, and near shore but also offshore, including over the deep ocean outfalls. Thus all offshore surface waters could be considered as possible routes of exposure and as issue boundaries.

The risk of swimming-associated acute gastroenteritis infection was estimated by the Southern California Association of Governments based on fiftieth and ninetieth percentile enterococcus levels applied to a model developed by the EPA, which assumes head immersion and 100 milliliters of water intake (MBC-AES 1988). For 17 stations between Topanga (near Malibu) and Malaga Cove (Palos Verdes), fiftieth percentile dry-weather acute gastroenteritis infection risk ranged from 0.2 per 1,000 persons (2×10^{-4}) at Malaga Cove to 11.3 per 1,000 persons (1.1×10^{-2}) at Pulga Canyon (north of Santa Monica) and Venice Beach; the highest risks (10^{-2}) were along the Santa Monica coast between these two points. During wet weather, risk increased sharply at the southern sites but only slightly at the Santa Monica area sites. Ninetieth percentile risks were one to ten times higher than the fiftieth percentile risks for dry weather and about twice those for wet weather. No estimates were made for respiratory illness risks or for risks associated with stormwater flows during storm events.

Viruses have been detected in stormwater in other areas at levels between 2.6 and 106 plaque-forming units per liter (PFU/L) (O'Shea and Field 1992). Taking the average level (12 PFU/L) and assuming it is similar to that of Santa Monica stormwater, estimates of infection risks could range from 2.7×10^{-1} to 2×10^{-2} when exposure occurs shortly after a storm event.

All these estimates are in general agreement that there is a seasonally-variable 1/100 to 1/5,000 chance of experiencing illness while bathing along Santa Monica Bay beaches. The enterococcus based model suggests that the risk, at least during dry weather, is ten times lower at beaches along the southern and northernmost shores than in the vicinity of Santa Monica and Venice.

Seafood Contamination

Because they are immobile, contaminated harvestable molluscan shellfish, such as mussels, clams, and rock scallops, are limited in areal extent, on the order of several meters to several kilometers from wastewater inputs. Mussels at several locations in Santa Monica Bay have in the past contained concentrations of DDT, PCBs, chlordane, and several metals approaching levels of concern (Phillips 1988). These data, however, have not yet been used in risk assessments. There are no data to support an assessment of risks from pathogens and toxicants in shellfish.

In contrast, finfish and mobile crustaceans, such as crab, lobster, and shrimp, have been the focus of intensive seafood consumption and risk studies. Several fish species of concern have potentially large distribution ranges that encompass tens of square kilometers. The highest levels of DDT and PCB contamination have been found in white croaker, which is among the most frequently caught and consumed fish. The white croaker effectively defines a boundary of concern that extends out to about 100 meters deep and from San Pedro to the central-northern shore of Santa Monica Bay.

The California Environmental Protection Agency Office of Environmental Health Hazard Assessment conducted a comprehensive study of chemical contamination of fish that was weighted by frequency of catch for various species (Pollock et al. 1991). They identified several chemical contaminants of concern: DDT and related compounds, PCBs, chlordane, mercury, and tributyltin. In contaminated areas, the white croaker, a bottom-feeder, was the most contaminated fish but corbina, queen fish, surfperch species, and sculpin (a.k.a. scorpion fish) were also relatively contaminated. Bonita, mackerel, halibut, sand dab, barracuda, opaleye, and halfmoon had the lowest level of contaminants. Cancer risks from PCBs and DDT and related compounds were the most significant health risks and ranged from 10^{-3} to 10^{-6} for a lifetime exposure to a particular species, depending on the location and species. Several general guidelines were issued as a result of the study including: eat a variety of fish species, consume fish caught from several locations, and trim fat (which concentrates DDTs and PCBs) from fillets. A number of site-specific advisories were issued recommending that anglers limit consumption or not eat certain fish species caught from specific locations.

Wetlands

Boundaries of wetlands are not limited to those in current existence but may expand to the natural historical boundaries, such as the wetland areas where Marina del Rey is now situated. The Santa Monica Bay Restoration Project also considers the potential for restoration or development of freshwater wetlands at historical sites, including within areas now occupied by storm drains (SMBRP 1992). Since wetlands contribute to the nourishment and reproduction of migratory birds, the boundary may reach as far as the extent of these migrations.

Ecological Health: Wetlands versus Marine Habitats versus Fisheries

How do we set priorities for marine ecosystems? One way is to evaluate current injury to each and then compare them on the basis of the fraction of habitat in need of restoration. Comparison of risks to various marine ecosystems might be made on several bases including total productivity of each, number of threatened or endangered species in each, or fraction of total habitat injured.

An attempt is made here to compare two habitats, the subtidal benthos and wetlands. About 11.4 square kilometers of the Santa Monica Bay sea floor are projected to incur changed benthic communities as a result of current inputs of suspended solids from sewage inputs. This area includes about 1.6 square kilometers, 8 kilometers offshore in the center of Santa Monica Bay shelf and 9.8 square kilometers located 3 kilometers offshore of the Whites Point area of the Palos Verdes shelf. Combined, this area affects about 1.7 percent of the 690-square-kilometer shelf of Santa Monica Bay. By contrast, there were historically about 9.2 square kilometers of wetland habitat at Ballona Creek and this has been highly modified by development (e.g., Marina del Rey) to less than 0.69 square kilometers, a reduction of over 92 percent. Therefore, on an areal basis, there is the possibility of comparing a 1.7 percent injury to the seafloor of the bay with a 92 percent injury to the major wetland of the bay (MBC-AES 1988).

In addition, there has been no documentation of a threatened or endangered marine invertebrate or fish in the bay, whereas the existing wetland is one of the few remaining habitats for the endangered Belding's Savannah Sparrow and possibly other terrestrial and shore species.

Develop and Compare Alternatives for Risk Management

The next step in integrated coastal management is to develop management alternatives for reducing the priority risks to coastal water quality, i.e., solving the problems. Most often the best solutions will involve a combination of actions, including engineering works (e.g., treatment plants and outfalls), source control for pollutants entering POTWs, or reduction of inputs from diffuse sources. This step involves engineering design of systems using the environmental-quality driven approach for both structural and nonstructural measures. To illustrate the engineering process this section focuses at first on a single major discharger of wastewater effluent from a POTW; then later in the section the discussion extends to other situations involving multiple point sources and diffuse sources. However, the same general concepts apply—i.e., working back from water and sediment quality objectives to find the optimum set of control measures needed.

The Need for Problem-Focused Management

Problems must be defined carefully and risks assessed as described in the preceding section. It is not enough to say "Clean up Boston Harbor." Rather, the dynamic planning process must be specific about exactly what pollution problems exist, what their severity and distribution are, and what the associated risks posed to ecosystems and human populations are. Close examination of all sources of a problem for coastal water bodies is required because one source may represent only a minor fraction of the problem and its tight control would be useless without a holistic approach that would address the bulk of the problem.

The purpose of this section is to explain the process of managing risks and developing alternative management strategies. Often, but not always, these risks can be managed by meeting numerical standards, such as water or sediment quality standards, concentration limits of pathogens in shellfish, or some other environmentally based standard.

After the precise nature of the problems and the sources that contribute to them are understood, the range of technical management options for their solution can be defined. Table 4.8 on pages 122-125 provides a list of possible problems that could be associated with a single major wastewater discharge and indicates what remedial actions may be taken for each. It illustrates in a descriptive way that different problems require different responses in order for pollution control efforts to be effective. There is no single approach to solve all problems even when they all are due to a single wastewater discharge. A similar table should be developed showing possible control measures and their effectiveness for all sources that the risk assessment process has shown to be of concern for a specific coastal management region. It is important to understand that it is easier to develop explicit control strategies as implied by Table 4.8 for sources such as outfalls from sewage treatment plants than for diffuse sources such as rivers, storm runoff channels, or atmospheric inputs.

Lessons from Existing Situations

In developing pollution control systems, much can be learned by studying the existing coastal pollution situation through detailed field observations and computer modeling. This effort will reveal the relative importance of different sources and information about the processes that are currently important in the coastal environment. For example, continuing with the hypothetical example discussed in Table 4.8, if the discharge point of an outfall for a POTW is moved to a new site much further offshore or the treatment is upgraded, then predictions can be made of the new ambient water quality to be achieved by modifying this one source while leaving unchanged all the other sources, such as CSOs, urban runoff, river discharges, atmospheric deposition, and other diffuse sources. Furthermore, when making water quality predictions, engineers usually have some past experience with discharges in the same region to provide data for model calibration and general guidance.

Ambient Water-Quality Objectives

The risk to human health and ecosystems and to our aesthetic appreciation of coastal water is linked directly to ambient water quality, which is often defined by measurable parameters such as concentrations of suspended solids, toxics, nutrients, bacteria and viruses, oil and grease, and dissolved oxygen. For instance, the risk management of pathogens has been systematized by setting standards for surrogates such as coliform bacteria or enterococci bacteria which can be monitored easily. Each individual discharger does not do a risk assessment for pathogens but uses a surrogate of fecal contamination. The deficiency in this approach is the inadequacy of the surrogate for predicting the presence (or absence) of various pathogens and not in the approach itself. (A more detailed discussion of this practice appears in Chapter 2 and Appendix B). Thus, bathing water standards can be set at various levels of bacteria or other appropriate organisms. Such limits constitute an explicit water quality standard used as an objective for engineers who design systems. Similarly, standards (or objectives) may be established to manage many risks. Setting objectives and standards is an important intermediate step for the design of management strategies and engineering systems.

Setting acceptable water-quality or sediment-quality standards for some

TABLE 4.8 Options for Improving Coastal Waters Affected by a Wastewater Treatment Plant Discharge (Brooks 1988)

Assumed present system: primary treatment, short outfall.

Assumed ocean environment density stratification due to temperature and/or salinity variation with depth during most seasons of the year.

Water Quality Problem	Effective Remedial Action	Comments
1. Pathogens/coliform counts too high at target areas (swimming areas, shellfish beds)	• Longer diffuser	• Higher initial dilution, possible submerged plume
	• Longer outfall (discharge farther offshore)	• Increases travel time back to shore and die-off
	• Disinfection	• Some methods (e.g., chlorine) are injurious to marine organisms
2. Undesirable biostimulation due to excessive nutrients (e.g., algal blooms)	• Longer outfall	• Improved dispersal to open ocean (nutrients less or not harmful)
	• Keep nutrients below photic zone	• Submerged effluent plume in naturally turbid waters
	• Nutrient removal	• With some methods, increased sludge volume, requires larger disposal site
	x Secondary treatment (not a remedy)	• Nutrients not significantly removed but may be more available • Increased light penetration due to particle removal
3. Excessive turbidity or light extinction	• None, if turbidity is natural	
	• Increased removal of particles by chemical coagulation and/or secondary biological treatment	• Increased light transmission may increase biological growth • Increased sludge production

4. Excessive oil and grease (slicks, grease balls, shoreline deposits)	• Better diffuser • Source control for sewer system • Better removal of oil and grease in primary tanks • Secondary treatment	• Increased dilution • Check other marine and land sources
5. Trace contaminants in water column; bioconcentration in fish and birds	• Source control • Better diffuser	
	• Move discharge farther away from biologically sensitive areas	• Higher dilution (more spread around, no decrease in total mass)
	• Improved treatment (e.g., secondary or advanced primary)	• Many toxicants transferred to sludge
6. Benthic accumulation of organic matter in sensitive areas	• Advanced primary	• Removes more particles by polymer (or other chemical) addition; makes more sludge
	• Physical-chemical treatment	• Further particle removal (makes even more sludge)
	• Secondary treatment (part or all of wastewater flow)	• Increased energy requirements; increased sludge production with increased handling, transportation, and disposal costs
	• Move discharge away from sensitive areas	• Only really effective approach
7. Trace contaminants in bottom sediments (effects on benthic organisms)	• Industrial pretreatment	• May increase sludge and toxics disposal by industry elsewhere
	• Pollution prevention • Greater particle removal (advanced primary or secondary)	
	• Move discharge further offshore	• Away from sensitive areas and to places with better natural dispersion of water and sediments

continued

TABLE 4.8 *Continued*

Assumed present system: primary treatment, short outfall.

Assumed ocean environment density stratification due to temperature and/or salinity variation with depth during most seasons of the year.

Water Quality Problem	Effective Remedial Action	Comments
8. Benthic release of contaminants deposited in earlier years	• Cover up with additional natural sediments • Continued wastewater particle discharge with control of toxics at source	• May require disruption in the area of the new sediment source • Delay implementing secondary treatment (if at all). Source control of toxics is essential
	• Excavation and removal	• Resuspension may cause rerelease • Need new disposal site • Temporary increase in release of toxics into water column
9. Depression of dissolved oxygen:	• Better diffuser	• Higher initial dilution to a lower concentration of BOD
A. Due directly to wastewater discharge	• Relocate discharge to area of better flushing • Secondary treatment • Secondary treatment with nitrification	• More dissolved oxygen resource tapped • Lowers oxygen demand • Lowers oxygen demand
B. Due to excessive plankton growth and then decay in deeper waters	• Reduce biostimulation	• See #2 above

10. Other aesthetic problems		
A. Floating solids of wastewater origin	• Better removal in sedimentation tanks • Effluent screening • Combined sewer overflow control	• Need to control floatables in wet weather overflows
B. Refuse	• Better cleaning of streets and paved areas • Public education	• Need to control other sources of refuse (e.g., solid waste from shipping traffic)
11. Occasional breakdown of treatment plants; toxic spills	• Long outfall • Good diffuser • Improved maintenance • In-line sewer and treatment plant flow equalization	

water uses (such as protection of ecosystems) is not easy and is subject to continual improvement as we learn more about various risks. For example, with respect to total nitrogen, we may have only some generalized brackets rather than specific numbers, but these too can be very useful as targets. When the water and sediment quality objectives are uncertain, an engineer normally would devise a system which will limit ambient pollutant concentrations at the lower end of the band of uncertainty to provide a safety margin. At later times, when more is understood and standards are changed or made more certain, then wastewater disposal systems and other activities can be adjusted to be most effective in achieving the goals.

Finally, the expression of water quality or environmental objectives may be stated in a way that is totally devoid of numerical description. A classic example is the one contained in the language of Section 301(h) of the Clean Water Act which requires that a "balanced indigenous population (of flora and fauna) be maintained." The California Ocean Plan also requires that "Marine communities, including vertebrate, invertebrate, and plant species, shall not be degraded" (CSWRCB 1990). A description of the California Ocean Plan is provided in Box 4.2. These qualitative standards require that the dischargers and regulators develop an agreement as to what these objectives mean on a case-by-case basis and how compliance will be evaluated on a local or regional basis.

Management of uncertainty is part of the design of environmental solutions. It is incorrect to say that if everything is not known, nothing can be done in terms of a wastewater disposal project. With sufficient empirical experience, which has been developed in many cases, successful projects can be built. The challenge of the future is to keep responding to new information so that as time goes on those things that are really necessary and worthwhile are continued and those things that have little or no benefit according to our developing knowledge base can be terminated.

The Environmental-Quality Driven Approach

The next step in risk management is to design engineering control systems that may be used to achieve compliance with the various water-quality and sediment-quality objectives that are established to manage risks to human health and ecosystems. In using the term *environmental-quality driven approach*, the implication is that other media effects will be included even though water and sediment are central. Sediment quality objectives are recent developments, as it is now well understood that contaminated sediments are significant contributors to exposure to the ecosystem and humans (through shellfish and benthic fish).

The design of engineering systems based on water and sediment quality objectives is shown schematically in Figure 4.4. The figure and the follow-

Box 4.2 THE CALIFORNIA OCEAN PLAN

The California Ocean Plan is the state of California's Water Quality Control Plan for Ocean Waters established and administered by the State Water Resources Control Board under the California Water Code, and approved by the U.S. Environmental Protection Agency under Section 303(c)(1) of the Federal Clean Water Act. It provides the basis for regulation of point and nonpoint discharges to California's ocean waters.

Periodic Review

First promulgated on July 6, 1972, the plan has a built-in triennial review process to accommodate advances in scientific understanding and to respond to public inputs on the criteria for protection of various beneficial uses. The plan was updated in 1978, 1983, 1988, and 1990; and is currently undergoing another review.

Use of an Environmental Quality Driven Approach

The Ocean Plan establishes water quality objectives for California's ocean waters. These objectives are designed to ensure that more than 66 physical, biological, and chemical parameters are within acceptable ranges, that designated beneficial uses are protected, and that human health is protected from potential risks associated with water contact and fish and shellfish consumption.

Toxics

The Ocean Plan was one of the earliest water quality plans to address toxic pollutant inputs by establishing water quality requirements based on protection of marine biota, preceding federal efforts to promulgate applicable marine water-quality standards. For toxic pollutants, the plan was unique in that it did not mandate technology per se, but established ambient water-quality standards that could be met through a variety of management or technologic controls. Effluent limitations for toxics are established for each discharger so that ambient water-quality requirements will be met after initial dilution. Dischargers can then set the most effective combination of controls (e.g., initial dilution, treatment, or source control) to meet requirements.

Pathogens

The plan establishes an ambient total and fecal coliform standard for receiving waters and a requirement to monitor enterococcus levels. Disinfection for control of pathogens is not required and is in fact discouraged because of the effluent toxicity associated with most disinfectants. One of the plan's general requirements for locating an outfall discharge point is that "Waste that contains pathogenic organisms or viruses should be discharged a sufficient distance

continued

Box 4.2—*continued*

from shellfishing and water-contact sports areas to maintain applicable bacterial standards without disinfection." If it is not possible to locate the discharge a sufficient distance away, disinfection is required.

Dissolved Oxygen

The plan establishes an ambient water-quality standard that limits reduction in dissolved oxygen levels to 10 percent of the natural value. It does not establish a standard for BOD in effluent nor does it establish treatment requirements for BOD.

Effluent Limitations

The plan establishes some standards without regard to initial dilution or receiving water conditions. Specific effluent limitations are established for suspended solids (60 mg/l or 75 percent removal, whichever is less restrictive), settleable solids, turbidity, oil and grease, pH, acute toxicity, and radioactivity.

Additional Highlights of the Ocean Plan

- Monitoring requirements. The plan requires the establishment of monitoring programs. Monitoring programs are reviewed and supervised by the Regional Water Quality Control Board.
- Recognition of naturally occurring variations. Naturally occurring variations in effluent quality and ocean conditions are accommodated through the use of statistical distributions for setting standards rather than setting rigid limits.
- Flexibility. Under certain conditions, the Regional Water Quality Control Board can establish more or less stringent requirements. Exceptions to any requirement must meet the Board's determination that it "will not compromise protection of ocean waters for beneficial uses, and the public interest will be served."
- Special protection. Discharges to areas of special biological significance are prohibited.

The Ocean Plan is Characterized by Its

- periodic (triennial) updates based on current scientific knowledge,
- public input on use attainability and beneficial uses,
- flexibility to deal with site-specific conditions,
- emphasis on nonconventional toxic pollutants removable by source control efforts, and
- comprehensive statewide coverage for varying situations.

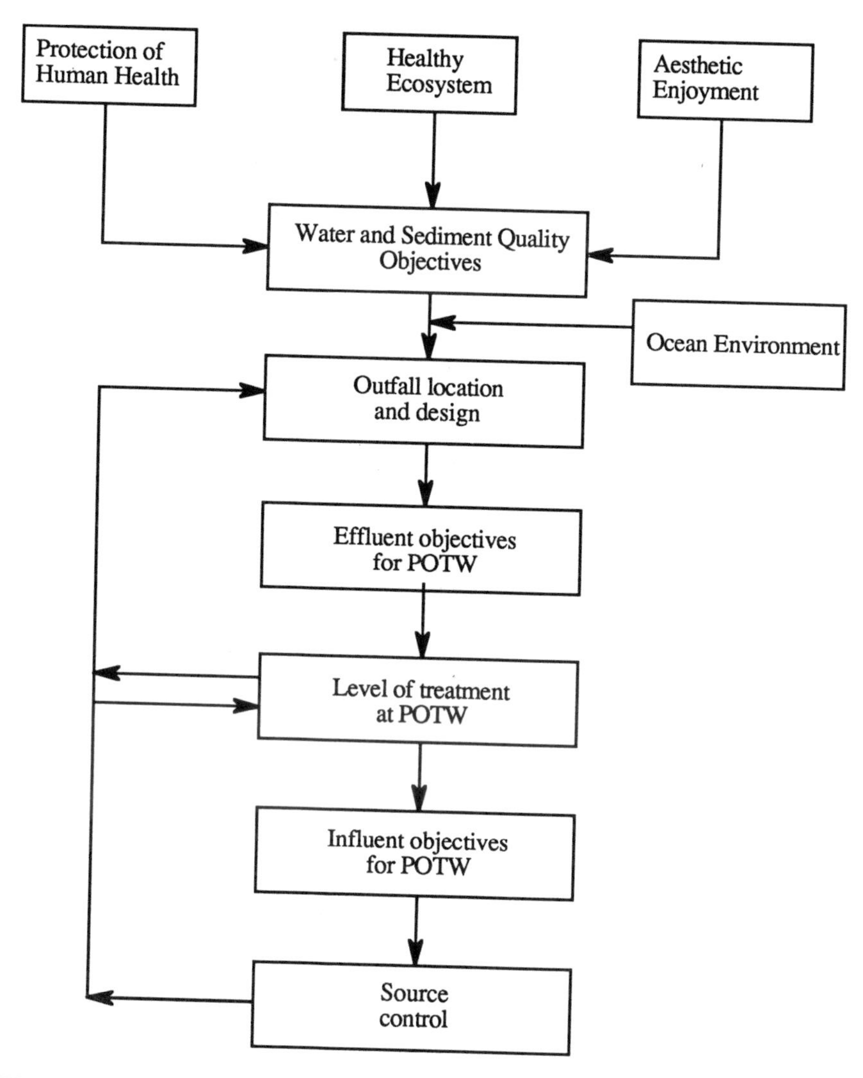

FIGURE 4.4 Overview of water- and sediment-quality driven approach for design of municipal wastewater disposal system. The water-quality/sediment-quality driven approach involves working backward (compared to direction of flow) to establish what components are needed.

ing discussion refer to a municipal wastewater disposal system in order to illustrate the concepts; however, the same concepts and procedures can be extended to water bodies with multiple inputs from point and diffuse sources.

The first step (top of Figure 4.4) is the expression of the environmental objectives, in terms of water quality objectives or standards, from which the engineering design may proceed. The chart then displays what appears to be a *backwards* calculation: from these standards is derived the combination of engineering measures that is best suited to reach the prescribed water quality. There are three basic types of engineering components for a municipal wastewater system:

1. the *outfall(s)*, including location and characteristics of multiport diffusers used for high dilution (see Appendix C);
2. the *treatment works*, with various possible types and levels of treatment (see Appendix D); and
3. *source control* (or source reduction) to limit the amount of toxic substances or other pollutants entering the sewer system and the treatment plant (see Appendix D).

These three parts constitute a system in which changes in one part will change the need for the others. For instance, a long outfall with high initial dilution generally reduces the need for secondary treatment; or better source control reduces the need for toxics removal at the POTW and simplifies the sludge disposal problem. Only by considering all three components and their associated environmental and financial costs and benefits at once can the optimum combination be found.

The conceptual design then proceeds with a trial choice of outfall and related treatment levels and source control programs necessary to meet the ambient water quality standards. The process involves complex modeling of the transport and fates of contaminants in the ocean after initial dilution. Given a certain outfall configuration, effluent limits needed at the POTW can be determined. To meet these limits, appropriate treatment components and upstream source control measures are then selected.

As a practical matter, such modeling must be done in the *forward* sense (sources → treatment → outfall → transport and fate → effects), but with iterations it is conceptually the same as Figure 4.4 with the reversed order of steps. However, with experience it is not difficult to work backwards from water-quality and sediment-quality standards to get approximate solutions for the three system components, which are then used as the initial iteration for the detailed forward water quality modeling. The use of modeling as a means for implementing a water-quality driven approach is discussed in further detail in the following section and Appendix C.

System Components

Source Control. Source control is the collective term used to describe all the ways that POTWs regulate the inputs of toxic materials into their sewer systems. These ways include prescribed effluent limits for industrial dischargers, effluent charges, and outright prohibitions. Source control programs implemented by many POTWs have been very effective in reducing the toxicity of both effluents and sludge. In the 1970s, regulators stated that one of the reasons for mandatory secondary treatment was the reduction in toxics in effluent, even though the toxics were then captured in the digested sludge. But in the 1980s, concern over the toxicity of sludge provided the impetus for implementation of stringent source control programs.

Treatment Plants. The various types of treatment plants are summarized briefly in Chapter 2 and in more detail in Appendix D. A study of performance and costs for various treatment components and combinations has been undertaken by the Committee on Wastewater Management for Coastal Urban Areas, and the results are presented in Appendix D.

For marine discharge, the usual configuration is full secondary treatment except for POTWs that have 301(h) waivers and operate at primary, or chemically-enhanced primary, or partial secondary treatment levels. In chemically enhanced primary treatment, sedimentation is enhanced by the addition of polymers or other coagulants that can increase the removal rate for suspended solids from about 55 to 65 percent for the traditional primary sedimentation up to 75 to 85 percent, or even higher, for the enhanced process. It is interesting to note that 85 percent removal of suspended solids, even though in a primary tank, is equivalent to the federal requirement for suspended solids removal for secondary treatment (for influent suspended solids less than 200 mg/l).

The environmental-quality driven approach will encourage further innovation in wastewater treatment focused on controlling water and sediment quality as needed.

Outfalls. An ocean outfall is a pipeline that conveys liquid effluent from a treatment plant to the receiving water. For the discharge end of the outfall, engineering practice has progressed in the last four decades from a simple open-ended pipe near the shore to the use of large multiple-port diffusers discharging the effluent in deep water far offshore. An example of such an outfall is the 27,400-foot one of the Orange County Sanitation Districts of California. As shown in the plan and profile in Figure 4.5, the outfall includes a 6,000-foot-long diffuser with 503 discharge ports and terminates at a water depth of 200 feet. Figure 4.6 is a schematic drawing of the discharge of buoyant effluent into a density-stratified receiving water

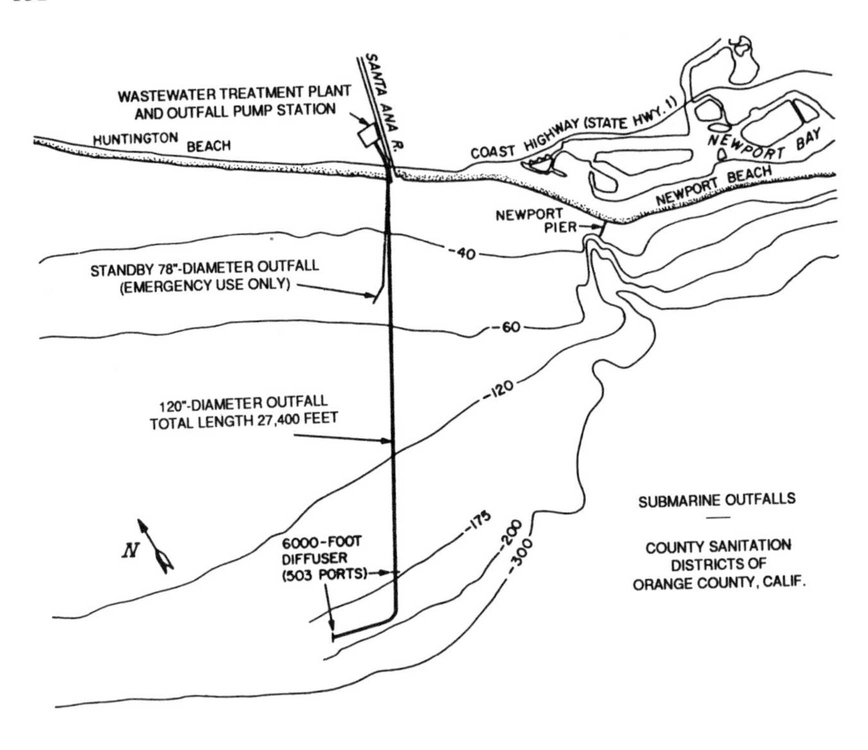

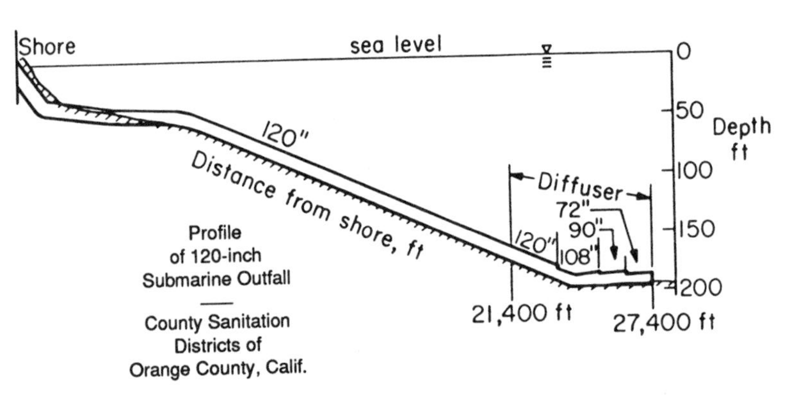

FIGURE 4.5 Schematic plan and profile of the 120-inch outfall, County Sanitation Districts of Orange County, California. (In metric units, the overall length is 8.35 km, the diffuser length is 1.83 km, the diffuser depth is 53-60 m, and pipe diameter is 3.05 m.) (Source: Koh and Brooks 1975. Reproduced, with permission, from the Annual Review of Fluid Mechanics, Vol. 7, © 1975 by Annual Reviews Inc.)

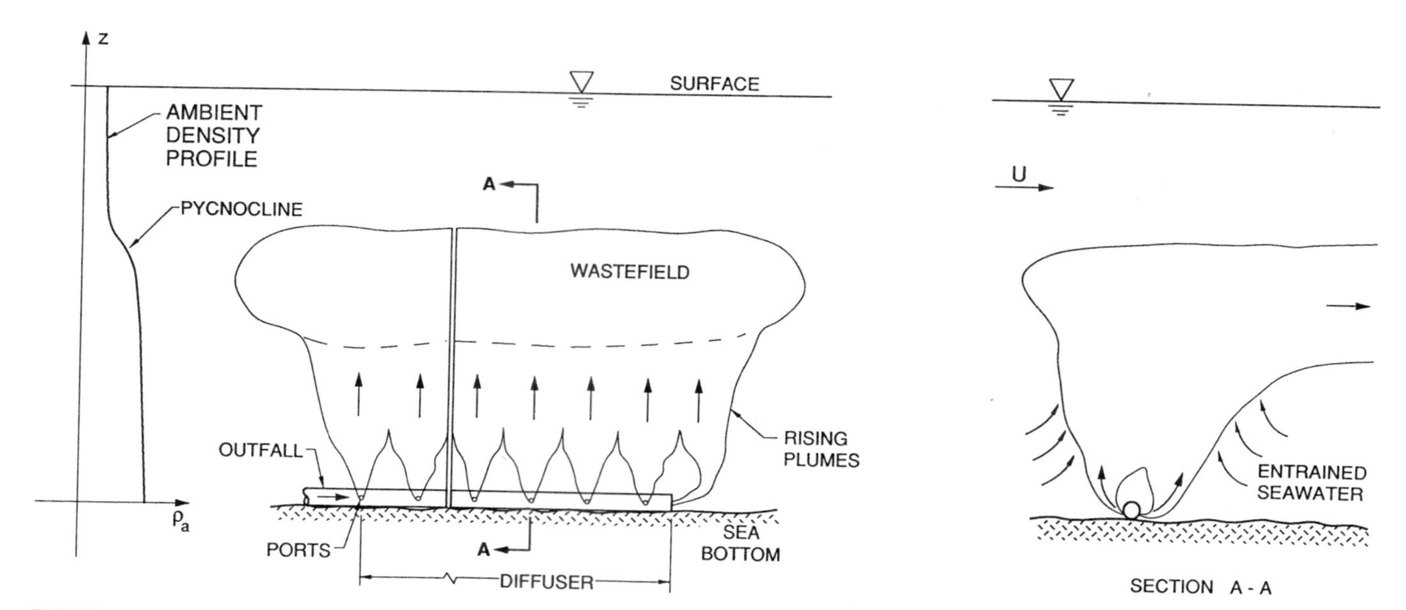

FIGURE 4.6 Formation of a submerged waste field over a multiport diffuser in a stratified ocean with a current perpendicular to the diffuser. For clarity, only a few ports are shown, as typically there are hundreds for a large outfall.

from a typical multiport diffuser at the end of an ocean outfall. Wastewater effluent, being effectively fresh water, rises in the ocean and mixes rapidly with the receiving water. Since the ocean is usually density-stratified due to surface heating or freshwater river discharges, the effluent mixes first with denser bottom water. The resulting mixture can become neutrally buoyant before the rising plume reaches the ocean surface. Neutral buoyancy leads to the formation of a submerged waste field or plume, which is then advected horizontally by the prevailing currents.

The mixing occurring in the rising plume is affected by the buoyancy and momentum of the discharge and is referred to as initial dilution. It is typically completed within a matter of minutes. This phase of the mixing process is under some control by the design engineer since it depends on the diffuser details such as length, jet diameter, jet spacings, and discharge depth. The initial dilution is also controlled partially by nature since it depends on the density stratification and currents in the receiving water.

A typical, large discharge diffuser (for a flow of 5 m^3/s) might be a kilometer in length and located in 60 m water depth at a distance of 10 km offshore. There might be several hundred discharge jets (typical diameter 10 cm) spaced along the diffuser. The initial dilution obtainable for such a diffuser would be expected to be in the hundreds to a thousand depending on details (mainly flow rate and density stratification). The initial dilution and waste field submergence can be estimated with a fair degree of confidence, thanks to three decades of engineering research on the mixing processes in buoyant jets and plumes. A number of computer models exist in the literature, each of which can provide predictions of sufficient reliability to make rational decisions on design choices.

In more complicated situations the water-quality driven approach can also include systems for CSOs controls and measures for abating diffuse sources.

Transport and Fates Modeling: Predicting Ambient Water and Sediment Quality

Mathematical and conceptual models are extremely useful in the development of risk management alternatives. Models are used to explain observed processes which disperse and modify pollutants in the ocean and to predict how various management options can be expected to perform. Various submodels may be combined to produce an overall model that relates pollutant inputs to water and sediment quality for single and multiple sources. These models are fundamental to the water-quality driven approach for system design because the limits on emission for any discharge or nonpoint source may be back-calculated using the models. Because of the various length and time scales associated with different problems (Table 4.9), a variety of models is needed—one mathematical model cannot provide answers to all questions. These models are analogous to the emissions-to-air-quality models used in developing air pollution control programs.

To be successful, there must be good predictive capability for the dominant factors that determine the engineering choices for satisfying the standards. These factors are determined by sensitivity analyses as well as the experience of the modeler. Thus for engineering purposes it is not necessary to understand every process if more knowledge would have no effect on the choice of control strategy. For example, it is not necessary to understand the behavior of a certain pollutant at a location where the input is far below any possible threshold value of concern. Or another example, if the range of uncertainty of a biological effect is a factor of 10, the effect may be of no importance if the upper end of the range is fully acceptable.

The preceding discussion relates to the wastewater disposal system for a municipality including source control programs, a treatment plant, and outfall. The three kinds of information required apply equally well to all other types of pollutant sources and the approach to devising an engineering system. In the case of CSOs, for example, it is necessary to have a set of water quality objectives; some knowledge of the environment; and information on the amount, quality, frequency, and distribution of existing CSOs.

TABLE 4.9 Typical Length and Time Scales of Effects Associated with Typical Pollutant Problems for Coastal Wastewater Discharges

Pollutant	General Length Scale	General Time Scale
Ammonia toxicity (if any)	One or two kilometers	Few hours
Other acute toxicity (if any)	Few kilometers	Less than 1 day
Coliforms	Up to 10 kilometers	Up to a week
Bacterial pathogens	Up to 10 kilometers	Up to a week
Viral pathogens	Up to 10 kilometers	Up to 4 months[1]
Protozoan pathogens	Up to 10 kilometers	Up to 3 months[2]
Deposition of organic matter	10 kilometers	Few days
Oil and grease (wastewater origin)	Few kilometers	Few days
BOD-caused DO decrease	Few kilometers	Few days
Nutrients (nitrogen)	Up to 100 kilometers	Months to a few years
Regional hypoxia	Up to 100 kilometers	Months to a few years
Heavy metals (sediments)	Few kilometers	Years to decades
Synthetic organics (sediments)	10s of kilometers	Decades

[1]Melnick and Gerba 1980, Goyal et al. 1984.
[2]G. Vessey, MacQuarie University, Sydney, Australia, personal communication, 1992.

If there are multiple sources contributing to the water quality impairment, then environmental modeling must integrate the effect of all sources and develop scenarios for different degrees of control and handling of different sources. Sources of the same kinds can be combined into classes to simplify the modeling, such as one for a large number of small POTWs all affecting the water quality of a large body of water such as Long Island Sound.

Predictive models have a number of uncertainties and need improvement, but nonetheless appropriate engineering systems for wastewater disposal and diffuse source control can be designed to meet prescribed water- and sediment-quality objectives. Since modeling for design of a management plan for pollution control always has some uncertainty covered by safety factors, it is cost-effective to implement a system (such as a waste treatment plants and an outfall) in a stepwise, flexible manner to allow for continuous feedback of operating experience and observed impacts on the coastal waters. In fact, there are few situations where there is not already an existing discharge that serves as a prototype to study before and during upgrading the system. For example, the full effect of upgrading primary treatment on coastal water quality might well be observed before proceeding to more advanced treatment levels. Or source control efforts for specific chemicals can be focused on dischargers whose concentrations are observed to be too high. This approach is always self-correcting, as the discharger is committed to take as many steps as necessary to solve any problem.

Diffuse Sources—Modeling and Control

An integrated management approach must also address the impacts of diffuse sources. The term *diffuse sources* describes the types of inputs to coastal water bodies other than municipal and industrial wastewater. These sources are diffuse in origin, such as urban storm runoff, and are usually intermittent. Current laws and regulations have redefined some of these discharges as point sources for which permits are now required. Herein the functional term *diffuse sources* is used for describing the nature of the sources, without making the legal distinctions.

Combined Sewer Overflows and Stormwater Outlets. Nearly all outlets for stormwater flows and CSOs discharge almost at the shoreline through either open channels or pipes. Because of the intermittent nature of the flows, often at discharge rates far greater than dry-weather sewage flows, and scattered locations, it is not feasible to use long outfalls or multiple-port diffusers, as is the practice for wastewater effluents from POTWs. In addition, because of the low use factor, such pipes tend to fill up with sand

and marine detritus to such an extent that the hydraulic capacity may be lost. Tide gates may be used to reduce saltwater and sand intrusion, but they must be readily accessible for maintenance rather than deep below the water surface.

Due to the general lack of feasible options to increase the initial dilution significantly at the discharge points, it is necessary to impose corrective measures on land, such as the banning of toxics for dispersed urban use (e.g., the elimination of lead from gasoline), better street cleaning, detention basins, skimming devices for floatables, and storage and diversions to treatment plants during dry weather. These measures may apply to either stormwater outlets or CSOs.

At one time, sewer separation was a common technique for abating CSOs as well as localized flooding. This practice was soon realized to be very expensive and unacceptably disruptive. Furthermore, in many areas it is impractical to separate all stormwater connections from sanitary wastestreams because a sizeable number of these take place on private property. Thus abatement efforts shifted toward maximizing the amount of wet weather flow treated at the wastewater treatment plant. Such efforts include enhancing in-line storage coupled with near surface or deep tunnel storage capacity. High-rate satellite treatment in a stand alone mode (e.g., vortex solids separators, and mechanical screening) has been used in a number of applications across the country over the last 20 years to decrease abatement costs; however, performance has often been disappointing. The most recent example is the large swirl concentrator installation in Washington, D.C. that is achieving suspended solids removal rates of only 25 percent and for which no measurable water-quality improvements have been found (Nemura and Pontikakis-Coyne 1991). In the last five years, high rate treatment has been packaged with satellite treatment concepts creating compact efficient solutions, but performance of these systems is not yet known. Also, during this time period, the cost of deep rock tunneling has decreased which makes the more efficient and dependable deep tunnel storage option increasingly attractive (see Appendix D for more details).

Natural Streams and Rivers. Coastal cities are often located where major streams discharge to the ocean directly or through estuaries. In these cases, the urban coastal waters may receive a large dose of pollutants washed from the entire watershed down the drainage system. These pollutants, which must be controlled at their sources, include nutrients and pesticides from farms, forests, and urban areas; dry and wet weather acid deposition on the watershed; products of chemical weathering; sediments; refuse; animal excretions; and other natural organic matter.

Various stream modification techniques, sometimes with adverse environmental consequences, can help to make the water flow easily into the ocean, and jetties may guide the flow slightly offshore, although they are usually built for sand control and stabilization of the outlet position. After discharge into an estuary or the ocean, the initial dilution is determined by the dynamics of the density-stratified flow in which a freshwater layer spreads laterally over the denser brackish water or seawater. Dilutions achieved in the first few minutes or fraction of an hour will remain low (i.e., less than 10). Subsequent mixing occurs at the density interface until it disappears. If the river water quality is managed to meet ambient water requirements, then initial dilution should not be an issue. Only long-term problems, such as eutrophication or toxics in sediments, will drive the planning of control measures, such as reduction of nutrients flowing from farmlands or mines into rivers. Source reduction measures have to be applied to the whole watershed with the tightness of control depending on what is needed to avoid problems of bioaccumulation of toxics in the food chain and eutrophication of semi-enclosed water bodies.

Fine sediments (i.e., silt and clay) in river discharges may be beneficial (if uncontaminated) in diluting or covering up previously contaminated deposits from other sources. Sediments carried by rivers replace beach material eroded by wave action. Also, increased turbidity due to finer (colloidal) suspended sediments reduces light penetration, a factor which may help to control excess growth in water bodies high in nutrients. However, the turbidity may be aesthetically unacceptable to the populace.

Atmospheric Deposition. Atmospheric deposition of pollutants directly onto a coastal water body in the form of precipitation or dry deposition is by its nature widely dispersed. The initial dilution is thus not relevant, and only the regional circulation and sedimentation processes determine how much becomes trapped in sediments versus the net transport to the deep ocean. If the residence time of a body of coastal water is short (days to months), the capture of airborne pollutants in the sediments will tend to be less than for lakes where the residence time is of the order of years, as is the case for the Great Lakes, where at present the principal new source of PCBs is atmospheric deposition.

The modeling of the transport and fate of airborne pollutants is similar to that for other input pathways, except that the first step will be vertical diffusion into the upper mixed layer of the ocean, typically to a depth of 10 meters. Whether such inputs are important must be evaluated on a case-by-case basis in comparison to the magnitude of the other sources. It may be found in some cases, as for nitrate in Chesapeake Bay, that as the strengths of other sources are reduced, the remaining depositional source may become relatively significant and must be included in the development of overall control strategies. Of course, control of atmospheric deposition can only be done back at the sources of emission of the pollutant in question to the atmosphere; the motivation for such control will usually come from other environmental problems first (such as acid deposition on forests and fresh waters), except in cases where the marine problem may be more acute than any land-based ecological or health risks. For example, there may be more bioconcentration of pesticides and PCB's in marine food chains than on land; on the other hand, acid deposition per se is of no consequence in coastal waters because of the huge buffering capacity of the sea.

In summary, for most pollutants direct deposition on coastal waters is a small term in their overall environmental mass balance; there are cases, however, in which direct deposition is a significant factor. Deposition on land areas and subsequent delivery to the shore by storm channels and rivers is often more significant. For example, before unleaded gasoline, the main pathway for lead transport to the ocean in southern California was urban storm runoff (Huntzicker et al. 1975).

Ocean Sludge Disposal

Disposal of digested sludge to the ocean, prohibited by federal law in the United States, is still practiced in other parts of the world. Digested sludge is a suspension of very small organic particles (averaging about 10 to 20 microns) that are similar to small particles still remaining in wastewater effluents. In recent years, because of source control, the concentration of contaminants in digested sludge has been steadily decreasing. Reduction in toxicants is required for safe land disposal and incineration as well. For discussion of other methods of sludge disposal see Appendix D.

The modeling of a possible sludge discharge (typically 100 times less in volume, but much more concentrated in particulate matter than effluent), whether by pipeline or by ship, proceeds by the same kind of transport and fate modeling as described above. The water- and sediment-quality driven approach applies for the development of whether, and, if so, how, sludge can be safely disposed of in the ocean. The principal uncertainty in the transport and fates modeling is the rate at which sludge particles will sink to the bottom. In other words, the size of the deposition footprint is in question. Much can be learned from study of existing sites such as The 106 Mile Dump Site off the New York Bight, where solids tended to reach the bottom somewhat faster than predicted, which resulted in a higher deposition rate near the source than expected.

The primary environmental impact associated with ocean disposal of digested sludge, if it is done far offshore, appears to be possible changes in sediments (Van Dover et al. 1992). On one hand, contaminants may be deposited on the ocean floor and enter the food web. On the other, in some circumstances, sludge disposal may provide beneficial nutrients to the ocean. It is reasonable that the question of sludge disposal to the sea be re-examined taking into account improved understanding of sludge particles' behavior and sediment quality criteria now being developed.

Digested sludge may be a beneficial resource; when it can be recycled feasibly in a beneficial way, it certainly should be. Sometimes, however, the extra costs of energy and facilities and the cross-media impacts of dewatering the sludge and transporting it to places where it can be used beneficially may not outweigh the environmental costs of ocean disposal.

Cross-Media Considerations

All of the foregoing discussion focused primarily on water and sediment quality and the development of appropriate engineering works and management strategies to produce the desired water and sediment quality while still disposing of wastewater and storm water. The integrated approach obviously requires that full consideration be given to the cross-media impacts of various actions for disposal of wastewater with respect to the land and the atmosphere. Some examples of these cross-media consequences are presented in Table 4.10. There are, of course, other examples as well, but the points are illustrative.

One general point is worth more attention. The building and operation of treatment plants, including the facilities for the processing of sludge, consume materials in their construction and use energy to operate. Materials extraction and production (such as steel for reinforcing bars or copper for windings in motors) have significant environmental costs and consume an important resource base. Energy to run plants requires the extraction of fuels and contributes CO_2 to the atmosphere, which may affect global warming. Digested sludge produces methane but this energy is not free. If the treatment plant, through energy conservation, were to have a surplus of electric power generated from the methane, then it could be put into the power grid and substituted for other sources. Secondary treatment is an energy-intensive process, the full environmental impacts of which should be considered along with the benefits and costs.

As a final example, the city of Los Angeles was recently required to stop discharging digested sludge to the ocean via a seven-mile sludge outfall and instead now has an EPA-approved disposal system consisting of dewatering and incineration even though the Los Angeles basin is one of the most severely impacted air pollution regions in the country. The option of moving the ocean discharge point much further offshore to water over 1,000-feet-deep has been predicted to have low cost and few offshore impacts. It was rejected by the EPA because the Clean Water Act prohibited such a project regardless of the environmental consequences of alternative options (LA/OMA 1980, Brooks 1983).

TABLE 4.10 Examples of Possible Cross-Media Consequences of System Choices for Ocean Disposal of Wastewater

Item	Potential Impacts or Hazards
Chlorination	• Hazardous manufacturing, transport, and storage and handling at the treatment plant • Impact on marine organisms
Sludge, prohibited for ocean disposal	• More energy used for dewatering for disposal elsewhere to reduce transportation costs • Centrate treatment and disposal • Ground water, land, and food contamination from land disposal, unless pathogens, metals, and toxics are controlled • Air pollutant emissions from incineration, including toxics, metals, and greenhouse warming gases
Increasing levels of treatment	• More sludge for disposal • More energy required for treatment and sludge processing • Associated impacts of extracting, refining, and burning fuels
Watershed controls on nitrogen release	• Reduces need for advanced treatment nitrogen removal
Source control of metals and toxics for reduction in effluent	• Reduces toxics in sludge

Costs and Feasibility

It is important to describe the costs associated not only with each of the alternatives for the solution of particular problems but also for the whole range of problems. A good understanding of the financial cost of a particular management option results from an analysis of the capital and operating costs of the project over time. The analysis also must take into account the cost of financing capital expenditures.

Consideration must be given to determining which sector of society will bear what portion of the costs. The simplest example of the impact of this consideration is seen in the construction of municipal treatment plants. In these cases, the end of federal grant support means that customers now bear a high proportion of, or all of, both construction and operating costs in their sewer rates. Where costs can be borne initially by the private sector, there may be opportunities to consider possible financial incentives or disincentives to induce the private sector to assume the burden. If examination of a particular risk suggests that suburban development requires effective management of stormwater, then a choice is posed about whether the necessary infrastructure should be provided by government or by developers.

These financial considerations need to be displayed in as great detail as are the technical solutions to the range of problems being considered. The comparative analysis of the cost factors associated with particular solutions will further define the practical achievability of any particular technical management option. It is noted that sometimes a more expensive solution will be favored because it is characteristically more reliable than less costly alternatives.

Finally, managers must consider the feasibility of each option within the range of solutions which have been developed. Feasibility is a function of a wide range of factors, including:

- public acceptability,
- legal authority, and
- institutional capacity.

One illustration will demonstrate the impact of these factors. In some situations, the engineering and cost analysis may suggest that the most technically practical solution to a wastewater management problem is discharge of advanced primary effluent through a very long outfall. However, this solution might be unacceptable to the public for nontechnical reasons, or may be illegal under existing statutory authorities.

Expectations and Benefits in Relation to Costs

After the various risks are analyzed with respect to possible technical solutions, costs, and feasibility, it is necessary to consider the public's evaluation of benefits in relation to costs for various possible solutions. For example, in a particular urban setting (e.g., Point Loma, San Diego), the public may desire to have swimmable water in adjacent kelp beds even though the cost may be much higher than if these benefits were forsaken. If maintaining a certain use is considered to be too expensive, then expectations may be scaled back. The question then is a matter of degree and choice: how much preservation of natural ecosystems and other beneficial uses is the public willing to pay for? In different places, the balance can and will be struck in different ways.

For environmental managers, the challenge is often to find ways to provide the most environmental protection for the money; i.e., engineering and other solutions must be cost effective and be perceived by the public to have produced significant long-term benefits in relation to the cost. At the same time, public expectations at a particular moment when management decisions must be made may lead to a choice that is either inadequate to protect the environment or that is quite costly with only marginal environmental benefits.

In summary, residuals must be managed and returned to the natural ecosystem by some procedure; the ocean disposal option cannot be considered in an absolute sense, and its risks, costs, and benefits must be evaluated in comparison to the other management and disposal options.

Summary

The main points in risk management for wastewater impacts on urban coastal waters are:

1. Integrated coastal management requires the use of the environmental-quality driven approach to model and manage the effects of single or multiple discharges and diffuse pollution sources and to make effective regional control strategies. The primary environmental objectives to be met are water-quality and sediment-quality standards for the receiving waters.

2. Predictive models have a number of uncertainties and need improvement, but, nonetheless, appropriate engineering systems for wastewater disposal and diffuse source control can be designed to meet prescribed water and sediment quality objectives.

3. There is much to be learned from existing discharge situations that is useful to support modeling and engineering efforts for designing new or upgraded facilities.

4. The ability to develop and use mathematical and conceptual models is ahead of the field of confirming the accuracy of models. More effort is needed to study prototype systems after construction to evaluate the preconstruction modeling and analysis.

5. A continuous, responsive approach is needed for future management of major discharge areas. This approach includes ongoing ocean studies and flexibility of management to modify the discharge system as needed in response to new research findings, new problems, or new environmental objectives. The corollary of this step is the need to proceed in a stepwise fashion, implementing the most effective steps first.

6. Coastal water quality management must be site or region specific because of widely varying conditions along the coastline of the United States.

7. Because of the wide range of spatial and time scales of various ocean processes and the various time scales of different water quality problems, different modeling approaches are required for different pollutants.

SELECTION, POLICY, AND INSTITUTIONS

Institutional Arrangements

The institutional setting for regulating wastewater discharges in the coastal environment is complex, fragmented, and compartmentalized. While the federal Clean Water Act provides a basic regulatory framework, the interplay of natural resource management, regional and local planning, programs for nonpoint and point source discharges, preventive and remedial actions, and multiple levels of government leads to an institutional setting far more diverse and extensive than the act alone would imply. Having a clear picture of this institutional setting is essential to any attempt to improve wastewater management.

Institutions are fragmented in at least three different ways: hierarchically, geographically, and functionally. Hierarchical fragmentation occurs when responsibilities are divided among two or more levels of government. Multiple political jurisdictions lead to geographic fragmentation. The further division of programs according to function adds still another layer of fragmentation. Within any particular watershed or region, the presence of multiple agencies, each with separate but relevant programs, requirements, and responsibilities, creates enormous complexity.

The Clean Water Act itself involves multiple levels of government and considerable complexity because it incorporates sewage treatment requirements, stormwater management, toxicant regulation, estuary management, and funding mechanisms, among other things. While the basic scheme of uniform federal requirements implemented through delegated state programs is straightforward, it leaves coastal states to struggle with some of the most difficult issues. Each state and local jurisdiction must decide how to mesh growth and land-use planning with water quality objectives; it must regulate toxicants in discharges in the absence of comprehensive federal regulations; and funding must be found for upgrading wastewater treatment in the absence of federal construction grants.

As the need for effective pretreatment and pollution prevention programs is recognized, municipal wastewater treatment requires a significantly increased regulatory role for local governments. Despite fragmentation and jurisdictional complexity throughout the nation, it is possible to discern an emerging trend toward better integration of planning, resource management, cross-media issues, and interjurisdictional concerns. Many coastal states provide promising examples of attempts to manage growth, to link land use and water resource management, and to control sources of water pollution in an overall context. Considerable integration and alignment of institutional arrangements has been achieved in some places. Signs of progress include the development of estuary protection strategies (e.g., for the Chesapeake

Bay), regional water resource agencies, and integrated growth management and land-use strategies.

There is no one *right* approach to better integration of environmental decision making. What is doable in one setting may be out of the question in another. Flexibility in approach and implementation is as necessary as leadership and long-term commitment of funding and political will.

Improving the institutional framework for water quality protection is a significant challenge. Increased integration of planning and implementation is desirable both within water resource functions and between water and other related resource decisions. Efforts at integrated environmental decision making are hopeful signs, but only a beginning. Most water pollution control decisions of this decade will in all likelihood continue to take place in a complex and fragmented institutional setting. Yet greater awareness of the broader context should assist both ad hoc and structural efforts to achieve better integration.

Development, Selection, and Implementation

Other sections of this report discuss goals and objectives for coastal wastewater management, sources of pollutants, and various strategies for the control of these pollutants. Application of the tools of risk management within a dynamic planning process, as described above, leads to identification of one or more strategies that meet the stated goals and requirements. Each strategy may consist of a variety of pollution reduction and pollution control initiatives, as well as other actions or modified behavior. The final steps in the planning process are to turn each strategy into a management plan, then to choose among alternative plans.

A management plan 1) describes a desired outcome as defined by the risk management strategy, 2) specifies the policies and actions necessary to achieve that outcome, and 3) assigns responsibility for implementing specific actions. Just as the process of risk management takes account of a wide range of possible actions and interventions in seeking the most appropriate strategy, it is important to consider all feasible management tools in devising management plans. But coastal wastewater management takes place in a fragmented, multi-institutional setting, where no one entity has the means or the mandate to set overall objectives or compel specific actions by others. A management plan in this situation takes the form of a set of tools that—directly or indirectly, through compulsion, incentive, motivation, or other means—cause necessary actions to be taken. The actions include various kinds of pollution prevention, pretreatment, wastewater reuse, treatment, and disposal practices.

Effective management plans do not rely on a single tool to achieve all required actions. Each kind of tool has unique advantages and disadvantages and may be more suited for one type of application than another. A comprehensive plan should incorporate a range of regulatory mechanisms and choose individual techniques to fit individual problems. The application of these tools can be overlapping—more than one policy can apply to a given activity. Command-and-control regulations, for example, might provide a *regulatory floor* that applies to all waste dischargers, while a combination of economic incentives, education initiatives, and other instruments of voluntary compliance are used to achieve progress beyond the command and control *floor*. The following sections outline the command and control approach as it is now practiced for wastewater management. Other kinds of regulatory tools are described and evaluated, using command and control as the basis of comparison.

No wastewater management strategy should be considered in isolation. An intricate network of laws, regulations, and policies is already in place and must be recognized in the development of any new policy. Improvements in the flexibility in these requirements would do much to facilitate effective management action. The same caution holds for existing institutions. Attempts should always be made to improve and rationalize institutions, but the end result of such efforts may fall short of the optimum. A management plan should, therefore, build on existing elements when it is reasonable to do so, superseding and rejecting only when such changes are imperative.

Tools for Management

Environmental quality in coastal areas is the result of numerous individual decisions by residential and business organizations, and governments regarding the use of resources and the discharge of pollutants. If environmental quality is to be protected, those individual decisions must reflect consideration of environmental consequences. One way to accomplish this is to set standards and compel individuals and organizations to observe them on threat of sanction. This command-and-control approach, the most common management strategy in the United States, is embodied in the Clean Water Act and other environmental legislation.

Many other management tools are available. They include economic incentives structured to induce voluntary behavior consistent with environmental quality goals; growth management, used to restrain human activity in sensitive coastal zones; education, which seeks to provide the basis for informed individual action; and financing mechanisms designed to facilitate the implementation of desired actions.

Economic incentives offer important advantages over the more familiar command-and-control instruments (see Appendix E). Properly designed and implemented, they tend to minimize the total cost of meeting environmental objectives. They promote technological innovation and improvement, both in pollution control and pollution prevention. Even though today's command-and-control instruments are more effective and flexible than their counterparts of twenty years ago, economic incentives may be designed to address sources beyond the reach of conventional regulation. However, experience to date suggests that economic instruments are often used most effectively in conjunction with command-and-control measures (Boland 1989).

Growth management is a comprehensive and integrative approach to planning for balancing resource protection and economic development, focused on implementing strategies. Growth management strategies offer some control over the total quantity of pollutants generated in sensitive environments. They also present opportunities for integration across environmental media. Education is another means of reducing pollution at the source. Carefully designed, interactive education programs can produce significant long-range reductions in the quantity and toxicity of materials discharged to the environment. A more detailed discussion of education initiatives is contained in Appendix E.

Wastewater management programs can be financed in numerous ways. These ways include the use of general tax revenue, dedicated tax revenue, user charges, intergovernmental transfers, and debt (long term or short term). Some of the characteristics of particular methods can be regarded as purely financial, such as the ability to provide sufficient revenue, the stability of the revenue stream, or the associated administrative details. Other properties of financing methods have important implications for wastewater management (see Appendix E for more details).

Differences in financing methods result in differences in the total cost burden imposed by wastewater management. More importantly, different financing methods produce very different incidences of cost—across sectors of the economy, across political jurisdictions, and across periods of time. The choice of financing method affects incentives for efficient management, as well. A more detailed presentation of these issues is contained in Appendix E.

The ultimate importance of these considerations derives from the willingness of the public, and various sectors of the public, to pay for improved wastewater management. If the cost of proposed programs is too high, or perceived as unfairly allocated, public and political support will be eroded or lost.

Selection

The planning process described in this report is intended to produce a number of competing management plans. These arise in part from the existence of alternative strategies for wastewater management (different assessments of relative risk, different allocations of abatement effort, etc.). Alternatives may also reflect consideration of different kinds and types of management tools. Each alternative plan has a unique set of likely consequences. While all alternatives are designed to protect the fundamental functions of the ecosystem, they may do so in somewhat different ways. All alternatives are intended to maintain important human uses of the resource, yet they may not all result in exactly the same set of protected uses. Economic costs will differ, as will such critically important characteristics as impacts on the distribution of income, political and social acceptability, etc. Some plans may be flexible and easily modified, while others are slow to adapt to changing circumstances. Some may promote innovation and individual initiative, while others lock in existing methods and technology. For these reasons and more, selection of the appropriate plan is only partly the responsibility of *experts*.

It is the task of planners and analysts to insure that all plans meet the following tests:

Adequacy—Each plan must satisfy the primary goals of wastewater management: to protect the fundamental functions and biological richness of the ecosystem and to maintain important human uses.

Integration—Each alternative must consider the full range of human activities linked to wastewater generation, as well as the full range of environmental effects resulting from wastewater management actions. All of this should be undertaken for a geographic area large enough to minimize important inter-area impacts.

Comprehensiveness—In developing the final plan, it is essential that all significant alternatives be considered, with respect to both control strategy, management tools, and costs. This consideration should encompass, among other things, alternatives that illustrate selected tradeoffs among the objectives (i.e., improved ecosystem protection at the expense of human activities, and increased human use at the expense of ecosystem protection).

Non-inferiority—No alternative should be put forward for selection that is, in all important respects, inferior to some other alternative. It follows, then, that each final candidate is preferred to some other candidates in one or more respects. The final set of alternatives are, in this sense, the best of all possible alternatives.

Properly applied, the planning process should give rise to novel and innovative solutions, as well as conventional and familiar approaches. Each will have various advantages and disadvantages. As wastewater management requirements become more severe, it can be expected that unconventional regulatory techniques will come to seem more conventional. Whether conventional or unconventional, each plan that meets the criteria must be fully described together with its expected consequences so that political leadership, regulatory agencies, and the public can make responsible choices.

MONITORING, INFORMATION MANAGEMENT, AND RESEARCH

The information developed through research, monitoring, and data management activities drives the planning process and keeps it dynamic. Information on the status of the coastal environment is obtained through monitoring. Data management allows for effective use of monitoring and research information. Research provides knowledge about the significance of various environmental changes, importance of different impacts, and potential for various technologies and management controls to be effective in mitigating impacts and protecting the environment. Research activities can also contribute new insights to the understanding of human expectations for the coastal environment and the economic impacts of alternative management strategies.

One important strength of the ICM process is that it is responsive to new information inputs and can respond to trends in coastal problems or new scientific findings with less delay than required by federal statutory changes and regulations. The linkage between the planning process and research, monitoring, and data management activities is critical to the success of a continuing, iterative ICM program.

Monitoring

The National Research Council report *Managing Troubled Waters: The Role of Marine Environmental Monitoring* defines monitoring in the marine environment as "a range of activities needed to provide management information about environmental conditions or contaminants." Monitoring is generally conducted to gather information about compliance with regulations and permit requirements, model verification, and trends (NRC 1990). The report concluded that monitoring can strengthen environmental management in several ways: 1) defining the extent and severity of problems, evaluating actions, and detecting emerging problems; 2) when coupled with research and predictive modeling, supporting integrated decision making; and 3) guiding the setting of priorities for management programs. It also concluded that comprehensive monitoring of regional and national trends was needed to better assess the extent of pollution problems and address broader public concerns (NRC 1990).

Since the release of that report, the U.S. Environmental Protection Agency has begun to implement a national and regional monitoring program called the Environmental Monitoring and Assessment Program (EMAP). EMAP will tend to provide a framework and a set of uniform data and quality objectives for regional programs. However, there is region-specific monitoring needed in many areas to examine identified issues, close data gaps found in the planning process, and monitor the performance of risk management strategies chosen.

In addition to the recent work of the EPA, the National Oceanic and Atmospheric Administration (NOAA) has had a significant marine monitoring program in place since 1984. NOAA's National Status and Trends (NS&T) Program monitors for selected metals and organic compounds in sediments and benthic organisms at nearly 300 coastal locations in the United States. The fundamental objective of the NS&T Program is to assess long-term trends in the concentrations of these toxic materials.

An effective ICM system requires a monitoring approach significantly different from approaches used in the past. As the National Research Council found in 1990,

> . . . monitoring designed principally to meet regulatory compliance needs generally does not adequately answer questions about the regional and national risks of pollutant inputs to public health, coastal environmental quality, or living resources. The reason is that compliance monitoring typically does not address potential effects removed from specific discharge points, including overall responses of the ecosystem to anthropogenic and natural stresses (NRC 1990).

Data analyses should be available in forms that not only address the current concerns but also allow for identification of new trends. In this manner, monitoring becomes the vehicle through which the process is accountable to the public and useful to practitioners.

Reliance on technology-based standards has led to universal monitoring of effluent parameters such as total suspended solids, BOD, and chemical oxygen demand, while far field and specific effects remain unexamined in many cases. For ecosystem effects, most useful information has been generated by special research projects in specific areas, with hardly any contribution from routine compliance-driven monitoring programs. For health effects, there is monitoring for pathogens and for toxic chemicals. These monitoring programs are designed to enhance risk management strategies that may include posting warnings, fishing limits, and shellfish bed and beach closure strategies, as well as to examine long-term trends.

Information Management

The ICM process is designed to make the fullest use possible of information relating to coastal systems and their management. It is an information intensive process that requires effective data and information management. Information and data should be collected and maintained in forms that are accessible to users and compatible with other data in the system. All too often monitoring data are collected and stored, but in forms that are of relatively little use for purposes of analysis (NRC 1990).

The results of scientific analyses should be presented in *user friendly* formats to all parties involved in the process ranging from scientists and engineers to politicians and the general public. Advances in computing technologies now allow for the display of information in a wide variety of graphical and pictorial formats. Monitoring information and modeling results can be manipulated in a variety of innovative ways to display information about coastal systems including three dimensional presentations of modeling scenarios.

Research

Areas where additional information in needed should be identified in the course of the dynamic planning process. For example, in the course of assessing and comparing risks, it should become clear where there are important uncertainties and data gaps. Research aimed at reducing these uncertainties would improve understanding and provide insight on how to manage the risks. The survey of anglers now under way to develop better estimates of fish consumption in Santa Monica Bay is a good example of a research effort aimed at refining information on risks. In developing management options, research to determine the potential efficacy of various management and control measures will be needed. The need for information should drive the areas of research targeted in an ICM program.

In addition to the specific kinds of research identified in the course of the ICM process, there is an array of more basic research needs in areas where greater understanding is required for fundamental understanding of environmental processes, ecological systems, human health, technological and engineering issues, and economic and policy effects. More specific research needs are identified in the discussion of relevant topics throughout this report and its appendices.

SUMMARY

This chapter has provided a description of how the ICM process should be applied to coastal areas. Several examples describe cases where some ICM concepts are being applied already.

The complexity of the ICM process reflects that of coastal systems including the physical and ecological interactions in the coastal zone and the human expectations and actions that affect them. ICM involves an iterative and dynamic planning process in which scientific information, engineering expertise, economic analyses, and public participation are combined to identify management alternatives that meet coastal objectives. The selection and implementation step of the ICM process is structured to select the most appropriate management plan and develop adequate institutional arrangements to ensure effective implementation. Research, monitoring, and data management are the activities that provide feedback on how well management controls are working, bring new information into the planning process, and further knowledge about coastal systems.

The following chapter, Benefits, Barriers, Solutions, and Implementation, examines issues relating to the application of ICM to urban coastal areas in the United States and provides recommendations for implementing ICM.

REFERENCES

Boland, J.J. 1989. Environmental Control Through Economic Incentive: A Survey of Recent Experience. Presented at the Prince Bertil Symposium on Economic Instruments in Environmental Control, Stockholm School of Economics, Stockholm, Sweden, June 12-14.

Brooks, N.H. 1983. Evaluation of key issues and alternative strategies. Pp. 709-759 in Ocean Disposal of Municipal Wastewater: Impacts on the Coastal Environment, E.P. Myers and E.T. Haring, eds. MITSG 83-33. Massachusetts: MIT Sea Grant College Program.

Brooks, N.H. 1988. The Experience with Ocean Outfalls to Dispose of Primary and Secondary Treated Sewage. Freeman Lecture, Boston Society of Civil Engineers, April 1988. Presented at Massachusetts Institute of Technology.

Cabelli, V.J., A.P. Dufour, L.J. McCabe, and M.A. Levin. 1983. A marine recreational water quality criterion consistent with indicator concepts and risk analysis. Journal of the Water Pollution Control Federation 55:1306-1314.

Capuzzo, J.M. 1981. Predicting Pollution Effects in the Marine Environment. Oceanus 24(1):25-33.

Carry, C.W., and J.R. Redner. 1970. Pesticides and Heavy Metals. Whittier, California: County Sanitation Districts of Los Angeles County.

CBP (Chesapeake Bay Program). 1992. Progress Report of the Baywide Nutrient Reduction Reevaluation. Washington, D.C.: U.S. Environmental Protection Agency.

Cosper, E.M., C. Lee, and E.J. Carpenter. 1990. Novel "brown tide" bloom in Long Island embayments: A search for the causes. In Toxic Marine Phytoplankton, E. Graneli, B. Sundsttrom, L. Edler, and D.M. Anderson, eds. New York: Elsevier.

CSWRCB (California State Water Resources Control Board). 1990. California Ocean Plan, Water Quality Control Plan for Ocean Waters of California. Sacramento, California: CSWRCB.

Doering, P.H., C.A. Oviatt, L.L. Beatty, V.F. Banzon, R. Rice, S.P. Kelly, B.K. Sulivan, and J.B. Frithsen. 1989. Structure and Function in a Model Coastal Ecosystem: Silicon, the Benthos and Eutrophication. Mar. Ecol. Prog. Ser. 52:287-299.

Dubos, R. 1965. Man Adapting. New Haven, Connecticut: Yale University Press.

EPA (U.S. Environmental Protection Agency). 1986. Ambient Water Quality Criteria Document for Bacteria. EPA A440/5-84-002. Washington, D.C.: U.S. Environmental Protection Agency.

EPA (U.S. Environmental Protection Agency). 1989. Sediment Classification Methods Compendium, Draft Final Report. U.S. Environmental Protection Agency, Watershed Protection Division.

EPA (U.S. Environmental Protection Agency). 1990. Reducing Risk: Setting Priorities and Strategies for Environmental Protection. Washington, D.C.: Science Advisory Board.

EPA (U.S. Environmental Protection Agency). 1991. Ecological Risk Assessment Guidelines, Strategic Planning Workshop. Miami, Florida. April 30 - May 2, 1991.

EPA (U.S. Environmental Protection Agency). 1992a. Environmental Equity: Reducing Risk for all Communities, Vol. 1: Workgroup Report to the Administrator. EPA230-R-92-008. Washington, D.C.: U.S. EPA, Policy, Planning, and Evaluation.

EPA (U.S. Environmental Protection Agency). 1992b. Environmental Equity: Reducing Risk for All Communities, Vol. 2: Supporting Document. EPA230-R-92-008A. Washington, D.C.: U.S. EPA, Policy, Planning, and Evaluation.

Fleisher, J.M. 1991. A reanalysis of data supporting U.S. federal bacteriological water quality criteria governing marine recreational waters. Research Journal of the Water Pollution Control Federation 63:259-265.

Gibb, H., and C. Chen. 1989. Is Inhaled Arsenic Carcinogenic for Sites Other than the Lung? PB 90-130675. Washington, D.C.: U.S. EPA, Office of Health and Environmenal Assessment.

Goyal, S.M., W.N. Adams, M.L. O'Malley, and D.W. Lear. 1984. Human pathogenic viruses at sewage sludge disposal sites in the middle Atlantic region. Appl. Environ. Microbiol. 48:758-763.

Graneli, E., H. Persson, and L. Edler. 1986. Connection between trace metals, chelators, and red tide blooms in the Laholm Bay, SE Kattagat—An experimental approach. Mar. Env. Res. 18:61-78.

Howarth, R.W. 1988. Nutrient limitation of net primary production in marine ecosystems. Annual Review of Ecology & Systematics 19:89-110.

Howarth, R.W. 1991. Comparative response of aquatic ecosystems to toxic chemical stress. Pp. 169-195 in Comparative Analysis of Ecosystems: Patterns, Mechanisms, and Theories, J.J. Cole, G. Lovett, and S. Findlay, eds. New York: Springer Verlag.

Huntzicker, J.J., S.K. Friedlander, and C.I. Davidson. 1975. A material balance for automobile emitted lead in the Los Angeles basin. Pp. 33-77 in Air-Water-Land Relationships for Selected Pollutants in Southern California, final report to the Rockefeller Foundation, J.J. Huntzicker, S.K. Friedlander, and C.I. Davidson, eds. Pasadena, California: W.M. Keck Laboratories, California Institute of Technology.

IRIS (Integrated Risk Information System). 1993. Cadmium. CASRN 7440-43-9. U.S. Environmental Protection Agency.

Jackson, J.B.C., J.D. Cubit, B.D. Keller, V. Batista, K. Burns, H.M. Caffey, R.L. Caldwell, S.D. Garrily, C.D. Getter, C. Gonzalelz, H.M. Guzman, K.W. Kaufmann, A.H. Knap, S.C. Levings, M.J. Marholl, R. Steger, R.C. Thompson, and E. Weil. 1989. Ecological effects of a major spill on Panamanian coastal marine communities. Science 243:37-44.

Koh, R.C.Y., and N.H. Brooks. 1975. Fluid mechanics of waste-water disposal in the ocean. Ann. Rev. Fluid Mechanics 7:187-211.

LA/OMA. 1980. Los Angeles/Orange County Metropolitan Area Regional Wastewater Solids Management Program. Final Facilities Plan/Program and Summary of Final EIS/EIR. Whittier, California.

Life Systems, Inc. 1989. Toxicological Profile for Cadmium. Oak Ridge, Tennessee: Agency for Toxic Substances and Disease Registry, Oak Ridge National Laboratory.

MBC-AES (MBC Applied Environmental Sciences). 1988. The State of Santa Monica Bay: Part One: Assessment of Conditions and Pollution Impacts. Los Angeles, California: Southern California Association of Governments.

Mearns, A.J., and J.Q. Word. 1982. Forecasting the effects of sewage solids on marine benthic communities. Pp. 495-512 in Ecological Stress in the New York Bight; Science and Management, G.F. Mayer, ed. Columbia, South Carolina: Estuarine Research Federation.

Mearns, A.J., M. Matta, G. Shigenaka, D. MacDonald, M. Buchman, H. Harris, J. Golas, and G. Lavenstein. 1991. Contaminant Trends in the Southern California Bight: Inventory and Assessment. NOAA Technical Memo NOS ORCA 62. Seattle, Washington: National Oceanic and Atmospheric Administration.

Melnick, J.L., and C.P. Gerba. 1980. The ecology of enteroviruses in natural waters. CRC Crit. Rev. Environ. Control 10:65-93.

Nemura, A., and E. Pontikakis-Coyne. 1991. Water Quality Benefits of CSO Abatement in the Tidal Anacostia River. Washington, D.C.: Metropolitan Washington Council of Governments.

Nixon, S.W., C. Oviatt, J. Frithsen, and B. Sullivan. 1986. Nutrients and the productivity of coastal marine ecosystems. J. Limnol. Soc. S. Africa 12:43-71.

NRC (National Research Council). 1983. Risk Assessment in the Federal Government: Managing the Process. Washington, D.C.: National Academy Press.

NRC (National Research Council). 1989a. Improving Risk Communication. Washington, D.C.: National Academy Press.

NRC (National Research Council). 1989b. Contaminated Sediments. Washington, D.C.: National Academy Press.

NRC (National Research Council). 1990. Managing Troubled Waters: The Role of Marine Environmental Monitoring. Washington, D.C.: National Academy Press.

NRC (National Research Council). 1993. Issues in Risk Assessment. Washington, D.C.: National Academy Press.

Officer, C.B., and J.H. Ryther. 1980. The possible importance of silicon in marine eutrophication. Mar. Ecol. Prog. Ser. 3:83-91.

O'Shea, M.L., and R. Field. 1992. Detection and disinfection of pathogens in storm-generated flows. Can. J. Microbiol. 38:267-276.

Phillips, P.T. 1988. California State Mussel Watch Ten Year Data Summary 1977-1987. Water Quality Monitoring Report No. 87-3. Sacramento, California: State Water Resources Control Board.

Pollock, G.A., I.J. Uhaa, A.M. Fan, J.A. Wisniewslei, and I. Withenell. 1991. A Study of Chemical Contamination of Marine Fish of Southern California. II. Comprehensive Study. Sacramento, California: Office of Environmental Health Hazard Assessment, California Environmental Protection Agency.

Rand, G.M., and S.R. Petrocelli (eds). 1985. Fundamentals of Aquatic Toxicology Methods and Applications. Washington, D.C.: Hemisphere Publishing Corporation. Pp. 651-657.

Rose, J.B., and C.P. Gerba. 1991. Use of risk assessment for development of microbial standards. Water Science Technology 24:29-34.

Schindler, D.W. 1987. Determining ecosystem responses to anthropogenic stress. Can. J. Fish. Aquat. Sci. 44:6-25.

Smayda, T.J. 1989. Primary production and the global epidemic of phytoplankton blooms in the sea: A linkage? Pp. 449-483 in Novel Phytoplankton Blooms: Causes and Impacts of Recurrent Brown Tides and Other Unusual Blooms. Lecture Notes on Coastal and Estuarine Studies, E.M. Cosper, E.J. Carpenter, and V.M. Bricelj, eds. Berlin: Springer-Verlag.

SMBRP (Santa Monica Bay Restoration Project). 1992. Progress Update: Santa Monica Bay Restoration Project 1991-92. Monterey Park, California: Santa Monica Bay Restoration Project.

SMBRP (Santa Monica Bay Restoration Project). 1993. Santa Monica Bay Characterization Report. Monterey Park, California: Santa Monica Bay Restoration Project.

Van Dover, C.L., J.F. Grassle, B. Fry, R.H. Garritt, and V.R. Starczak. 1992. Stable isotope evidence for entry of sewage-derived organic material into a deep-sea food web. Nature 360:153-155.

WHO (World Health Organization). 1948. Preamble, Constitution. Geneva: World Health Organization.

Younger, L.K., and K. Hodge. 1992. 1991 International Coastal Cleanup Results. R. Bierce and K. O'Hara, eds. Washington, D.C.: Center for Marine Conservation.

5

Benefits, Barriers, Solutions, and Implementation

INTRODUCTION

Prescriptive regulations derived from nationally applied technology-based performance standards have been the primary vehicle for achieving clean water objectives over the last twenty years. For industrial dischargers, specific requirements were developed for groups of industries, that tended to standardize treatment technology within each group. In the case of municipal wastewater dischargers, the performance standards include a minimum requirement of full secondary treatment for all but a small number of publicly owned treatment works (POTWs). The cost of constructing these treatment plants has been the single largest component of federal, state, and local expenditures for clean water over this time.

Targeting municipal wastewater plant sources over the last twenty years made considerable practical sense. First, the municipalities had the institutional framework in place to expend large amounts of public funds to construct the needed infrastructure. The issues of fiscal management, professional services procurement, contract bidding, construction management, and facility operations were familiar subjects to most local governments. Not unlike the transportation, water supply, and flood control construction programs preceding them, wastewater construction could be managed by municipalities.

Second, many municipal dischargers were major sources of poorly managed sanitary sewage and industrial waste. These discharges were, in turn, responsible for significant local and regional water pollution.

Third, because these effects were so visible and reversible, it was obvious to Congress that the return on investment from an environmental, public, and political viewpoint would be great. The costs, though considerable, would be met with widespread public support. Beaches and fisheries could be improved. The public would see results—dramatic results, it was hoped—that would give this new public policy tangible value.

Finally, even though understanding of water pollution causes and effects was somewhat limited, it was time to get on with the task of improving the country's waters. Previous attempts by the federal government, beginning with the Federal Water Pollution Control Act of 1948 and the uneven and sometimes half-hearted attempts of the states to improve water quality in the 1950s and 1960s, had left a patchwork quilt of water quality problems. Directing massive amounts of federal, state, and local dollars to the problem, though arguably inefficient, would result in significant improvements in the most adversely impacted water bodies of the country. Thus Congress converted the high hopes of "fishable and swimmable"[1] waters into a national mandate and a six-year plan.

Undercontrol and Overcontrol

As sensible as the strategy appeared in 1972, and as understandable as it looks today, its goals have not been universally achieved. For example, it appears that the act has led to frequent instances of *undercontrol* and *overcontrol*. Undercontrol occurs when mandated levels of treatment do not or are not expected to meet water quality goals. Although the act permits the Environmental Protection Agency to set more stringent standards in cases of undercontrol, solving these problems has proved far more daunting than the relatively simple task of implementing uniform technology-based regulations. In fact, significant portions of the coastal zone remain adversely affected and still short of being fishable and swimmable.

Available evidence suggests cases of overcontrol as well, but the Clean Water Act has no present cure for this problem. Overcontrol means that some part of mandated treatment is not considered justified by the resulting environmental improvement. This implies that spending some of the money elsewhere could produce a larger overall environmental improvement. Identification of overcontrol requires careful analysis of incremental ecosystem effects as well as value judgements on social, and economic effects, e.g., are the environmental quality improvements too small? are social goals still satisfied? is the cost too great? Ideally, these judgements should be

[1]One of the goals declared by Congress is to achieve ". . .water quality which provides for the protection and propagation of fish, shellfish, and wildlife and provides for recreation in and on the water be achieved. . ." (Federal Water Pollution Control Act, as amended, Section 101[a][2]).

made in the context of an integrated planning process, such as Integrated Coastal Management (ICM).

Overcontrol is particularly likely along the ocean coasts, where dispersion of partially treated wastes in deep ocean water may produce few, if any, adverse effects. Nevertheless, full secondary treatment is required in every such case, regardless of cost or lack of benefits. For a time, such issues could be addressed in the waiver process provided in Section 301(h) of the Clean Water Act, but the provision to apply for an initial waiver expired at the end of 1982. Dischargers who have a waiver can request a renewal. Some initial applications are still pending.

An example of undercontrol is provided by the case of Long Island Sound. Despite major improvements in treatment of municipal wastewater, the sound continues to experience hypoxia in the summer because of marine algae blooms caused by nitrogen enrichment. Contributions of nitrogen to the sound come from varied sources. While an estimated 28 percent of the nitrogen input comes from sewage treatment plants discharging directly into the sound, the remaining 72 percent comes from widespread point and nonpoint sources, some of which are upstream of the sound (LISS 1990).

Those sewage treatment plants discharging to the sound and not yet at full secondary treatment are taking action to complete upgrades to full secondary treatment and some are adding nitrogen removal technology. Nevertheless, under the most optimistic projections, plants with nitrogen removal capability will control only three-quarters of their current nitrogen input and only one-fifth of the total input to the sound. Therefore, in the absence of controls for the other nitrogen sources, the hypoxia problems, while decreased, will not be eliminated by the future actions of the POTWs discharging directly into the sound. An integrated nitrogen management plan that manages all sources is envisioned in the Long Island Sound Study (LISS 1990). Such a plan must be fully implemented to solve the hypoxia problem.

Long Island Sound is, therefore, a case of good news and bad news. The bad news is that the original national technology-based, end-of-pipe approach of the 1972 Clean Water Act will leave the hypoxia problems of the sound unresolved. The good news is that the trend toward future collaborative efforts of the states, counties, and cities to address nitrogen enrichment holistically indicate a great potential for success. The promise of an integrated approach to solving the hypoxia problem of Long Island Sound provides a model for a national approach to water quality objectives during the third decade of the Clean Water Act.

Other cases of undercontrol can be found throughout the country. Overcontrol is less easy to demonstrate, since the necessary value judgements have usually not been made. Also, water quality conditions in the absence of some part of existing or improved treatment are sometimes not known with any great confidence. Questions have been raised in the press and elsewhere, about the need for massive investment in secondary treatment for the Boston metropolitan area, or the similarly large investment by San Diego in biological secondary treatment, to replace chemically enhanced primary treatment. At issue are not only the appropriate levels of treatment but how to set priorities for addressing related problems, such as combined sewer overflows in Boston and water reclamation and stormwater discharges in San Diego.

BENEFITS

Chapters 3 and 4 describe a proposal for managing wastewater in coastal areas within the framework of integrated coastal management. A dynamic planning process is described, that permits management strategies and the means for implementing them to be selected from among a wide range of alternatives. The approach is fundamentally different from the command-and-control, technology-based strategy of the Clean Water Act. It is comprehensive rather than single-issue in nature; it expands the range of alternative strategies rather than promoting uniform responses; and it gives equal attention to industrial pretreatment and pollution prevention, rather than focusing on end-of-pipe measures. This approach would allow regulators to move beyond the limitations and inefficiencies of current practice without sacrificing any of the accomplishments of the Clean Water Act. Some specific benefits of integrated coastal management are discussed below.

Clear Goals

Wastewater management has two overall objectives: to restore and maintain the integrity of coastal ecosystems and to maintain important human uses of the coastal resource. The first step of the process is to transform these general objectives into specific goals that reflect the most important conflicts and values in the planning area. Since all problems will never be solved, priorities must be developed and accepted. The resulting ecosystem-specific goals constitute a clear and unambiguous statement of the purpose of wastewater management while providing a coherent set of criteria against which to measure progress.

Improved Ability to Achieve Objectives

Under integrated coastal management, all management actions would be devised in order to meet the stated goals. Put another way, regulation is water- and sediment-quality driven, not technology driven. It is also ecosystem-specific, not standardized across the country.

Goals are translated into water, sediment, and environmental quality requirements, then management actions are designed to meet those requirements. This approach would make effective use of existing scientific knowledge and engineering skills while facilitating the incorporation of new knowledge as it becomes available. Moreover, actions that fail to perform could be quickly identified and corrected. Since planning and regulation would be carried out on an ecosystem-specific basis, the likelihood of meeting goals—and of avoiding damaging undercontrol as well as costly overcontrol—would be significantly better than under the existing policy.

Cost-Effective Solutions

Current regulatory practice provides only limited scope for minimizing the cost of compliance. In most important respects, cost-minimizing behavior is discouraged or prohibited. Dischargers are regulated individually, using the same standards, regardless of differences in abatement cost. Standardized end-of-pipe treatment is required regardless of opportunities for pollution reduction elsewhere in the system. Regulatory actions, even when they achieve wastewater management objectives, may often impose higher than necessary costs on government and industry. Excessive costs, in turn, slow environmental progress and divert funds from other important activities.

Integrated coastal management, on the other hand, would imply considerable flexibility in the way in which management objectives are met. New and innovative solutions would be encouraged, as would alternative forms of familiar approaches. Cost-effectiveness would be an important consideration in selecting the final management approaches for preventing a pollutant from entering the waste stream. If economic incentives could be used to reallocate treatment requirements to achieve a lower cost while still meeting the management objectives, then that strategy would be preferable to uniform discharge standards.

Improved Local Support and Commitment

One of the most striking features of the integrated coastal management proposal presented here is its high degree of dependence on local initiative. The necessary planning and decision making can only be accomplished by local organizations in a local setting. Federal agencies lack the information and the authority to do this type of planning, or to make the required choices. Integrated coastal management can only happen where local and regional leaders make a conscious decision to seek more effective and efficient wastewater management. Integrated coastal management would shift the culture of water quality protection from federal mandate to local and state empowerment and responsibility.

The advantage of encouraging local initiative is that it produces competent local and regional agencies fully committed to the effective management of wastewater. If the planning process has been properly conducted, businesses, community leaders, and many ordinary citizens will be aware of what is being done and why it is necessary and desirable to do it in this way. Even where the selected strategies are innovative or require some level of voluntary compliance or cooperation, widespread support can be expected.

BARRIERS AND SOLUTIONS

Barriers to an integrated management approach can be found everywhere. A persistent lack of open communication among federal agencies, states, local governments, and private stakeholders contributes to an air of distrust and limited cooperation. Leaving the inefficiencies of the status quo for the uncertainties and risks of a new approach may be difficult for all and nearly impossible for some. Suspicion regarding the motives of others may make the development of an integrated coastal management strategy difficult to accomplish. Consensus building, commitment, and sign-off may require finesse, time, patience, and risk. If a region is unwilling to approach integrated management with its collective eyes open to this reality, then successful adoption will be elusive.

Integrated coastal management requires answers to a number of questions:

- What are the boundaries of an integrated coastal management region?
- What is the authority of the various agencies involved in the region?
- What are the objectives and approaches of an integrated management plan?
- Who will implement and pay for the management plan?
- Who will monitor progress toward achieving the objectives of the management plan?
- Who will enforce the commitments enumerated in the management plan?
- What contingencies will be provided for?

In each application, the answers to these questions will determine both the benefits achieved from integrated management and the barriers to its implementation.

Definition of the Coastal Management Region

Definition of a coastal management region requires a compromise between contradictory criteria: the definition should be comprehensive, holistic, and all-encompassing on the one hand, but also comprehensible, manageable, and focused on the other. Boundary lines implied by geographical, hydrological, and ecological factors may bear little resemblance to political jurisdictional limits. A river, for example, may serve as an excellent political boundary but make little sense as a dividing line between two coastal reaches that are ecologically interrelated. Adherence to a strict ecological definition of regions may place two political jurisdictions in the same management area, making the solution of their conflicting interests a continuing challenge.

All authorities and all sources (including those now unregulated or underregulated) are subject to inclusion in a management plan. With all relevant entities identified and involved, those responsible for overseeing the management plan would have more complete information and opportunity for addressing the needs of the region. Monitoring and analytical resources could be broadly applied to a coastal region to identify and evaluate various sources of stressors. For this reason, it may be difficult to develop consensus on boundaries. The inclusiveness of the process will have tremendous implications for government jurisdictions and private stakeholders. Pressures to be included or excluded may be significant.

Assignment of Authority

Resistance to an integrated management approach may begin with the perception that it cannot possibly work. Previous regional water-quality planning failures, including some of those mandated under Section 208 of the 1972 act, may be pointed to as examples of how difficult it is to produce working regional consensus and action. Completed in the late 1970s, most Section 208 regional wastewater plans were not implemented. Inaction occurred for a number of reasons. First, the 208 program was conducted in parallel with a very aggressive POTW construction program, which meant that action often preceded planning. Thus, local government had little motivation to devote effort to 208 planning since resulting recommendations were likely to become moot. Second, no federal funding was provided to support the staff work needed to maintain and update the plans, and state and local agencies had no motivation to provide such funding. Third, some of the lead agencies responsible for creating and approving 208 plans were not vested with the authority to monitor or enforce the plans' provisions. Most such agencies (e.g., regional land-use planning agencies) had few or no environmental responsibilities beyond the completion of the 208 plan. In the absence of an authority to monitor and enforce the plans, they had no caretaker and agencies soon became irrelevant to ongoing planning, construction, and enforcement activities in the regions.

As described earlier, integrated coastal management is a process in which each step is dynamic and iterative. Implementation at the regional scale requires appropriate institutionalization. While centralized regional agencies appear, in concept, to be the ideal locations for integrated management, they may not be the best alternatives. In many areas, existing arrangements, intergovernmental relations, or other situations will preclude the establishment or the use of such an organization. It is important to recognize that many different multiple organizational and interorganizational forms can be effective. An institutional model that is successful in one region may not be appropriate for another.

For integrated coastal management to work effectively, the implementing institution(s) must be vested with sufficient responsibility, resources, and authority to be a viable force within a region. One approach might be to start with consolidating various governmental expectations into a single voice by creating a commission, joint powers agreement, or other organizational vehicle that would serve as the focal point for developing and implementing regional policy. Such a regional authority, not necessarily residing in a single agency or even a single jurisdiction, could provide the opportunity to develop cooperation between relevant international, federal, state, and local agencies. Existing resources could be combined and leveraged to maximize efficiency.

To be effective, the governmental authorities in the region would require sufficient scientific information to assess the condition of their coastal region; sufficient public input to assess the human and economic needs of the region; and sufficient empowerment to approve a plan, implement it, and then monitor and enforce it. Today's fragmented regulatory and jurisdictional setting makes planning and implementation difficult to achieve. Existing governmental mandates and charters may require modifications to allow for an integrated approach to coastal management. Some government entities that now have jurisdiction in a region may have to either divest themselves of some of their authority in favor of a regional body or participate as equal partners seeking consensus and moving forward with action based on the consensus.

Sharing those environmental and public health mandates now held exclusively by state and federal agencies with regional and local government may be difficult to achieve. Congress and federal agencies will need to be convinced that state and local government can make appropriate decisions and implement needed changes. A regional approach could provide the responsiveness necessary to use the most effective approaches for achieving site-specific management objectives of a region.

Goals and Approaches

A well-selected list of management goals will be a key to success. However, achieving consensus on the measurable goals of an integrated coastal plan may be difficult. Government agencies, environmental interest groups, industrial and agricultural trade associations, private stakeholders, taxpayer groups, and the general public will have conflicting objectives. Existing government jurisdictions and mandates may be in conflict with one another. Reconciling contradictory agency objectives is difficult. For this reason, consensus building is an important part of the selection of objectives.

If the goals are not sufficiently comprehensive, then the solutions to environmental problems may not be adequate. On the other hand, there may be concern that a comprehensive approach will cause the process to become overextended and unfocused. If the scope is too large there is a danger of trivializing everything and accomplishing little.

The approach to establishing objectives would begin with the assumption that scientifically-based assessments of environmental and public health status and trends would provide the best information for managing the coastal region and achieving long-term resource protection. Scientific research, environmental monitoring, and technical analysis are raised to a position of importance, and the prevailing conditions of the region can be assessed to provide the real problems and real priorities.

Barriers may be encountered here, however. Existing environmental and public health information may be inadequate, poorly integrated, or irrelevant to the questions posed during the planning process. Assembling available data into a form that can be assimilated into the process and understood by the public and policy makers will be a challenge. Developing consensus on the meaning and relevance of available data may be difficult.

In the absence of consensus, it is likely that some parties will object to the goals and approaches chosen. It is important, therefore, to provide an administrative appeal process, similar to those used by other local and regional programs of similar scope, such as land use planning. Those who still feel aggrieved would ultimately have access to the courts.

If these problems can be addressed, risk assessment and risk management can be used to incorporate cross-media and cross-program environmental and public health considerations into a management plan. Comparisons of various management options and the associated risks can facilitate the identification of the best overall solutions. A balanced approach that would maximize the overall environmental and public health objectives while restraining infrastructure costs, environmental impact, and energy usage would be sought.

Integrated coastal management planning could identify the most effective tactical and strategic tools to address the problems unique to a region. Best management practices for POTWs, combined sewer overflows, stormwater, and nonpoint sources could be incorporated into a management plan, and the opportunity to determine the affordability of potential management practices or engineering solutions could help establish the most practical approach.

Most important, a broad scientific and technical understanding of the issues of concern could enhance the region's opportunity to identify nondeleterious multiple uses. Local responsiveness could open up the process to early innovation and proactive solutions.

Finally, a regional plan could encourage more public awareness, involvement, and support. Popular support could be mustered to develop the political will to do the right thing. The public is the most important component in making the ideals of an integrated approach a reality.

Plan Implementation

The development of a viable integrated coastal management plan will take some time. This time will be perceived by some as a way to delay action, confuse objectives, and avoid responsibility. Timely implementation will be difficult to achieve if the scope or cost of an integrated plan is too large or is simply very comprehensive. Staging plans and related actions in incremental segments may be necessary to expedite implementation. An evolutionary approach that starts small and becomes large may be helpful. Setting priorities and staging early actions according to ease of implementation or urgency of need would be a logical approach.

Implementation may be difficult to achieve if the interests of the stakeholders in the region are affected adversely. Providing incentives (or the opportunity to avoid disincentives) may be a strategic measure to aid implementation. It may be unrealistic to assume that costly management or structural changes required of a local government entity or a private stakeholder will be affordable. Funding alternatives that include tax incentives, government loans, low-interest bonding alternatives, and other forms of innovative financing may be a necessary element of a successful integrated management plan.

Managing coastal zones to achieve environmental and public health objectives may require significant changes in the management of land use. However, local control issues and private land-use prerogatives may be in conflict with the objectives identified in a regional management plan. This will mean that divestiture of government control affects not only federal and state government but local government and private citizens as well. Without governmental commitment to striking new balances among competing and varied interests, the ability to implement a plan effectively will be limited.

Monitoring Progress

Monitoring of the performance of an implementation plan is an important ingredient in success. A reporting system that adequately informs governmental entities and other stakeholders will be necessary to communicate the status and trends of the coastal zone.

A baseline inventory of the conditions present in the region and a regularly scheduled status report could provide regulators with meaningful feedback on environmental and public health improvements. A system geared toward the display of environmental status information could replace today's dependence on benchmarks that use number of permits issued or number of enforcement actions as indices of progress. The original environmental quality objectives of the plan would be served best by demonstrating measurable and quantifiable results of the plan's implementation.

To help in measuring results, existing monitoring and assessment activities could be regionalized to save on costs and to best utilize the scientific resources of the region. If existing monitoring and assessment activities are limited, then regionalizing resources can help in the start-up of this component of an integrated plan.

Fulfilling Commitments

Enforcement of the commitments made in an implementation plan is another important factor in guaranteeing continuing progress. Public disclosure, arbitration, fees, fines, or other sanctions may have to be incorporated into a plan as enforcement options. These features must be clearly understood by all stakeholders before the plan is first implemented. Committing to the plan, knowing that others are committed to the plan, and fearing penalties if one does not deliver on the plan will provide momentum and pressure to move forward.

Contingencies

Population growth, economic development, and economic restructuring within the coastal zone may erode the gains made by an integrated plan. Unforeseen pressures caused by changing land use, market forces, and factors outside the boundary or charter of the region may be significant barriers to the short-term or long-term success of an integrated management plan. Therefore, the flexibility to modify the plan to account for unanticipated factors may be helpful. Localized control and flexibility in the plan could promote more rapid and effective rehabilitation of damaged ecosystems and response to unforeseen problems.

IMPLEMENTATION

Clearly, an argument can be made for proceeding on a new course to achieve environmental and economic objectives in a more effective and efficient way through the use of methods of integrated coastal management. The potential benefits are large. Unfortunately, the barriers may seem large as well. For those coastal regions that are compelled to make changes, the concept of integrated coastal management may be relatively attractive. Such regions are likely candidates for early application.

For those regions with little reason for change, or those that have little institutional motivation to change, an integrated approach may not be feasible. Strongly motivated governments or other regional entities, willing to commit sufficient time and energy, are of paramount importance. An outside suggestion or mandate to move forward with an integrated plan may have little chance of success if local will is absent. For these reasons, integrated management can be expected to have varying levels of success or require varying periods of time for implementation.

At least two models embodying some of the basic concepts of integrated coastal management are operating today. These are the programs for the Chesapeake Bay and the Great Lakes. Frameworks for limited forms of integrated management exist within the National Coastal Zone Management Act (CZMA), the National Estuaries Program, and other programs. These examples have had varying scopes and degrees of success and commitment from the various entities they encompass. Implementation of ICM in the context of the Clean Water Act would be a strategy allowing more effective and efficient achievement of the nation's clean water objectives. This should not be confused with the wider range of planning activities provided for under the CZMA, including the requirement that federal activities be consistent with state plans. Plans under the CZMA, for example, are more directed toward coordination of a broad range of land and water plans, while the Clean Water Act is more focused on water quality management per se. The fact is that few regions of the country have developed comprehensive, integrated, or coordinated management strategies. This prior experience illustrates both the potential for integrated wastewater management and the various barriers that may be encountered.

It is not immediately clear how the promise of integrated coastal management can be achieved within the strictures of existing legislation, institutions, and expectations. In fact, universal implementation will require dramatic and courageous changes in all of these things. Such changes may begin immediately but still require a number of years to complete. There is, however, much that can be done now, and assuming a commitment to better wastewater management, more can be done at each step of the way.

The following paragraphs discuss possible changes in federal legislation. Nothing in this section should be taken to imply an expanded federal role in ICM. In fact, the opposite is true. A successful application of ICM depends upon the initiative of local, regional, and state agencies, and on the willingness of the federal agencies to grant them the necessary autonomy and freedom of action. How this can be done within the web of detailed environmental laws and regulations now administered by the federal government is a key question. Although there is the opportunity to undertake some aspects of ICM under various provisions of the existing Clean Water Act, some changes would be required to allow for implementation of the full range of alternatives that may be identified through the ICM process. Additionally, legislative changes could provide a framework in which regionally-based ICM initiatives could be fostered. The purpose of this section is merely to suggest some federal actions that can be taken immediately to provide latitude for local governments as they seek to achieve the benefits of a more integrated approach. This discussion is followed by brief description of some longer term strategies that could more fundamentally change the governance of the coast, substituting flexibility and local initiative for rigidity and detailed federal control.

Immediate Actions

The National Estuary Program

The National Estuary Program, administered by the Environmental Protection Agency pursuant to section 320 of the Clean Water Act, provides an excellent opportunity to practice integrated coastal management. Seventeen estuary programs have been approved and several more are in the designation process. Two features of the estuary program make it suitable for utilization of ICM. First, the program attempts to bring a wide range of interests to the decision-making process. Second, there is a requirement that a management strategy, known as the Comprehensive Conservation and Management Plan (CCMP), be developed for each estuary. The process of developing the CCMP is an excellent opportunity to practice ICM. ICM can assure that the full range of public values and all relevant scientific information are used, and that planning is conducted on the basis of comparative risk assessment and management.

Those estuary programs that have not yet developed CCMPs might be encouraged by the Environmental Protection Agency, through supplemental grants and technical assistance, to apply the concepts of ICM. At this juncture, maximum flexibility ought to be afforded each estuary program to expand planning beyond the existing framework, so as to achieve integration across jurisdictions, activities, stakeholders, and environmental media. ICM would take estuary programs several steps further by addressing tradeoffs between ecosystem protection and human uses and by formulating and comparing a range of management options designed to achieve the desired level of protection.

This application will provide actual field testing of the ICM concepts, which can form the basis for further refinements. Furthermore, rapid application of the principles of ICM to the development of the CCMP can help in bringing the estuary planning process to bear on real issues and choices, increasing the likelihood that effective actions follow studies. This application would do much to correct the principal shortcoming of the National Estuary Program, which is the absence of any real implementation.

Public Involvement

ICM is dependent on effective public involvement, both in early planning stages and throughout decision making and program implementation. Public values and expectations for the quality of the coastal environment can and should define the problems that must be solved and influence decisions on how resources should be allocated. Public involvement also builds and reinforces public interest, which is crucial to enhancing the political priority of coastal issues. Perhaps most important, developing a sense of individual responsibility for water quality, thus fostering behavior changes regarding the use and disposal of toxic products, transportation choices, and consumer purchases, can be the result of effective public involvement in coastal management. For example, a monitoring program in the Chesapeake Bay that uses citizens to help collect data has produced good quality data on which analyses of trends are based and has served to increase public awareness of the status of the bay and its resources.

Effective public involvement is rare in complex governmental decision-making processes. Public opinion is devalued by technical experts, yet public support is essential in political decisions. Lengthy planning processes are not often exciting enough to garner attention in the important early phases or conducive to meaningful participation by lay people who work at other jobs during the day. Yet proposals put out for public comment in their advanced stages can be greeted with skepticism and resentment. Public-awareness-building and education about water quality and individual actions are rarely seen as a major responsibility of any government entity.

Immediate actions to achieve effective public involvement would include 1) public funding for citizen organizations to participate in coastal planning and management processes, 2) significant agency budgets for outreach and public awareness-building, 3) environmental monitoring programs structured and funded with significant citizen volunteer and student/teacher components, 4) careful attention to expunging jargon and bureaucratic writing styles from all communications, and 5) clear lines of accountability established at the outset for carrying out decisions and implementing plans.

Science and Technical Information

A common criticism of current wastewater management concerns the use and misuse of scientific information. In one sense, ICM can be thought of as a process that allows the rational display and discussion of scientific information prior to decision making. The purpose of this exercise is to identify, in a clear way, that which is important and about which something can be done. The methodology of ICM suggests a number of actions that need to be followed to accomplish this objective, such as comparative analysis, adequate peer review by scientists outside of government, environmental monitoring, appropriate research, and creating access to information.

In a wide variety of ongoing coastal activities, the suggested improvements in the management of science can be implemented now. This would be an important step toward actually achieving a more integrated approach to coastal problems and toward assuring a wider base of public and political support for needed allocation of resources or other management responses.

Institutional Arrangements

An important feature of ICM is the identification of an official or agency that has the paramount responsibility for assuring wise management and protection of the coastal environment. While it might seem easiest to have this responsibility at the federal level, it is more important to fix responsibility at the state or regional level, where operational actions are much more concentrated (including those implementing federal programs). Without identifying a specific office or agency with this responsibility, there are several attributes that ideally ought to be associated with the assignment. These include the following: 1) operational responsibility for carrying out the ICM planning process, 2) oversight responsibility for assuring development and implementation of budgets that reflect management decisions, 3) responsibility for the design and general conduct of the monitoring program, and 4) responsibility as the focal point for public accountability.

These recommendations should not necessarily be taken as a call for agency consolidation. A coalition, commission, joint powers agreement, or some other arrangement of responsible public agencies could serve as the vehicle for designing and implementing an ICM program. If implementation of ICM were to depend on major reorganization of government, it could be doomed to failure. Reorganization is difficult, but, more importantly, in the coastal environment it is simply unrealistic to imagine that all of the disparate government functions could, or even ought to, be brought under one gigantic roof. Instead, there are some actions that, if taken, could facilitate the likelihood that integrated planning is done and that integrated plans are implemented. These include interagency and interjurisdictional agreements that allocate responsibility for specific activities; lead agency/subordinate agency relationships for hierarchical division of activity; temporary transfer of key persons or units to responsible agencies for portions of the planning activity; and use of contractors for selected aspects of the task. Whatever solution is selected by a region, it should recognize the long-term, iterative nature of ICM, and take care to preserve access to the skills and experience needed to continually improve the plan.

The Longer Term

Federal Legislation

A variety of federal statutes could be amended to provide greater opportunities to implement ICM and to provide incentives both for the process itself and to assure the implementation of its outcomes. Although the total number of legislative provisions that require change may be large, certain statutory provisions stand out for particular attention.

- Section 320 of the Clean Water Act could be modified to establish a National Coastal Quality Program as a supplement to the estuary program and that would address coastal areas beyond those in the National Estuary Program. In addition to the expanded geographic coverage, the National Coastal Quality Program and National Estuary Program would have the following new program elements: 1) implementation of an integrated coastal management process, 2) creation of an ICM permit which would subsume all existing discharge National Pollutant Discharge Elimination System (NPDES) permits and/or apply to a class of activities not otherwise regulated by the Clean Water Act, 3) establishment of an iterative action plan as a substitute for the CCMP in which first steps are taken based on the current knowledge base, and 4) the opportunity for trade-offs where consequences are economically advantageous and environmentally equivalent or benign.

The Committee did not explore whether the NPDES permit process should be extended to all ICM activities. Some Committee members believe that the federal government should only mandate the results in terms of environmental goals, and let the states manage the implementation of ICM in their regions. If there were an expanded ICM permit or other approval system, some Committee members fear that the imposition of federal rules and regulations leading to a final federal approval might stifle local initiative and innovation. This is considered a risk even in states that already have comprehensive programs for coastal water management (e.g., California's Ocean Plan).

- Both the Clean Water Act and the Coastal Zone Management Act could be amended to build a more integrated planning and management process between the two statutory systems. The basis for this evolution has been established with the 1990 amendments to the Coastal Zone Management Act which created a joint effort for nonpoint source controls. This cooperation could be expanded in the areas of monitoring, resource and land management, federal activities, and government organization (to statutorily authorize the ideas set forth in the previous section).

Section 301(h) of the Clean Water Act providing for waivers from secondary treatment continues as an example of site-specific management practices bounded by overall federal objectives. Section 301(h) differs from ICM in important ways, however. It permits no tradeoffs: waiver grantees must demonstrate equivalent or better environmental conditions irrespective of cost differences or of the relative importance of the affected ecosystems. Also, Section 301(h) requires monitoring and analysis from waiver grantees that are not required from those localities that install secondary treatment. As a result, the waiver approval process is regarded by some Committee members as unwieldy and burdensome. Still, it should be noted that ICM uses science and environmental monitoring in a way that is similar to that developed by the Environmental Protection Agency in its numerous guidance documents for implementing the 301(h) waiver process. The difference is that ICM covers a much broader range of activities and a larger spatial scale. Moreover, full implementation of the science- and value-based approach suggested by an ICM approach might require levels of treatment lower (and higher) than those currently required.

- In the event that the reauthorized Clean Water Act contains significant new funding authorizations to assist states and local communities meet with its provisions, such funding, especially in complex or high cost situations, should be tied to the conduct of an ICM process. If this linkage were made, one of the major reasons for the failure of the previous 208 program would be overcome. In addition, the linkage would ensure that federal funds were expended only after the necessary planning had been done to obtain the desired results.

Alternative Modes of Regulation

The accumulated mass of federal water quality legislation makes it clear that a command-and-control structure will remain a fundamental part of the nation's clean water programs for a long time to come. It is equally clear that the complexity of the goals that must be met in order to achieve and maintain healthy ecosystems suggests that measures beyond command-and-control regulation will be required. If all of the opportunities for addressing environmental problems that could be identified through integrated coastal management are to be pursued, it is especially important that flexible, easily deployed tools are available to induce fundamental changes in complex human behavior.

In particular, economic incentives may be used to minimize the cost of regulation or to reach activities or discharge levels not readily subjected to command-and-control instruments. Pollution prevention may be achieved through educational programs, economic incentives, or, in a few cases, by command and control. Growth management planning, implemented through economic incentives or zoning changes, can also be employed to modify the amount and location of wastes that must be managed.

Pollution Prevention

ICM is a system based on the rational application of resources to problems that have been deemed important by science and human values. It recognizes that knowledge will often be imperfect or incomplete and that resources are finite. Accordingly, some problems will not be identified, and on occasion resources will be unavailable for the control or management of some problems. In this context, it is important to recognize that ICM includes continuing efforts to reduce the quantity of potential pollutants at the source, especially for those materials for which understanding of the environmental and health effects is poor.

Social Science Dimensions

Successful application of ICM depends on a rich body of knowledge, a keen understanding of human expectations, and robust analytical capacity. Resources are growing rapidly in all of these areas; yet, some areas need greater attention. In particular, the capacity to use economic analysis in environmental management must be strengthened. The ability to predict the outcomes of complex social decisions also needs to be improved. Improved financial support for research in these areas could pay substantial dividends in the future.

WASTEWATER MANAGEMENT FOR THE NEXT CENTURY

This report proposes nothing less than a new paradigm for wastewater management in the United States, which must evolve from the existing structure of policy and practice. Existing policy focuses on setting national standards and writing individual permits for discharges. It is a top-down, command-and-control strategy that takes most aspects of wastewater generation and treatment as given and addresses any unique local ecosystem impacts only as second-order adjustments.

Instead, the central task of wastewater management should be to assess and compare risks and find the appropriate balance between restoring and protecting aquatic ecosystems and maintaining important human uses of the same environment. As argued in previous sections, this task demands integrated coastal management at the regional level. While general principles and constraints can and should be set at the national level, the planning and development of a management strategy must be conducted at the regional level. Management is, therefore, bottom-up rather than top-down.

Furthermore, integrated coastal management encompasses a far greater range of possible actions and policies than contemplated under existing practices. Instead of simply applying a few technologies at known discharge points, integrated coastal management considers all possible interventions, from growth management to pollution prevention to end-of-pipe treatment to ocean discharge options. Choices are water- and sediment-quality driven, rather than technology-constrained. Compliance may be sought through command-and-control methods as well as through economic incentives, educational programs, land-use zoning, or any combination of these. Finally, management is not seen as a once-and-for-all activity, but as an ongoing process, constantly adjusting to changing conditions and achieving goals through a process of continuing iteration and improvement.

REFERENCE

LISS (Long Island Sound Study). 1990. Long Island Sound Study, Status Report and Interim Actions for Hypoxia Management. Boston, Massachusetts: U.S. Environmental Protection Agency.

APPENDIXES

A

The Role of Nutrients in Coastal Waters

Urban wastewaters contain high concentrations of nutrients and as such, contribute significantly to the mass loadings of nitrogen and phosphorus to coastal waters. The various inorganic forms of nitrogen and phosphorus stimulate aquatic plant growth, and since they are relatively hydrophilic, their removal from the water column is more biologically mediated than for trace metals or hydrophobic organics. These characteristics are somewhat unique for urban wastewater contaminants. Therefore, nutrients will be considered here in detail.

Coastal waters receive large amounts of nutrients from wastewater treatment plants and nonpoint sources. In particular, estuaries receive more nutrient inputs per unit surface area than any other type of ecosystem. Many estuaries receive nutrient inputs per unit area that are more than 1,000-fold greater than those of heavily fertilized agricultural fields (Nixon et al. 1986). In moderation, nutrient inputs to estuaries and coastal seas can be considered beneficial. They result in increased production of phytoplankton (the microscopic algae floating in water), which in turn can lead to increased production of fish and shellfish (Nixon 1988, Hansson and Rudstam 1990, Rosenberg et al. 1990). However, excess nutrients can be highly damaging, leading to effects such as anoxia and hypoxia from eutrophication, nuisance algal blooms, dieback of seagrasses and corals, and reduced populations of fish and shellfish (Ryther 1954, 1989; Kirkman 1976; McComb et al. 1981; Kemp et al. 1983; Cambridge and McComb 1984; Gray and Paasche 1984; Officer et al. 1984; Larsson et al. 1985; Price et al. 1985; Rosenberg 1985; D'Elia 1987; Baden et al. 1990; Cederwall and Elmgren 1990; Hansson and

Rudstam 1990; Rosenberg et al. 1990; Parker and O'Reilly 1991; Lein and Ivanov 1992; Smayda 1992). Eutrophication also may change the plankton-based food web from one based on diatoms toward one based on flagellates or other phytoplankton, which are less desirable as food to organisms at higher trophic levels (Doering et al. 1989).

Whether or not nutrient inputs should be considered excessive depends in part upon the physics and ecological sensitivity of the receiving water body. In many parts of the world, estuaries and coastal seas clearly are receiving an excess of nutrients, and the resulting eutrophication is one of the major causes of decline of coastal waters. Nutrients should be considered a major ecological concern along with sewage disposal in many coastal urban areas and be regulated accordingly.

This section first briefly discusses the negative effects of eutrophication and nuisance algal blooms in coastal marine ecosystems. It then reviews in more detail the controls on eutrophication; discusses the issue of whether nitrogen or phosphorus is more limiting to eutrophication; presents dose-response information, which relates nitrogen to algal biomass and production; and reviews the information on controls of nuisance algal blooms.

ADVERSE CONSEQUENCES OF EUTROPHICATION AND NUISANCE ALGAE

Anoxia and Hypoxia

Anoxia is the complete removal of dissolved oxygen from the water column, an event which obviously causes widespread damage to aquatic plants and animals. Even mobile animals that can escape from anoxic waters can suffer population declines from the loss of habitat area. For example, in parts of the Baltic Sea, cod eggs laid in oxic surface waters sink into anoxic bottom waters where they die (Rosenberg et al. 1990). Oxygen concentrations in the bottom waters of the deep basins of the Baltic between 1969 and 1983 correlate negatively with codfish populations (Hansson and Rudstam 1990).

Oxygen need not be completely absent for damage to occur, and a lowering of oxygen to concentrations as low as 3 to 4.3 mg per liter can cause ecological harm in some estuaries and coastal seas (EPA 1990). Such a depletion of oxygen is termed hypoxia. Examples of ecological damage from hypoxia include lowered survival of larval fish, mortality of some species of benthic invertebrates, and loss of habitat for some mobile species of fish and shellfish that require higher concentrations of oxygen, such as lobster and codfish (Baden et al. 1990, EPA 1990). Significant mortalities of lobsters and population declines of both lobster and codfish have been observed in some Swedish coastal waters as a result of increased incidences of hypoxia (Baden et al. 1990).

Anoxia and hypoxia are major and growing problems in many estuaries and coastal seas. Over the past few decades, the volume of anoxic bottom waters has been increasing in the Chesapeake Bay (Officer et al. 1984, D'Elia 1987), the Baltic Sea (Larsson et al. 1985), and the Black Sea (Lein and Ivanov 1992). The apex of the New York Bight (an area of some 1,250 km^2) becomes hypoxic every year, and a large region of the Bight became anoxic in 1976 (Mearns et al. 1982). Hypoxic events appear to be becoming more common in waters such as the Long Island Sound (EPA 1990, Parker and O'Reilly 1991), the North Sea (Rosenberg 1985), and the Kattegat (the waters between Denmark and Sweden; Baden et al. 1990), although historical data on oxygen concentrations in coastal waters are often poor.

Anoxia and hypoxia result from oxygen consumption exceeding oxygen supply. Oxygen is consumed by the respiration of organisms, including animals, plants, and the decomposing activity of microorganisms. Oxygen is supplied to waters through the process of photosynthesis and through diffusion and surface entrapment from the atmosphere. Organic matter released in sewage effluent can contribute to anoxia and hypoxia by creating biochemical oxygen demand (BOD), the oxygen consumed during the microbial decomposition of this organic matter and chemical oxygen demand (COD), the oxygen consumed through the oxidation of ammonium and other inorganic reduced compounds. For example, altered structure, reduced diversity, and elevated biomass of benthic animal communities—related in part to high organic solids inputs—once characterized up to 95 square kilometers of the coastal shelf around major outfalls in San Diego, Orange, and Los Angeles Counties in southern California (Mearns and Word 1982). However, BOD inputs to most estuaries and coastal seas are well controlled and represent at most a localized problem. Of more concern in most estuaries and coastal marine ecosystems is the oxygen consumption that results from the decomposition of the excess phytoplankton production characteristic of eutrophication (Officer et al. 1984, Larsson et al. 1985, Jensen et al. 1990, Rydberg et al. 1990, EPA 1990, Parker and O'Reilly 1991, Lein and Ivanov 1992). Photosynthesis by phytoplankton produces oxygen, but much of the photosynthesis in eutrophic waters occurs near the surface, and oxygen readily escapes to the atmosphere. The majority of the phytoplankton material is decomposed deeper in the water column, consuming oxygen there. In contrast to the rather localized effects of BOD inputs, nutrient inputs can lead to eutrophication and anoxia or hypoxia far from the original source of the nutrient. In some cases, improved sewage treatment may aggravate this situation by resulting in more distant transport of nitrogen (Chesterikoff et al. 1992). Some evidence points to increasing hypoxia in the western basin of the Long Island Sound as being a result of improved sewage treatment in the East River in New York City with resulting increase of flows of nitrogen from the East River into the Sound (Parker and O'Reilly 1991).

Dieback of Seagrasses, Algal Beds, and Corals

In addition to anoxia and hypoxia, eutrophication can lead to the die-back of seagrass beds, which are important habitats and nursery grounds for a variety of fish and other animals. One mechanism for such die-back is a shading out of the grasses by the abundant phytoplankton in the overlying water, a process thought to have caused the die-back of macrophytes in the upper portions of the Chesapeake Bay (Kemp et al. 1983, Twilley et al. 1985, D'Elia 1987), in the Dutch Wadden Sea (Gieson et al. 1990), and of both tropical and temperate seagrasses in Australia (Kirkman 1976, Cambridge and McComb 1984, Cambridge et al. 1986). Die-back caused by such shading usually manifests itself in a rather gradual loss of the seagrasses (Robblee et al. 1991), although the occurrence of unusual nuisance algal blooms in 1985 and 1986 greatly reduced the abundance of seagrass beds near Long Island (Dennison et al. 1989). Shading by enhanced epiphytic (Twilley et al. 1985) and macroalgal (Valiela et al. 1990) growth may be an additional cause of seagrass die-back. Nutrient enrichment may also have a direct physiological response on seagrasses, with internal nutrient imbalances appearing to lead to reduced survival (Burkholder et al. 1992a).

Beds of attached macro-algae on bottom sediments or rocks can also be adversely affected by eutrophication. Nutrient enrichment of rocky intertidal areas typically leads to a reduction in the overall diversity of both attached algae (Borowitzka 1972, Littler and Murray 1978) and associated animals (Gappa et al. 1990). These nutrient-enriched areas tend to be dominated by opportunistic algae with rapid growth rates, such as *Cladophora* sp. and *Enteromorpha* sp., which can take advantage of the elevated nutrient levels and shade out other species (Littler and Murray 1975, 1978). This phenomenon is seen clearly along the Swedish coast of the Baltic Sea, where since the mid-1970s nuisance forms of filamentous algae (*Cladophora* and *Enteromorpha* species) have become more dominant, coinciding with a decline of the former dominant bladderwrack algae, *Fucus* sp. (Baden et al. 1990, Rosenberg et al. 1990). The bladderwrack is used by herring in spawning, and the change to dominance by filamentous macroalgae has led to decreased hatching of herring eggs (Rosenberg et al. 1990). Subtidal forests of giant kelp (*Macrocystis pyrifera*) died back and failed to reproduce during the 1960s and early 1970s along the Palos Verdes Peninsula near the Los Angeles County outfalls (Wilson et al. 1980). This may have been due to light limitation, perhaps a result of eutrophication, but was more likely due directly to solids discharged in poorly treated sewage. Toxic substances associated with these particle discharges may also have played a role. The kelp beds recovered following significant reductions in the solids emissions (Grigg 1978, Harris 1980, Wilson et al. 1980).

Reduced light levels from excessive phytoplankton and macroalgal growth in eutrophic environments can cause coral die-back. Shading affects coral growth by decreasing the productivity of the zooxanthallae, symbiotic algae in the coral tissue, which provides much of the coral's nutrition (Smith 1981). Excess nutrients can also cause a shift in the composition of the coral community as other species outcompete corals for space. Filter-feeding species such as sponges take advantage of the high phytoplankton productivity and become dominant, displacing the corals (Pastorak and Bilyard 1985). Sedimentation of decomposing phytoplankton may have an additional detrimental impact on coral growth (Smith 1981).

Nuisance Algal Blooms

Blooms of nuisance algae are characterized by very high abundances of one overwhelmingly dominant species in the phytoplankton. These blooms often result in noticeable color and are popularly named by this color: red tides, green tides, and brown tides. As with eutrophication generally, these blooms can result in anoxic or hypoxic conditions. In addition, many nuisance blooms produce substances toxic to aquatic organisms or humans (Cosper 1991). Green tides during the 1950s heavily damaged oyster populations on Long Island (Ryther 1954, 1989), and brown tides in 1985 and 1986 greatly reduced populations of bay scallops on Long Island (Cosper et al. 1987, Bricelj and Kuenstner 1989) and of blue mussels in Narragansett Bay (Tracey et al. 1989). These shellfish starved to death since they were unable to graze on the brown-tide algae. Blooms of some dinoflagellates (red tides) can result in the accumulation of toxins in shellfish, which, when eaten by humans, cause paralytic or diarrhetic shellfish poisoning (Smayda 1989). Frequent blooms of a gold-brown dinoflagellate in Northern Europe have caused extensive fish mortality since the mid-1960s (Smayda 1989). In 1991, toxins produced by a diatom bloom concentrated in anchovy and caused the death of pelicans that fed on these fish (Work et al. undated, as cited in Smayda 1992). Recently, Burkholder et al. (1992b) discovered a new toxic dinoflagellate that releases toxins only in the presence of fish and appears to be responsible for several fish kills in estuaries in North Carolina.

Nuisance-bloom tides have been known since biblical times (Cosper 1991), but blooms of many species appear to be occurring with greater frequency throughout the world (Hallegraeff et al. 1988; Anderson 1989; Smayda 1989, 1992; Robineau et al. 1991). Red-tide blooms of toxic dinoflagellates appear to be more frequent in many parts of the world (Anderson 1989, Smayda 1989, Wells et al. 1991), and blooms of cyanobacteria have become more prevalent in the less saline portions of the Chesapeake Bay (D'Elia 1987) and in the Baltic Sea and related waters over the past 10 to 20 years (Smayda 1989 and references therein). Many of the new toxic phytoplankton blooms are subpopulations of previously non-toxic species, which now occur at previously unseen abundances (Smayda 1989, 1992). Brown-tide blooms of *Aureococcus anophagefferens* were unknown before 1985 (Sieburth et al. 1988). As discussed below, the cause(s) of increased nuisance blooms is not known, but evidence points toward the importance of increased nutrient inputs (nitrogen, phosphorus, iron) to estuaries and coastal seas.

CONTROLS ON EUTROPHICATION AND NUISANCE BLOOMS IN COASTAL WATERS

Nutrient Limitation

Nutrients are elements essential for plant growth, such as nitrogen, phosphorus, silica, and sulfur. Phytoplankton production in most coastal marine ecosystems and estuaries is nutrient limited, and increased nutrient inputs lead to higher production and eutrophication (Ryther and Dunstan 1971; Graneli 1978, 1981, 1984; McComb et al. 1981; Boynton et al. 1982; Nixon and Pilson 1983; Smith 1984; Valiela 1984; D'Elia et al. 1986; Nixon et al. 1986; D'Elia 1987; Howarth 1988; Andersen et al. 1991). Unfortunately, the discussion of nutrient limitation in coastal marine waters has been surrounded by some confusion, in part because the term can have many different meanings and is often used quite loosely (Howarth 1988) and in part because of potential methodological problems in determining nutrient limitation (Hecky and Kilham 1988, Howarth 1988, Banse 1990). If one is concerned with eutrophication, then the appropriate definition of nutrient limitation is the regulation of the potential rate of net primary production by phytoplankton (Howarth 1988). Net primary production is defined as the total amount of photosynthesis minus the amount of plant respiration occurring in a given area (or volume) of water in a given amount of time. If an addition of nutrients would increase the rate of net primary production, even if this means a complete change in the species composition of the phytoplankton, then production is considered to be nutrient limited (Howarth 1988, Vitousek and Howarth 1991).

Factors other than nutrient input can also influence or partially control primary production. For instance, phytoplankton production in some estuaries (such as the Hudson River) is limited by light availability. This tends to occur in extremely turbid estuaries or in estuaries where moderate turbidity coexists with deep mixing of the water. The turbidity can result both from suspension of inorganic particles and from high phytoplankton biomass, and so light limitation often is a result of self-shading by the phytoplankton, as is the case in many lakes (Wetzel 1983). In estuaries where nutrient inputs are high and production is limited by light, the nutrients are simply transported further away from the source before being first assimilated by phytoplankton, as is seen in the transport of nutrients from the Hudson River and New York Harbor into the New York Bight (Malone 1982). This further transport may or may not provide sufficient dilution to avoid excessive eutrophication, which may just occur further afield from the nutrient source.

Through their grazing on phytoplankton, zooplankton and other animals can also influence the rate of primary production and the biomass of phytoplankton. This has received extensive study and discussion in freshwater ecosystems (Carpenter et al. 1985, Morin et al. 1991) and in offshore ocean ecosystems (Steele 1974, Banse 1990) although it is virtually unstudied in estuaries and coastal seas. Nonetheless, changes in grazing in estuaries may have serious effects on water quality. For instance, some researchers believe that at one time, oyster populations in the Chesapeake Bay were sufficiently high to filter the bay's entire water volume on average once every week; the currently declined oyster populations probably filter the water of the bay only once per year on average (Newell 1988). This lower grazing pressure on phytoplankton populations may be contributing to eutrophication of the Chesapeake. However, reductions in grazing pressure can result in algal blooms only where nutrient availabilities are high. Thus, nutrient supply should be viewed as the cause of eutrophication with grazing pressures being a secondary regulator.

Nitrogen Versus Phosphorus Limitation

Nitrogen is the element usually limiting to primary production by phytoplankton in most estuaries and coastal seas of the temperate zone (Ryther and Dunstan 1971; Vince and Valiela 1973; Smayda 1974; Norin 1977; Graneli 1978, 1981, 1984; Boynton et al. 1982; Nixon and Pilson 1983; Valiela 1984; D'Elia et al. 1986; Nixon et al. 1986; Frithsen et al. 1988; Howarth 1988; Rydberg et al. 1990; Vitousek and Howarth 1991; Nixon 1992), although some temperate estuaries such as the Apalachicola in the Gulf of Mexico may be phosphorus limited (Myers and Iverson 1981; Howarth 1988) and others such as parts of the Chesapeake Bay and the Baltic Sea may switch seasonally between nitrogen and phosphorus limitation (McComb et al. 1981, D'Elia et al. 1986, Graneli et al. 1990, Andersen et al. 1991). Many tropical estuarine lagoons may be phosphorus limited as well (Smith 1984, Smith and Atkinson 1984, Howarth 1988, Vitousek and Howarth 1991). Sewage often accounts for 50 percent or more of the nitrogen inputs to

TABLE A.1 Total Annual Inputs of Dissolved Inorganic Nitrogen to Various Estuaries and the Percent of this Input which comes from Sewage Treatment Plants (Source: Nixon and Pilson 1983. Reprinted, by permission, from Academic Press, 1983.)

	DIN input (mmole m^{-2} $year^{-1}$)	Sewage (%)
Kaneohe Bay	230	78
Long Island Sound	400	67
Chesapeake Bay	510	33
Apalachicola Bay	560	2
Barataria Bay	570	<1
Patuxent Estuary	600	48
Potomac Estuary	810	48
Pamlico Estuary	860	<1
Narragansett Bay	950	41
Mobile Bay	1,280	7
Delaware Bay	1,300	50
Raritan Bay	1,460	86
South San Francisco Bay	1,600	≈100
North San Francisco Bay	2,010	45
New York Bay	31,900	82

various estuaries and may be a more controllable input than those from agriculture and other nonpoint sources (Table A.1, Nixon and Pilson 1983).

That nitrogen limits primary production in most temperate-zone estuaries and coastal seas was much debated throughout the 1980s (D'Elia 1987, Howarth 1988, Nixon 1992). One argument against nitrogen limitation was that phosphorus is generally limiting in temperate-zone lakes (Edmondson 1970; Vollenwieder 1976, 1979; Schindler 1977, 1978; Wetzel 1983) and, until recently, there was little evidence that the biogeochemical processes regulating nutrient limitation were fundamentally different between freshwater and marine ecosystems (Schindler 1981, Smith 1984). Another argument was that much of the evidence for nitrogen limitation in marine ecosystems came from extremely short-term (generally a few days), small-scale enrichment experiments in flasks or bottles. It may not be possible to extrapolate the results of such short-term enrichment experiments to the whole ecosystem, in part because they only measure the physiological response of phytoplankton species that are present in the water and can respond in a significant way during the time of the experiment (Smith 1984, Hecky and Kilham 1988, Howarth 1988, Banse 1990, Marino et al. 1990).

In recent years, increasing evidence has accumulated both that nitrogen is limiting in many coastal marine ecosystems and that the biogeochemical processes regulating nutrient limitation do in fact vary between marine and freshwater ecosystems. The evidence for nitrogen limitation, in addition to the short-term enrichment experiments, consists of generally low concentrations of dissolved nitrogen compared to dissolved phosphorus (Boynton et al. 1982, Graneli 1984, Valiela 1984) and longer, larger scale enrichment experiments (D'Elia et al. 1986), including one mesocosm experiment of many months duration in the Marine Ecosystem Research Laboratory (MERL) facility (Frithsen et al. 1988, and unpublished data). While any one such piece of evidence may not be entirely convincing, the agreement among the several lines of evidence convincingly demonstrates nitrogen limitation (Howarth 1988, Vitousek and Howarth 1991).

What differences in biogeochemical cycles lead toward nitrogen limitation in temperate coastal marine ecosystems and toward phosphorus limitation in temperate lakes? At least three factors appear important (Figure A.1): the ratio of nitrogen to phosphorus in nutrient inputs to estuaries is

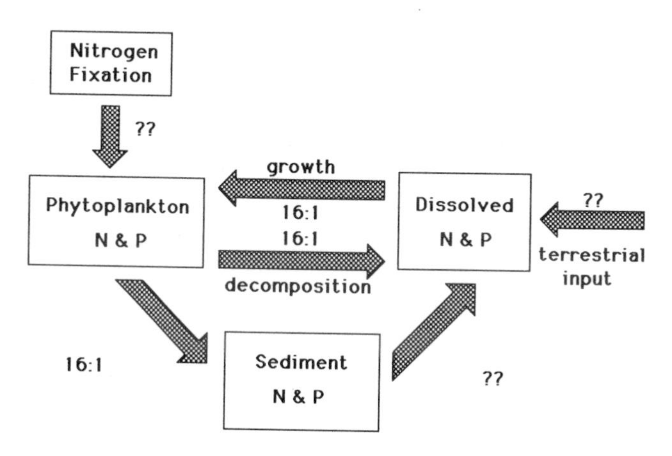

FIGURE A.1 Summary of biogeochemical processes controlling nutrient limitation in aquatic ecosystems. The ratio of nitrogen to phosphorus from terrestrial inputs varies greatly, generally being lower for ecosystems receiving more sewage inputs. Nitrogen and phosphorus are assimilated by phytoplankton in an approximate molar ratio of 16:1. The N:P ratio of nutrients released in the water column during zooplankton feeding and during decomposition frequently approximates 16:1 but can be lower. Conversely, the N:P ratio of sedimenting material also frequently approximated 16:1 but can be as high as 30:1 in oligotrophic waters. Nutrients released from marine and estuarine sediments back to the water column tend to have low N:P ratios, but such releases in lack sediments often have N:P ratios near 16:1 or higher. The extent of nitrogen fixation varies but tends to be much greater in lakes than in marine ecosystems. (Source: Howarth 1988. Reproduced, with permission, from the Annual Review of Ecology and Systematics, Vol. 19, © 1988 by Annual Reviews Inc.)

frequently less than for lakes, the sediments are often a more important sink of phosphorus in lakes than in marine ecosystems, and nitrogen fixation is a more prevalent process in the plankton of lakes (Howarth 1988). Each of these is discussed briefly below.

In both freshwater and marine ecosystems, the relative requirements of phytoplankton for nitrogen and phosphorus are fairly constant, with the two elements being assimilated in the approximate molar ratio of 16:1, the Redfield ratio (Redfield 1958). If there were no biogeochemical processes acting within a water body, the ratio of nitrogen to phosphorus in the nutrient inputs to the ecosystem would determine whether the system were nitrogen or phosphorus limited. Ratios below 16:1 would lead to nitrogen limitation and higher ratios would lead to phosphorus limitation (Howarth 1988). In fact, the N:P ratio in nutrient loadings to many (but by no means all) estuaries and coastal seas are below this ratio, while nutrient inputs to temperate lakes tend to have higher N:P ratios (Jaworski 1981, Kelly and Levin 1986, NOAA/EPA 1988). This difference in ratios probably reflects the relative importance of sewage, which tends to have a low N:P ratio, as a nutrient source to coastal waters.

Biogeochemical processes within sediments act to alter the relative abundance of nitrogen and phosphorus in an ecosystem. Denitrification, the bacterial reduction of nitrate to molecular nitrogen, removes nitrogen and tends to make coastal marine ecosystems more nitrogen limited (Nixon et al. 1980, Nixon and Pilson 1983). However, this process appears to be even more important in lakes than in estuaries and coastal seas; a higher percentage of the nitrogen mineralized during decomposition is denitrified in lake sediments than in estuarine sediments (Seitzinger 1988, Gardner et al. 1991, Seitzinger et al. 1991). Of more importance in explaining a tendency for nitrogen limitation in coastal marine ecosystems of the temperate zone, therefore, is the relatively high phosphorus flux from sediments; nutrient fluxes from these sediments have fairly low N:P ratios (Rowe et al. 1975, Boynton et al. 1980, Nixon et al. 1980). In many lakes, phosphorus is bound in the sediments (Schindler et al. 1977), although in others, phosphorus fluxes are comparable to marine sediments (Khalid et al. 1977). Nutrient fluxes from lake sediments can be either enriched or depleted in nitrogen relative to phosphorus (Kamp-Nielsen 1974). Caraco et al. (1989, 1990) have suggested that the abundance of sulfate in an ecosystem partially regulates the sediment flux of phosphorus. Phosphorus binding in sediments is greatest where sulfate concentrations are lowest, which is consistent with variable fluxes in lakes and higher fluxes in coastal marine ecosystems.

When the relative abundance of nitrogen to phosphorus is low in the water column of lakes, nitrogen-fixing species of cyanobacteria are favored since they can convert molecular nitrogen to ammonium or organic nitrogen. Under such nitrogen-depleted conditions in lakes, these cyanobacteria often are the dominant phytoplankton species and fix appreciable quantities of nitrogen. As a result, nitrogen deficits (relative to phosphorus) can be alleviated, and primary production in the lake is phosphorus limited (Schindler 1977, Flett et al. 1980, Howarth 1988, Howarth et al. 1988a). In contrast, nitrogen-fixing cyanobacteria are rare or absent from the plankton of most estuaries and coastal seas, which is a condition helping to maintain nitrogen limitation in these ecosystems (Howarth 1988, Howarth et al. 1988a). Exceptions are found in the Baltic Sea (Lindahl and Wallstrom 1985) and in the Harvey-Peel estuary in Australia (McComb et al. 1981) but are unknown in the waters of the United States. The explanation for the rarity of planktonic, nitrogen-fixing cyanobacteria in coastal marine waters is still subject to debate (Paerl et al. 1987, Howarth et al. 1988b, Paerl and Carlton 1988, Carpenter et al. 1990, Marino et al. 1990, Vitousek and Howarth 1991). Possible reasons include one or more of the following: a lower availability of iron and molybdenum, trace metals required for nitrogen fixation, in saline water (Howarth and Cole 1985, Howarth et al. 1988b, Marino et al. 1990); greater turbulence in coastal marine systems, allowing oxygen to poison the nitrogenase enzyme responsible for nitrogen fixation (Paerl et al. 1987, Paerl and Carlton 1988); greater grazing pressure on cyanobacteria in marine systems (Vitousek and Howarth 1991); and a lower light availability in estuaries and coastal waters due to higher turbidity and/or deeper mixed layers (Howarth and Marino 1990, Vitousek and Howarth 1991).

As noted above, many tropical estuaries and coastal systems may be phosphorus limited (Smith 1984, Smith and Atkinson 1984). Although the evidence for limitation of production by phytoplankton is not entirely clear in tropical systems (Howarth 1988), and production by seagrasses and attached macroalgae is sometimes nitrogen limited in tropical systems (Lapointe et al. 1987, McGlathery et al. 1992), primary production by seagrasses in many tropical areas is clearly limited by phosphorus (Short et al. 1985, 1990; Littler et al. 1988; Powell et al. 1989). Phosphorus limitation in these systems is probably the result both of a high degree of phosphorus adsorption in the calcium-carbonate sediments that dominate such tropical systems (Morse et al. 1985) and the high rates of nitrogen fixation associated with benthic algal mats and with symbionts of seagrasses in clear, relatively oligotrophic lagoons (Howarth 1988, Howarth et al. 1988a).

This report does not explicitly consider nutrient limitation in open ocean systems away from the coast because there is no reason to believe that such ecosystems can be significantly affected by wastewater inputs.

Dose-response Relationship: Nitrogen and Eutrophication

The relationship between phosphorus loadings, phosphorus concentrations, chlorophyll (a measure of phytoplankton biomass), and primary pro-

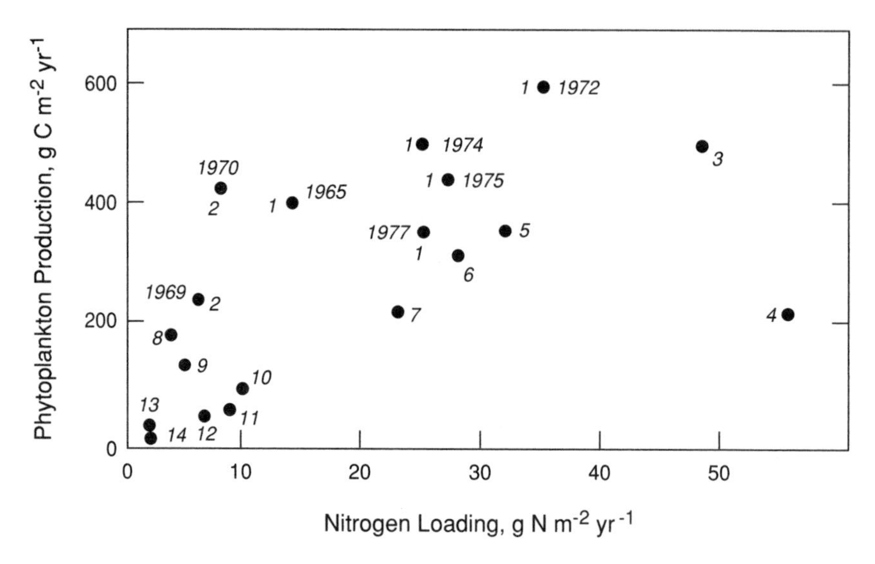

FIGURE A.2a Regression plot relating nitrogen loadings to annual phytoplankton production in a variety of estuarine ecosystems: 1) Chesapeake Bay, 2) Patuxent River, 3) Pamlico River, 4) Byfjord, 5) Apalachicola Bay, 6) Narragansett Bay, 7) San Francisco Bay, 8) St. Margarets Bay, 9) Long Island Sound, 10) Kungsbacka Fjord, 11) Loch Etive, 12) St. Lawrence River, 13) Baltic Sea, and 14) Kaneohe Bay. (Source: Boynton et al. 1982. Reprinted, by permission, from Academic Press, 1982.)

duction has been extensively studied through statistical analysis in freshwater lakes using large, multi-lake data sets (Vollenweider 1976, 1979; Schindler 1978; Smith 1979; Wetzel 1983; Molot and Dillon 1991; and references therein). Generally, a positive relationship is found between either phosphorus loadings or concentrations of phosphorus in the water column and either chlorophyll or primary production. These relationships have been used successfully to manage eutrophication in lakes (Wetzel 1983).

Although the available data sets are smaller, similar approaches have been applied to study nitrogen and eutrophication in estuaries and coastal marine ecosystems. Data compiled from a variety of estuaries show a positive relationship between loadings of either total nitrogen per unit area (Boynton et al. 1982) or dissolved inorganic nitrogen per unit volume (Nixon and Pilson 1983, Nixon 1992) to the estuary and rates of primary production (Figures A.2a and A.3). Interestingly, nitrogen loadings to estuaries are a better predictor of primary production than are phosphorus loadings (Figures A.2a and A.2b)—further evidence that a majority of these ecosystems are probably nitrogen limited (Boynton et al. 1982). Whether a particular

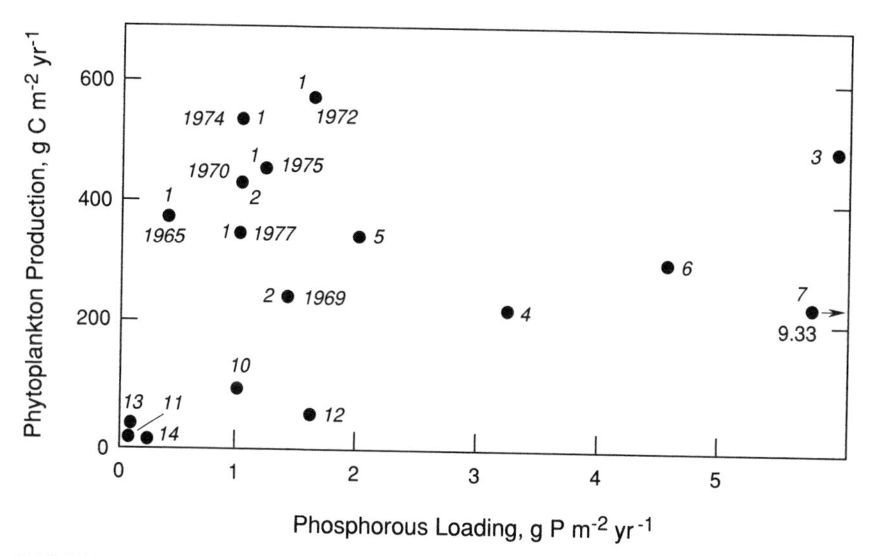

FIGURE A.2b Regression plot relating phosphorus loadings to annual phytoplankton production in a variety of estuarine ecosystems: 1) Chesapeake Bay, 2) Patuxent River, 3) Pamlico River, 4) Byfjord, 5) Apalachicola Bay, 6) Narragansett Bay, 7) San Francisco Bay, 8) St. Margarets Bay, 9) Long Island Sound, 10) Kungsbacka Fjord, 11) Loch Etive, 12) St. Lawrence River, 13) Baltic Sea, and 14) Kaneohe Bay. (Source: Boynton et al. 1982. Reprinted, by permission, from Academic Press, 1982.)

level of production should be considered excessive would depend upon the circulation of the water body and other factors affecting the supply of oxygen to the ecosystem.

Chlorophyll concentrations (phytoplankton biomass) are also correlated with inputs of dissolved inorganic nitrogen to various estuarine and coastal marine ecosystems (Nixon and Pilson 1983). Figure A.4a shows the relationship between annual inputs of dissolved inorganic nitrogen per unit volume and the average annual abundance of chlorophyll for a variety of estuaries. Data from an enrichment experiment at MERL are also included. Figure A.4b is similar but relates nitrogen inputs to average chlorophyll abundances during the summer season instead of giving annual values. The results from the MERL experiment compare favorably with the data from natural ecosystems, particularly at lower rates of nitrogen input. At the higher nutrient inputs, compared with natural ecosystems, the MERL mesocosms tend to have lower chlorophyll values during the summer and higher annual average chlorophyll values (Figures A.4a and A.4b).

No guidelines exist by which to determine whether coastal marine eco-

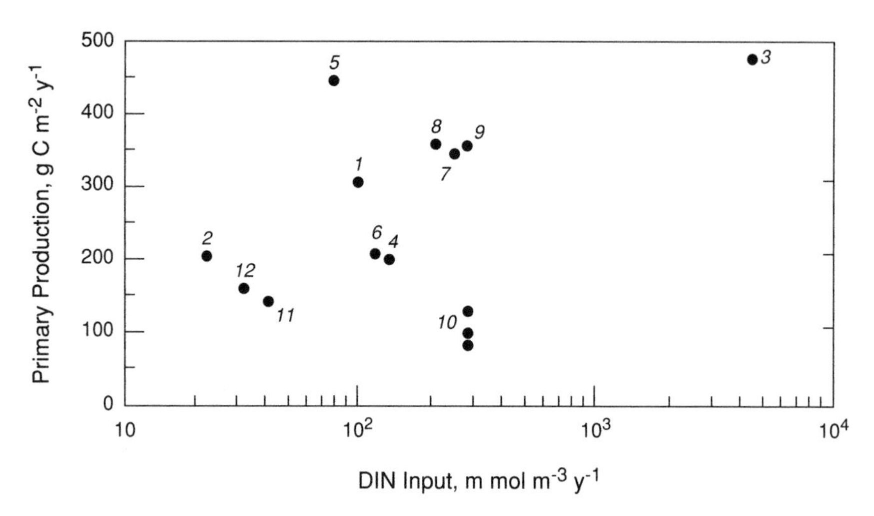

FIGURE A.3 Annual measurements of primary production as a function of the estimated annual input of dissolved inorganic nitrogen (DIN) in various estuaries. 1) Narragansett Bay, Rhode Island; 2) Long Island Sound; 3) Lower New York Bay; 4) Lower Delaware Bay; 5) Chesapeake Bay; 6) Patuxent estuary, Maryland; 7) Pamlico estuary, North Carolina; 8) Apalachicola Bay, Florida; 9) Barataria Bay, Louisiana; 10) North and South San Francisco Bay, California; 11) Kaneohe Bay, Hawaii. (Source: Nixon and Pilson 1983. Reprinted, by permission, from Academic Press, 1983.)

systems are in fact eutrophic. In lakes, however, over the past few decades, collective expert opinion has tended toward the following relationship of annual chlorophyll to trophic status (Wetzel 1983): oligotrophic lakes average 1.7 mg m^{-3} chlorophyll a (range = 0.3 to 4.5), mesotrophic lakes average 4.7 mg m^{-3} (range = 3 to 11), and eutrophic lakes average 14.3 mg m^{-3} (range = 3 to 78). There is no reason why these criteria should not be equally applicable to marine ecosystems, and the averages and ranges are drawn onto Figures A.4a and A.4b for reference. Using these delineations, few estuaries are oligotrophic, many are mesotrophic, and many are extremely eutrophic. Note that based on this interpretation of trophic status, Kaneohe Bay (Honolulu, Hawaii) is considered oligotrophic (average annual chlorophyll of 2 and 6 mg m^{-3}; Nixon and Pilson 1983), and Long Island Sound is considered mesotrophic (average annual chlorophyll of 6 mg m^{-3}; Nixon and Pilson 1983). Yet nutrient enrichment has severely modified the coral community of Kaneohe Bay (Smith 1981, Pastorak and Bilyard 1985), and eutrophication in Long Island Sound appears to be leading to increasingly frequent hypoxic events (EPA 1990). Thus, the applica-

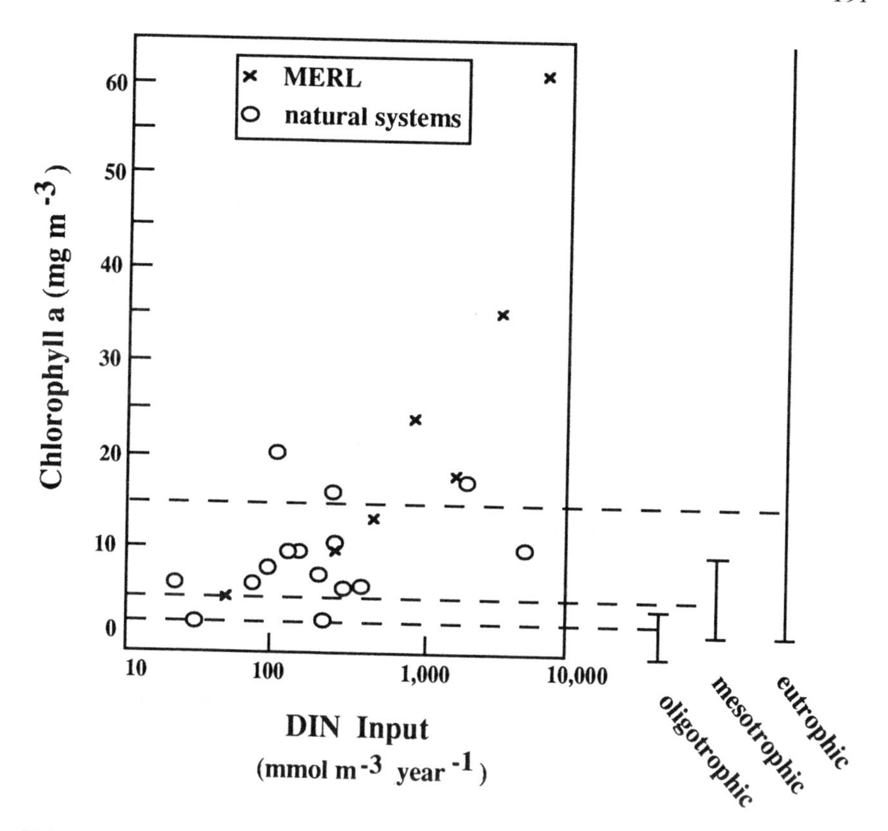

FIGURE A.4a Annual average concentration of chlorophyll compared to inorganic nitrogen inputs in a variety of natural estuaries and in experimental mesocosms at the Marine Ecosystem Research Laboratory, University of Rhode Island. The mesocosms were fertilized with nutrients. Data are from Nixon and Pilson (1983). Chlorophyll is a measure of phytoplankton abundance. Bars to the right indicate ranges of chlorophyll believed to characterize lakes as being oligotrophic, mesotrophic, or eutrophic. Dashed horizontal lines represent the mean chlorophyll values for each classification (Wetzel 1983). See text for a discussion of the applicability of this classification to estuaries.

tion of this lake-based approach to the trophic status of estuaries may not provide sufficient protection. This is the case, in part, because such trophic guidelines are based on phytoplankton response to nutrients yet problems such as those experienced in Kaneohe Bay are the result of benthic algae overgrowing the coral.

Jaworski (1981) proposed "permissible" nutrient inputs to protect shallow (4-meters to 9-meters deep) temperate-zone estuaries from eutrophica-

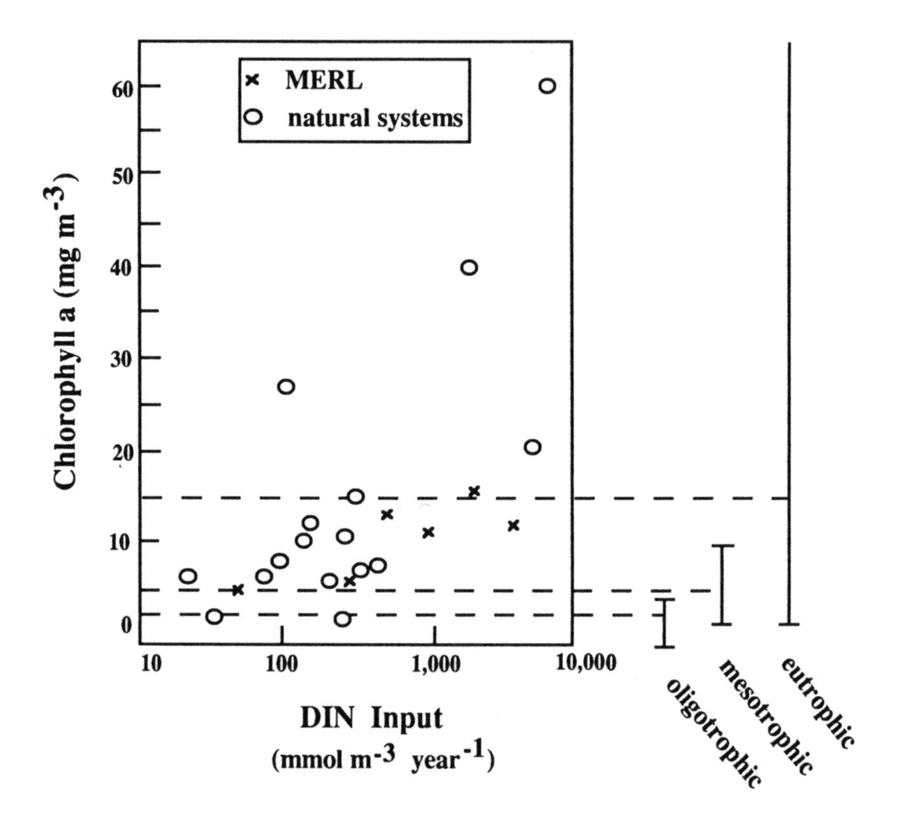

FIGURE A.4b Summer mean concentration of chlorophyll. The comparison shown here is the same as for Figure A.4a, except chlorophyll values represent the mean for the summer season rather than annual means.

tion. For phosphorus-limited estuaries, he suggested that inputs of phosphorus should be kept below 24 mmol P m^{-2} $year^{-1}$. For nitrogen-limited estuaries, Jaworski proposed that inputs be kept below 380 mmol N m^{-2} $year^{-1}$. For these shallow estuaries, this corresponds to a nitrogen loading of 40 to 95 mmol m^{-3} $year^{-1}$. From Figures A.4a and A.4b, it would appear that Jaworski's proposed nitrogen loadings would indeed keep most estuaries from becoming eutrophic. Note that numerical models of the Chesapeake Bay suggest that current nitrogen loadings need to be reduced by 40 percent to eliminate anoxia there (Butt 1992); this reduction corresponds to a decrease in nitrogen inputs from 510 mmol m^{-2} $year^{-1}$ (Nixon and Pilson 1983, and Table A.1) to some 300 mmol m^{-2} $year^{-1}$ or within the range suggested by Jaworski (1981) for nitrogen-limited estuaries.

Causes of Nuisance Algal Blooms

Smayda (1989) has compiled extensive evidence in support of the hypothesis that the worldwide increase in nuisance algal blooms is related to increased nutrient availability. For instance, a 2.5-fold increase in nutrient loadings accompanied an 8-fold increase in the annual number of red-tide blooms in a harbor in Hong Kong between 1976 and 1986. Increased nutrient concentrations in the North Sea, the Baltic Sea, and in waters between Denmark and Sweden (the Skagerrak and Kattegat) have co-occurred with increased primary production and increased incidence of blooms in these waters (Smayda 1989). The green-tides which occurred in the Great South Bay of Long Island in the 1950s were also clearly associated with nitrogen loading from duck farms there (Ryther 1954), and the reduction of nutrient loadings and the opening of a channel to increase water exchange between the bay and ocean have greatly reduced these blooms (Ryther 1989). Also, nuisance algal blooms are much more likely to occur in nutrient-rich estuarine waters than in more coastal or shelf waters (Cosper 1991, Prego 1992). The new dinoflagellate discovered by Burkholder et al. (1992b), which produces toxins only in the presence of fish, seems to be stimulated by phosphorus additions.

On the other hand, there is little if any evidence to show a direct connection between either nitrogen or phosphorus concentrations and blooms of most brown-tide or red-tide organisms (Cosper 1991, Wells et al. 1991). Red-tide blooms in Florida are not correlated with concentrations of any measured form of nitrogen or phosphorus (Rounsefell and Dragovich 1966). Similarly, the brown-tide blooms of the mid-1980s along the northeastern coast of the United States did not appear to be correlated with higher levels of nitrogen or phosphorus (Cosper et al. 1989, Cosper 1991). However, it is important to note that the concentration of a nutrient at any given point of time may not be correlated with its availability to phytoplankton (Howarth 1988), and phytoplankton can grow for long periods of time off of internally stored pools of nutrients (Andersen et al. 1991).

Perhaps more importantly, it may not be the availability of nitrogen or phosphorus alone that matters in controlling nuisance algal blooms but rather the relative availability of these nutrients in comparison to silicon (Officer and Ryther 1980, Smayda 1989). When Si:N and Si:P ratios are relatively high, silicon is relatively available, favoring the growth of diatoms, which have a high requirement for silicon. However, as the Si:N and Si:P ratios decrease, competition begins to favor other algae with no silicon requirement, such as the red-tide, green-tide, and brown-tide organisms. Most silicon comes from natural sources, and as inputs of nutrients from sewage increase, the Si:N and Si:P ratios decrease (Officer and Ryther 1980). Eutrophication itself can decrease the abundance of silicon by increasing sedimentation of phytoplankton, as has been demonstrated in the Baltic Sea (Wulff et al. 1990). Where long-term nutrient data are available, the increased occurrence of nuisance algal blooms has always been found to be correlated with a decrease in Si:N and Si:P ratios (Smayda 1989 and references therein). Net primary production probably remains controlled by nitrogen or phosphorus availability throughout the range of silicon availabilities (Howarth 1988), but the relative availability of silicon may well control the abundance of diatoms versus other phytoplankton species, thereby setting the stage for nuisance blooms (Smayda 1989).

Trace-metal availability may be another factor in the initiation of many nuisance algal blooms, with high iron availability favoring the bloom-forming species. A variety of physiological studies with pure cultures and short-term enrichment studies with natural populations have shown that the requirement for iron is high for red-tide dinoflagellates (Wilson 1966, Martin and Martin 1973, Graneli et al. 1986, Doucette and Harrison 1990), for the brown-tide algae (Cosper et al. 1990), and for cyanobacteria (Wurtsbaugh and Horne 1983, Howarth et al. 1988b). Red-tide outbreaks over a 25-year period in Florida have been correlated with iron inputs from rivers (Kim and Martin 1974) although inputs of a number of other substances are undoubtedly correlated with the iron inputs (Wells et al. 1991). However, most iron is not directly available to phytoplankton, and the available fraction of iron is not correlated with the outbreak of dinoflagellate blooms in Maine (Wells et al. 1991). Proof of a critical role for iron in initiating blooms must await further study.

CONCLUSIONS

Many estuaries and coastal marine ecosystems receive excessive inputs of nutrients, and sewage effluent is often a major component of this. These excessive inputs cause eutrophication, leading to anoxic and hypoxic conditions, loss of seagrass and algal beds, and damage to coral reefs. In many temperate estuaries and coastal seas, nitrogen is the primary nutrient of concern leading to increased eutrophication. Phosphorus controls eutrophication in some coastal marine ecosystems, at least during some seasons, and may be the primarily controlling element in tropical seas. To control coastal eutrophication, both nitrogen and phosphorus need to be controlled (D'Elia et al. 1986, Howarth 1988, Graneli et al. 1990). For phosphorus, the approach used in lakes—correlational models relating phosphorus to phytoplankton in lakes—should provide an adequate strategy for determining acceptable inputs to most temperate estuaries and coastal marine ecosystems as well. Acceptable inputs for nitrogen to marine ecosystems can be developed from similar models relating nitrogen inputs to phytoplankton production or biomass. However, such approaches may not adequately protect tropical and semi-tropical marine ecosystems that contain nutrient-sensitive corals or other ecosystems with sensitive benthic communities in shallow water.

No strong consensus exists on the extent to which excess nutrients contribute to the increasingly frequent occurrence of toxic algal blooms in many parts of the world's coastal oceans. However, to minimize problems from such blooms, it seems prudent to maintain availabilities of nitrogen and phosphorus at levels low compared to silicon availability, thereby favoring diatoms. Since increased iron availabilities may also contribute to the formation of toxic blooms, increases in the availability of iron in coastal waters should also be avoided. Factors that increase iron availability include erosion from land and increased reducing conditions in sediments from eutrophication.

REFERENCES

Andersen, T., A.K.L. Schartau, and E. Paasche. 1991. Quantifying external and internal nitrogen and phosphorus pools, as well as nitrogen and phosphorus supplied through remineralization, in coastal marine plankton by means of a dilution technique. Mar. Ecol. Progr. Ser. 69:67-80.

Anderson, D.M. 1989. Toxic algal blooms and red tides: A global perspective. In Red Tides: Biology, Environmental Science and Toxicology, T. Okaichi, D.M. Anderson, and T. Nemoto, eds. New York: Elsevier.

Baden, S.P., L.O. Loo, L. Pihl, and R. Rosenberg. 1990. Effects of eutrophication on benthic communities including fish: Swedish west coast. Ambio 19:113-122.

Banse, K. 1990. Does iron really limit phytoplankton production in the offshore subarctic Pacific? Limnol. Oceanogr. 35:772-775.

Borowitzka, M.A. 1972. Intertidal algal species diversity and the effect of pollution. Australian Journal of Marine and Freshwater Science 23:73-84.

Boynton, W.R., W.M. Kemp., and C.G. Osborne. 1980. Nutrient fluxes across the sediment-water interface in the turbid zone of a coastal plain estuary. In Estuarine Perspectives, V.S. Kennedy, ed. New York: Academic Press.

Boynton, W.R., W.M. Kemp, and C.W. Keefe. 1982. A comparative analysis of nutrients and other factors influencing estuarine phytoplankton production. In Estuarine Comparisons, V.S. Kennedy, ed. New York: Academic Press.

Bricelj, M., and S. Kuenstner. 1989. The feeding physiology and growth of bay scallops and mussels. In Novel Phytoplankton Blooms: Causes and Impacts of Recurrent Brown Tides and Other Unusual Blooms. Lecture Notes on Coastal and Estuarine Studies, E.M. Cosper, E.J. Carpenter, and V.M. Bricelj, eds. Berlin: Springer-Verlag.

Burkholder, J.M., K.M. Mason, and H.B. Glasgow. 1992a. Water-column nitrate enrichment promotes decline of eelgrass *Zostera marina*: Evidence from seasonal mesocosm experiments. Mar. Ecol. Prog. Ser. 81:163-178.

Burkholder, J.M., E.J. Noga, C.H. Hobbs, and H.B. Glasgow. 1992b. Now "phantom" dinoflagellate is the causative agent of major estuarine fish kills. Nature 358:407-410.

Butt, A.J. 1992. Numerical models and nutrient reduction strategies in Virginia. Coastal Management 20:25-36.

Cambridge, M.L., and A.J. McComb. 1984. The loss of seagrasses in Cockburn Sound, Western Australia. I. The time course and magnitude of seagrass decline in relation to industrial development. Aquatic Botany 20:229-242.

Cambridge, M.C., A.W. Chaffings, C. Brittan, L. Moore, and A.J. McComb. 1986. The loss of seagrass in Cockburn Sound, Western Australia. II. Possible causes of seagrass decline. Aquatic Botany 24:269-285.

Caraco, N., J.J. Cole, and G. E. Likens. 1989. Evidence for sulfate-controlled phosphorus release from sediments of aquatic systems. Nature 341:316-318.

Caraco, N., J.J. Cole, and G.E. Likens. 1990. A comparison of phosphorus immobilization in sediments of freshwater and coastal marine systems. Biogeochemistry 9:277-290.

Carpenter, E.J., J. Chang, M. Cottrell, J. Schubauer, H.W. Paerl, B.M. Bebout, and D.G. Capone. 1990. Re-evaluation of nitrogenase oxygen-protective mechanisms in the planktonic marine cyanobacterium *Trichodesmium*. Mar. Ecol. Prog. Ser. 65:151-158.

Carpenter, S.R., J.F. Kitchell, and J.R. Hodgson. 1985. Cascading trophic interactions and lake productivity. BioScience 35:634-639.

Carpenter, S.R., J.R. Kitchell, and J.R. Hodgson. 1987. Regulation of lake primary production by food web structure. Ecology 68:1863-1876.

Cederwall, H., and R. Elmgren. 1990. Biological effects of eutrophication in the Baltic Sea, particularly the coastal zone. Ambio 19(3):109-112.

Chesterikoff, A., B. Garban, G. Billen, and M. Poulin. 1992. Inorganic nitrogen dynamics in the River Seine downstream from Paris (France). Biogeochemistry 17(3):147-164.

Cosper, E.M. 1991. Recent and Historical Novel Algal Blooms. Monospecific Blooms Occurred Along Northeast Coast in 1980s. Waste Management Research Report (SUNY Buffalo, SUNY Stony Brook, and Cornell Univ.) 3(2):3-6.

Cosper, E.M., C. Lee, and E.J. Carpenter. 1990. Novel "brown tide" bloom in Long Island embayments: A search for the causes. In Toxic Marine Phytoplankton, E. Graneli, B. Sundsttrom, L. Edler, and D.M. Anderson, eds. New York: Elsevier.

Cosper, E.M., W.C. Dennison, E.J. Carpenter, V.M. Bricelj, J.G. Mitchell, S.H. Kuenstner, D.C. Colflesh, and M. Dewey. 1987. Recurrent and persistent "brown tide" blooms perturb coastal marine ecosystem. Estuaries 10:284-290.

Cosper, E.M., W. Dennison, A. Milligan, E.J. Carpenter, C. Lee, J. Holzapfel, and L. Milanese. 1989. An examination of the environmental factors important to initiating and sustaining "brown tide" blooms. In Novel Phytoplankton Blooms: Causes and Impacts of Recurrent Brown Tides and Other Unusual Blooms. Lecture Notes on Coastal and Estuarine Studies, E.M. Cosper, E.J. Carpenter, and V.M. Bricelj, ed. Berlin: Springer-Verlag.

D'Elia, C.F. 1987. Nutrient enrichment of the Chesapeake Bay—Too much of a good thing. Environment 29:6-33.

D'Elia, C.F., J.G. Sanders, and W.R. Boynton. 1986. Nutrient enrichment studies in a coastal plain estuary: Phytoplankton growth in large-scale, continuous cultures. Can. J. Fish. Aquat. Sci. 43:397-406.

Dennison, W.C., G.J. Marshall, and C. Wigand. 1989. Effect of "brown tide" shading on eelgrass *(Zostera marina* L.) distributions. In Novel Phytoplankton Blooms: Causes and Impacts of Recurrent Brown Tides and Other Unusual Blooms. Lecture Notes on Coastal and Estuarine Studies, E.M. Cosper, E.J. Carpenter, and V.M. Bricelj, eds. Berlin: Springer-Verlag.

Doering, P.H., C.A. Oviatt, L.L. Beatty, V.F. Banzon, R. Rice, S.P. Kelly, B.K. Sullivan, and J.B. Frithsen. 1989. Structure and function in a model coastal ecosystem: Silicon, the benthos and eutrophication. Mar. Ecol. Prog. Ser. 52:287-299.

Doucette, G.J., and P.J. Harrison. 1990. Some effects of iron and nitrogen stress on the red tide dinoflagellate *Gymnodinium sanguineum*. Mar. Ecol. Prog. Ser. 62:293-306.

Edmondson, W.T. 1970. Phosphorus, nitrogen, and algae in Lake Washington after diversion of sewage. Science 169:690-691.

EPA (U.S. Environmental Protection Agency). 1990. Long Island Sound Study, Status Report and Interim Actions for Hypoxia Management. New York: U.S. Environmental Protection Agency.

Flett, R.J., D.W. Schindler, R.D. Hamilton, and N.E.R. Campbell. 1980. Nitrogen fixation in Canadian precambrian shield lakes. Can. J. Fish. Aquat. Sci. 37:494-505.

Frithsen, J.B., C.A. Oviatt, M.E.Q. Pilson, R.W. Howarth, and J.J. Cole. 1988. A comparison of nitrogen vs. phosphorus limitation of production in coastal marine ecosystems. EOS 69(44):1100.

Gappa, J., J. Lopez, A. Tablado, and N.H. Magaldi. 1990. Influence of sewage pollution on a rocky intertidal community dominated by the mytilid *Brachidontes rodriguezi*. Mar. Ecol. Prog. Ser. 63:163-175.

Gardner, W.S., S.P. Seitzinger, and J.M. Malczyk. 1991. The effects of sea salts on the forms of nitrogen released from estuarine and freshwater sediments: Does ion pairing affect ammonium flux? Estuaries 14:157-166.

Gieson, W.B.J.T., M.M. van Katwijk, and C. den Hartog. 1990. Eelgrass condition and turbidity in the Dutch Wadden Sea. Aquatic Botany 37:71-85.

Graneli, E. 1978. Algal assay of limiting nutrients for phytoplankton production in the Oresund. Vatten 2:117-128.

Graneli, E. 1981. Bioassay experiments in the Falsterbo Channel—nutrients added daily. Kieler Meeresforsch. Sonderh. 5:82-90.

Graneli, E. 1984. Algal growth potential and limiting nutrients for phytoplankton production in Oresund water of Baltic and Kattegat origin. Limnologica (Berlin) 15:563-569.

Graneli, E., H. Persson, and L. Edler. 1986. Connection between trace metals, chelators, and red tide blooms in the Laholm Bay, SE Kattegat—An experimental approach. Mar. Env. Res. 18:61-78.

Graneli, E., K. Wallstrom, U. Larsson, W. Graneli, and R. Elmgren. 1990. Nutrient limitation of primary production in the Baltic Sea area. Ambio 19:142-151.

Gray, J.S., and E. Paasche. 1984. On marine eutrophication. Mar. Pollut. Bull. 15:349-350.

Grigg, R.W. 1978. Long-term changes in rocky bottom communities of Palos Verdes. Pp. 157-184 in Coastal Water Research Project Annual Report, 1978, W. Bascom, ed. El Segundo, California: Southern California Coastal Water Research Project.

Hallegraeff, G.M., D.A. Steffensen, and R. Wetherbee. 1988. Three estuarine dinoflagellates that can produce paralytic shellfish toxins. Journal of Plankton Research 10:533-541.

Hansson, S., and L.G. Rudstam. 1990. Eutrophication and Baltic fish communities. Ambio 19:123-125.

Harris, L. 1980. Changes in intertidal algae at Palos Verdes. Pp. 35-73 in Coastal Water Research Project Biennial Report 1979-1980, W. Bascom, ed. Long Beach, California: Southern California Coastal Water Research Project.

Hecky, P.E., and P. Kilham. 1988. Nutrient limitation of phytoplankton in freshwater and marine environments: A review of recent evidence on the effects of enrichment. Limnol. Oceanogr. 33:796-822.

Howarth, R.W. 1988. Nutrient limitation of net primary production in marine ecosystems. Annual Review of Ecology & Systematics 19:89-110.

Howarth, R.W., and J.J. Cole. 1985. Molybdenum availability, nitrogen limitation, and phytoplankton growth in natural waters. Science 229:653-655.

Howarth, R.W., and R. Marino. 1990. Nitrogen-fixing cyanobacteria in the plankton of lakes and estuaries: A reply to the comment by Smith. Limnol. Oceanogr. 35:1859-1863.

Howarth, R.W., R. Marino, J. Lane, and J.J. Cole. 1988a. Nitrogen fixation in freshwater, estuarine, and marine ecosystems. 1. Rates and importance. Limnol. Oceanogr. 33:669-687.

Howarth, R.W., R. Marino and J.J. Cole. 1988b. Nitrogen fixation in freshwater, estuarine, and marine ecosystems. 2. Biogeochemical controls. Limnol. Oceanogr. 33:688-701.

Jaworski, N.B. 1981. Sources of nutrients and the scale of eutrophication problems in estuaries. In Estuaries and Nutrients, B.J. Neilson and L.E. Cronin, eds. New York: Humana.

Jensen, L.M., K. Sand-Jensen, S. Marcher, and M. Hansen. 1990. Plankton community respiration along a nutrient gradient in a shallow Danish estuary. Mar. Ecol. Prog. Ser. 61:75-85.

Kamp-Nielsen, L. 1974. Mud-water exchange of phosphorus and other ions in undisturbed sediment cores and factors affecting the exchange rate. Arch. Hydrobiol. 13:218-237.

Khalid, R.A., W.H. Patrick, and R.D. DeLaune. 1977. Phosphorus sorption characteristics of flooded soils. Soil Sci. Soc. Am. J. 41:305.

Kelly, J., and S. Levin. 1986. A comparison of aquatic and terrestrial nutrient cycling and production processes in natural ecosystems, with reference to ecological concepts of relevance to some waste disposal issues. In The Role of Oceans as a Waste Disposal Option, G. Kullenber, ed. Amsterdam: Reidel.

Kemp, W.M., R.R. Twilley, J.C. Stevenson, W.R. Boynton, and J.C. Means. 1983. The decline of submerged vascular plants in upper Chesapeake Bay: Summary of results concerning possible causes. J. Mar. Technol. Soc. 17:78-85.

Kim, Y.S., and D.F. Martin. 1974. Interrelationship of Peace River parameters as a basis of the iron index: A predictive guide to the Florida red tide. Wat. Res. 8:607-616.

Kirkman, R.H. 1976. A Review of the Literature on Seagrass Related to its Decline in Moreton Bay, Qld. CSIRO Report no. 64. CSIRO.

Lapointe, B.E., M.M. Littler, and D.S. Littler. 1987. A comparison of nutrient-limited productivity in macroalgae from a Caribbean barrier reef and from a mangrove ecosystem. Aquatic Botany 28:243-255.

Larsson, U.R., R. Elmgren, and F. Wulff. 1985. Eutrophication and the Baltic Sea: Causes and consequences. Ambio 14:10-14.

Lein, A.Y., and M.V. Ivanov. 1992. Interaction of carbon, sulphur, and oxygen cycles in continental and marginal seas. In Sulphur Cycling on the Continents: Wetlands, Terrestrial Ecosystems, and Associated Water Bodies, R.W. Howarth, J.W.B. Stewart, and M.V. Ivanov, eds. Chichester, U.K.: Wiley & Sons.

Lindahl, G., and K. Wallstrom. 1985. Nitrogen fixation (acetylene reduction) in planktonic cyanobacteria in Oregrundsgrepen, SW Bothnian Sea. Arch. Hydrobiol. 104:193-204.

Littler, M.M., and S.N. Murray. 1975. Impact of sewage on the distribution, abundance and community structure of rocky intertidal macro-organisms. Mar. Biol. 30:277-291.

Littler, M.M., and S.N. Murray. 1978. Influence of domestic wastes on energetic pathways in rocky intertidal communities. J. Appl. Ecol. 15:583-596.

Littler, M.M., D.S. Littler, and B.E. Lapointe. 1988. A comparison of nutrient- and light-limited photosynthesis in psarnmophyfic versus epilithic forms of *Halimeda* (caulerpales, halimedaceae) from the Bahamas. Coral Reefs 6:219-225.

Malone, T.C. 1982. Factors influencing the fate of sewage-derived nutrients in the lower Hudson estuary and New York bight. In Ecological Stress and the New York Bight: Science and Management, G.F. Mayer, ed. Columbia, South Carolina: Estuarine Research Federation.

Marino, R., R.W. Howarth, J. Shamess, and E.E. Prepas. 1990. Molybdenum and sulfate as controls on the abundance of nitrogen-fixing cyanobacteria in saline lakes in Alberta. Liminol. Oceanogr. 35:245-259.

Martin, D.F., and B.B. Martin. 1973. Implications of metal organic interactions in red tide outbreaks. In Trace Metals and Metal-organic Interactions in Natural Waters, P.C. Singer, ed. Ann Arbor, Michigan: Ann Arbor Science.

McComb, A.J., R.P. Atkins, P.B. Birch, D.M. Gordon, and R.J. Luketelich. 1981. Eutrophication in the Peel-Harvey Estuarine System, Western Australia. In Estuaries and Nutrients, B.J. Nielson and L.E. Cronin, eds. New York: Humana.

McGlathery, K.J., R.W. Howarth, and R. Marino. 1992. Nutrient limitation of the macroalga,

Penicillus capitatus, associated with subtropical seagrass meadows in Bermuda. Estuaries 15:18-25.

Mearns, A.J., and J.Q. Word. 1982. Forecasting the effects of sewage solids on marine benthic communities. Pp. 495 in Ecological Stress in the New York Bight: Science and Management, G.F. Mayer, ed. Columbia, South Carolina: Estuarine Research Federation.

Mearns, A.J., E. Haines, G.S. Klepple, R.A. McGrath, J.J.A. McLaughlin, D.A. Segar, J.H. Sharp, J.J. Walsh, J.Q. Word, D.K. Young, and M.W. Young. 1982. Effects of nutrients and carbon loadings on communities and ecosystems. In Ecological Stress and the New York Bight: Science and Management, G.F. Mayer, ed. Columbia, South Carolina: Estuarine Research Federation.

Molot, L.A., and P.J. Dillon. 1991. Nitrogen/phosphorus ratios and the prediction of chlorophyll in phosphorus-limited lakes in central Ontario. Can. J. Fish. Aquat. Sci. 48:140-145.

Morin, A., K.D. Hambright, N.G. Hairston, D.M. Sherman, and R.W. Howarth. 1991. Consumer control of gross primary production in replicate freshwater ponds. International Vereinigung fuer Theoretische und Angewandte Limnologie. Verhandlungen IVTLAP 24(3):1512-1516.

Morse, J.W., J.J. Zullig, L.D. Bernstein, F.J. Millero, P. Milne, A. Mucci, and G.R. Choppin. 1985. Chemistry of calcium carbonate-rich shallow water sediments in the Bahamas. Am. J. Sci. 285:147-185.

Myers, V.B., and R.I. Iverson. 1981. Phosphorus and nitrogen limited phytoplankton productivity in northeastern Gulf of Mexico coastal estuaries. In Estuaries and Nutrients, B.J. Nielson, and L.E. Cronin, eds. New York: Humana.

Newell, R.I.E. 1988. Ecological changes in Chesapeake Bay: Are they the result of over harvesting the American oyster, *Crassotrea virginica*? Pp. 29-31 in proceedings of Understanding the Estuary: Advances in Chesapeake Bay Research, March 1988. Publication 129. Baltimore, Maryland: Chesapeake Research Consortium.

Nixon, S.W. 1988. Physical energy inputs and the comparative ecology of lake and marine ecosystems. Limnol. Oceanogr. 33:1005-1025.

Nixon, S.W. 1992. Quantifying the relationship between nitrogen input and the productivity of marine ecosystems. Adv. Mar. Techn. Conf. 5:57-83.

Nixon, S.W., and M.E.Q. Pilson. 1983. Nitrogen in estuarine and coastal marine ecosystems. In Nitrogen in the Marine Environment, E.J. Carpenter and D.G. Capone, eds. New York: Academic Press.

Nixon, S.W., J.R. Kelly, B.N. Fumas, C.A. Oviatt, and S.S. Hale. 1980. Phosphorus regeneration and the metabolism of coastal marine bottom communities. In Marine Benthic Dynamics, K.R. Tenore and B.C. Coull, eds. Columbia, South Carolina: University of South Carolina Press.

Nixon, S.W., C. Oviatt, J. Frithsen, and B. Sullivan. 1986. Nutrients and productivity of estuaries and coastal marine ecosystems. J. Limnol. Soc. S. Afr. 12:43-71.

National Oceanographic and Atmospheric Administration and the U.S. Environmental Protection Agency (NOAA/EPA). 1988. Strategic Assessment of Near Coastal Waters: Northeast Case Study. Susceptibility and Status of Northeast Estuaries to Nutrient Discharges. Rockville, Maryland: National Oceanic and Atmospheric Administration.

Norin, L.L. 1977. ^{14}C-bioassays with the natural phytoplankton in the Stockholm archipelago. Ambio Spec. Rep. 5:15-21.

Officer, C.B., and J.H. Ryther. 1980. The possible importance of silicon in marine eutrophication. Mar. Ecol. Prog. Ser. 3:83-91.

Officer, C.B., R.B. Biggs, J. Taft, L.E. Cronin, M.A. Tyler, and W.R. Boynton. 1984. Chesapeake Bay anoxia: Origin, development, and significance. Science 223:22-27.

Paerl, H.W., and R.C. Carlton. 1988. Control of nitrogen fixation by oxygen depletion in surface-associated microzones. Nature 332:260-262.

Paerl, H.W., K.M. Crocker, and L.E. Prufert. 1987. Limitation of N_2 fixation in coastal marine waters: Relative importance of molybdenum, iron, phosphorous, and organic matter availability. Limnol. Oceanogr. 32:525-536.

Parker, C.A., and J.E. O'Reilly. 1991. Oxygen depletion in Long Island Sound: A historical perspective. Estuaries 14:248-264.

Pastorak, R.A., and G.R. Bilyard. 1985. Effects of sewage pollution on coral-reef communities. Mar. Ecol. Prog. Ser. 21:175-189.

Prego, R. 1992. Flows and budgets of nutrient salts and organic carbon in relation to a red tide in the Ria of Vigo (NW Spain). Marine Ecology Progress Series 79:289-302.

Price, K.S., D.A. Flemer, J.L. Taft, and G.B. Mackiernnan. 1985. Nutrient enrichment of Chesapeake Bay and its impact on the habitat of striped bass: A speculative hypothesis. Trans. Am. Fish. Soc. 114:97-106.

Powell, G.V.N., W.J. Kenworthy, and J.F. Fourqurean. 1989. Experimental evidence for nutrient limitation of seagrass growth in a tropical estuary with restricted circulation. Bull. Mar. Sci. 44:324-340.

Redfield, A.C. 1958. The biological control of chemical factors in the environment. Am. Sci. 46:205-221.

Robblee, M.B., T.R. Barber, P.R. Carlson, M.J. Durako, J.W. Fourqurean, L.K.Muehlstein, D. Porter, L.A. Yarbro, R.T. Zieman, and J.C. Zieman. 1991. Mass mortality of the tropical seagrass *Thalassia testudinum* in Florida Bay (USA). Mar. Ecol. Prog. Ser. 71:297-299.

Robineau, B., J.A. Gagne, L. Fortier, and A.D. Cembella. 1991. Potential impact of a toxic dinoflagellate (*Alexandrium excovatum*) bloom on survival of fish and crustacean larvae. Marine Biology 108:293-301.

Rosenberg, R. 1985. Eutrophication—The future marine coastal nuisance? Mar. Poll. Bull. 16:227-231.

Rosenberg, R., R. Elmgren, S. Fleischer, P. Jonsson, G. Persson, and H. Dahlin. 1990. Marine eutrophication case studies in Sweden. Ambio 19:102-108.

Rounsefell, G.A., and A. Dragovich. 1966. Correlation between oceanographic Rfactors and abundance of the Florida redtide *(Gymnodiniwn breve* Davis), 1954-1961. Bull. Mar. Sci. 16:402.

Rowe, G.T., C.H. Cliffer, K.L. Smith, and P.L. Hamilton. 1975. Benthic nutrient regeneration and its coupling to primary productivity in coastal waters. Nature 225:215-217.

Rydberg, L., L. Edler, S. Floderus, and W. Graneli. 1990. Interaction between supply of nutrients, primary production, sedimentation and oxygen consumption in SE Kattegat. Ambio 19:134-141.

Ryther, J.H. 1954. The ecology of phytoplankton blooms in Moriches Bay and Great South Bay, Long Island, New York. Biol. Bull. 106:198-209.

Ryther, J.H. 1989. Historical perspective of phytoplankton blooms on Long Island and the green tides of the 1950's. In Novel Phytoplankton Blooms: Causes and Impacts of Recurrent Brown Tides and Other Unusual Blooms. Lecture Notes on Coastal and Estuarine Studies, E.M. Cosper, E.J. Carpenter, and V.M. Bricelj, eds. Berlin: Springer-Verlag.

Ryther, J.H., and W.M. Dunstan. 1971. Nitrogen, phosphorus and eutrophication in the coastal marine environment. Science 171:1008-1012.

Schindler, D.W. 1977. Evolution of phosphorus limitation in lakes. Science 195: 260-262.

Schindler, D.W. 1978. Factors regulating phytoplankton production and standing crop in the world's freshwaters. Limnol. Oceanogr. 23:478-486.

Schindler, D.W. 1981. Studies of eutrophication in lakes and their relevance to the estuarine environment. In Estuaries and Nutrients, B.J. Neilson and L.E. Cronin, eds. New York: Humana.

Schindler, D.W., R. Hesslein, and G. Kipphut. 1977. Interactions between sediments and overlying waters in an experimentally eutrophied pre-cambrian shield lake. In Interactions Between Sediments and Fresh Water, H.L. Golttemian, ed. Junk, The Hague.

Seitzinger, S.P. 1988. Denitrification in freshwater and marine ecosystems: Ecological and geochemical significance. Limnol. Oceanog. 33:702-724.

Seitzinger, S.P., W.S. Gardner, and A.K. Spratt. 1991. The effect of salinity on ammonium sorption in aquatic sediments: Implications for benthic nutrient cycling. Estuaries 14:167-174.

Short, F.T., W.C. Dennison, and D.G. Cappone. 1990. Phosphorus-limited growth of the tropical seagrass *Syringodiumfiliforme* in carbonate sediments. Mar. Ecol. Prog. Ser. 62:169-174.

Short, F.T., M. W. Davis, R.A. Gibson, and C.F. Zimmerman. 1985. Evidence for phosphorus limitation in carbonate sediments of the seagrass *Syringodiumfiliforme*. Estuarine, Coastal and Shelf Science. 20:419-430.

Sieburth, J., P.W. Johnson, and P.E. Hargraves. 1988. Ultrastructure and ecology of *Aureococcus anophagefferens* gen. et sp. nov. (Chrysophyceae); the dominant picoplankter during a bloom in Narragansett Bay, Rhode Island, summer 1985. J. Phycol. 24:416-425.

Smayda, T.J. 1974. Bioassay of the growth potential of the surface water of lower Narragansett Bay over an annual cycle using the diatom *Thalassiosira pseudonana* (oceanic clone, 13-1). Limnol. Oceanogr. 19:889-901.

Smayda, T.J. 1989. Primary production and the global epidemic of phytoplankton blooms in the sea: A linkage? Pp. 449-483 in Novel Phytoplankton Blooms: Causes and Impacts of Recurrent Brown Tides and Other Unusual Blooms. Lecture Notes on Coastal and Estuarine Studies, E.M. Cosper, E.J. Carpenter, and V.M. Bricelj, ed. Berlin: Springer-Verlag.

Smayda, T.J. 1992. A phantom of the ocean. Nature 358:374-375.

Smith, V.H. 1979. Nutrient dependence of primary productivity in lakes. Limnol. Oceanogr. 24:1051-1064.

Smith, S.V. 1981. Responses of Kaneohe Bay, Hawaii, to relaxation of sewage stress. In Estuaries and Nutrients, B.J. Neilson and L.E. Cronin, eds. New York: Humana.

Smith, S.V. 1984. Phosphorus vs. nitrogen limitation in the marine environment. Limnol. Oceanogr. 29:1149-1160.

Smith, S.V., and M.J. Atkinson. 1984. Phosphorus limitation of net production in a confined aquatic ecosystem. Nature 207:626-627.

Steele, J.H. 1974. The Structure of Marine Ecosystems. Cambridge, Massachusetts: Harvard University Press.

Tracey, G.A., R.L. Steele, J. Gatzke, D.K. Phelps, R. Nuzzi, M. Waters, and D.M. Anderson. 1989. Testing and application of biomonitoring methods for assessing environmental effects of noxious algal blooms. In Novel Phytoplankton Blooms: Causes and Impacts of Recurrent Brown Tides and Other Unusual Blooms. Lecture Notes on Coastal and Estuarine Studies, E.M. Cosper, E.J. Carpenter, and V.M. Bricelj, eds. Berlin: Springer-Verlag.

Twilley, R.R., W.M. Kemp, K.W. Staver, J.C. Stevenson, and W.R. Boynton. 1985. Nutrient enrichment of estuarine submerged vascular plant communities. 1. Algal growth and effects on production of plants and associated communities. Mar. Ecol. Prog. Ser. 23:179-191.

Valiela, I. 1984. Marine Ecological Processes. New York: Springer-Verlag.

Valiela, I., Costa, J., Foreman, K., Teal, J.M., Howes, B., and Aubrey, B. 1990. Transport of groundwater-borne nutrients from watersheds and their effects on coastal waters. Biogeochemistry 10:177-197.

Vince, S., and I. Valiela. 1973. The effects of ammonium and phosphate enrichment on chlorophyll a, a pigment ratio and species composition of phytoplankton of Vineyard Sound. Mar. Biol. 19:69-73.

Vitousek, P.M., and R.W. Howarth. 1991. Nitrogen limitation on land and in the sea: How can it occur? Biogeochemistry 13:87-115.

Vollenweider, R.A. 1976. Advances in defining critical loading levels for phosphorus in lake eutrophication. Mem. 1st. Ital. ldrobiol. 33:53-83.

Vollenweider, R.A. 1979. Das Nahrstoffbelastungskonzept als Grundlage fur den externen Eingriff in den Eutrophierungsprozess stehender Gewasser und Talsperren. Z. Wasser-u. Abwasser-Forschung 12:46-56.

Wells, M.L., L.M. Mayer, and R.R.L. Guillard. 1991. Evaluation of iron as a triggering factor for red tide blooms. Mar. Ecol. Prog. Ser. 69:93-102.

Wetzel, R.G. 1983. Limnology. Philadelphia, Pennsylvania: Saunders.

Wilson, W.B. 1966. The Suitability of Seawater for the Survival and Growth of Gymnodinium breve Davis, and Some Effects of Phosphorus and Nitrogen on its Growth. Professional Paper Series, Marine Laboratory Florida 7:1-42.

Wilson, K., A.J. Mearns, and J.J. Grant. 1980. Changes in kelp forests at Palos Verdes. Pp. 77-92 in Coastal Water Research Project Biennieal Report 1979-1980, W. Bascom, ed. Long Beach, CA: Southern California Coastal Water Research Project.

Wulff, F., A. Stigebrandt, and L. Rahm. 1990. Nutrient dynamics of the Baltic Sea. Ambio 19:126-133.

Wurtsbaugh, W.A., and A.J. Horne. 1983. Iron in eutrophic Clear Lake, California: Its importance for algal nitrogen fixation and growth. Can. J. Fish. Aquat. Sci. 40:1419-1429.

B

Microbial Pathogens in Coastal Waters

Bacterial diseases associated with polluted recreational waters and shellfish have been documented for over 100 years. Typhoid, for example, was first documented in association with recreational waters as early as 1888 (Craun 1986). Transmission of viral disease via recreational exposure to sewage contaminated waters was first documented as early as the 1950s, and is now well established (Stevenson 1953, Balarajan et al. 1991, Alexander et al. 1992, Fewtrell et al. 1992). Transmission of typhoid and cholera associated with the consumption of contaminated seafood has long been recognized, and by 1956 the risk of viral diseases, specifically hepatitis, was documented (Roos 1956). Disease occurs through two pathways of exposure: swimming in contaminated waters or eating contaminated fish or shellfish. Bathing in contaminated water can result in accidental swallowing or aspiration of infective pathogens. Ingestion of contaminated seafood can cause infection by pathogens or toxicity from toxins elaborated by microorganisms or algae. The effects of microbial infections can range from infection without overt disease to acute, self-limited respiratory, skin, gastrointestinal, and ear infections to extreme gastrointestinal and liver disorders and even to death.

MICROBIOLOGIC AGENTS ASSOCIATED WITH WASTEWATER

Over 100 different enteric pathogens may be found in sewage. These includes viruses, parasites, and bacteria, all of which may be associated with waterborne disease.

Viruses

Enteric viruses are obligate human pathogens. That is, they replicate only when within the human host. Their structures may allow prolonged survival outside the human body, in the environment. There are over 120 enteric viruses that may be found in sewage. Table B.1 lists some of the better described viruses, including the enteroviruses (polio-, echo-, and coxasackieviruses), hepatitis A virus, rotavirus, and Norwalk virus, and the annual incidence of disease and case mortality rates for all sources of exposure (Bennett et al. 1987). Virus numbers reported in sewage vary greatly and reflect the variation in infection in the population excreting the agent, the season of the year (outbreaks of viral disease are often seasonal), and methods used for their recovery and detection. Table B.1 shows virus numbers that have been reported in sewage. Treatment reduces but does not eliminate viral contamination (Melnick and Gerba 1980, Rose and Gerba 1990, Asano et al. 1992).

Over 100 outbreaks of hepatitis and viral gastroenteritis have been associated with the consumption of sewage contaminated shellfish in the United States (Richards 1985). The reported outbreaks have increased from less than 10 in the years 1966-1970 to more than 50 in the years 1981-1985. Although this apparent increase could be due to reporting artifacts, the number reported most certainly represents a great underestimate because of the long incubation period for hepatitis A and the difficulty in tracing the source. From 1983-1989, the incidence of hepatitis A increased 58 percent with 14.5 cases in 100,000 in the United States. An estimated 10 percent of these cases may be due to foodborne transmission, including shellfish (CDC 1990).

Viral outbreaks due to recreational exposure to contaminated waters have been documented in the United States. Between the years 1986 and 1988, 41 percent of these were an undefined gastroenteritis and likely of a viral etiology (CDC 1990). *Shigella* and *Giardia* were also predominant causes of recreational outbreaks of disease. Viruses (entero- and rotaviruses) have been isolated from recreational waters in the absence of any discharge from a wastewater treatment plant (Rose et al. 1987).

Parasites

The parasites of primary public health concern for wastewater exposure are the protozoa and helminths. The helminths include roundworms (*Ascaris*), hookworms, tapeworms, and whipworms. These organisms are endemic in areas where there is inadequate hygiene and their transmission is generally associated with untreated sewage, untreated sludges, and night soil, with very little documentation of waterborne transmission.

TABLE B.1 Characteristics of Enteric Viruses (Bennett et al. 1987)

Virus Group	1985 Reported Cases	Mortality Rates (%)	Levels in Sewage/L	Diseases
Enterovirus:	6,000,000	0.001	182-92,000	
Poliovirus	7	10		Paralysis Aseptic meningitis
Coxasackievirus				
A				Herpangina Aseptic meningitis Respiratory illness
B				Paralysis fever Pleurodynia Aseptic meningitis Pericarditis Myocarditis Congenial heart anomalies Nephritis
Echovirus:				Respiratory infection Aseptic meningitis Diarrhea Pericarditis Myocarditis fever, rash
Hepatitis A Virus	48,000	0.6	5101	Infectious hepatitis
Reovirus			1-1,247	Respiratory disease Gastroenteritis
Adenovirus	10,000,000	0.01	100-100,000	Acute conjunctivitis Diarrhea Respiratory illness Eye infection
Rotavirus	8,000,000	0.01	401	Infantile gastroenteritis
Norwalk agent (probably a calcivirus)	6,000,000	0.0001		Gastroenteritis
Astrovirus				Gastroenteritis
Calcivirus				Gastroenteritis
Snow Mt. Agent (probably a calcivirus)				Gastroenteritis
Norwalk-like virus				Gastroenteritis
Non-A, Non-B Hepatitis	50,000	0.4		Hepatitis

The pathogenic enteric protozoa *Giardia lamblia*, *Cryptosporidium*, and *Entamoeba histolytica* are listed in Table B.2. These enteric protozoa are important waterborne pathogens and are a cause of acute and chronic diarrhea. They replicate only within their host. The cyst or oocyst that is excreted in the feces is the infective form, able to survive in the environment and 10 to 1,000 times more resistant to water disinfection than the bacteria (Jarroll 1988, Korick et al. 1990). Filtration is a more effective means of removal. Ingestion of small numbers of cysts (between 1 and 10) are capable of initiating an infection (Rose et al. 1991a). Therefore, as for viruses, low levels are of greater public health concern than low levels of bacterial contamination.

Entamoeba histolytica infects only humans. Waterborne transmission is usually from raw sewage contamination of the water. Although only one

TABLE B.2 Characterization of Pathogenic Protozoa in Relationship to Waterborne Diseases

	Giardia	Cryptosporidium	Entamoeba	Isospora
Type of Protozoan	Obligate enteric amoebae	Obligate enteric coccidian	Obligate enteric amoebae	Obligate enteric coccidian
Transmission Routes	Fecal-oral by cysts	Fecal-oral by oocysts	Fecal-oral by cysts	Fecal-oral by oocysts
Reservoirs of Infection for Man	Infected animals and man, Chronic human carriers	Infected animals and man, Chronic human carriers	Infected animals and man, Chronic human carriers	Infected animals and man, Chronic human carriers
Documented Waterborne Disease in the U.S.	106 outbreaks 1965-1988 >26,010 cases	3 outbreaks 1980-1988 > 13,117 cases	8 outbreaks[1] 1920-1988 1,495 cases	None
Type of Illness	Acute (5-30d.) and chronic (months) infections of diarrhea	Self-limiting diarrhea, Cholera-like 7-10 day	Diarrhea, Liver abscesses, Mortality 0.02-6%	Diarrhea
Levels in Sewage	530-100,000/L	10-1,000/L	28-52/L	Unknown

[1]Only 1 since 1971.

waterborne outbreak has been documented since 1971 in the United States (Craun 1991), this organism is an important cause of morbidity worldwide.

Giardia lamblia is the most common protozoan infection in the United States and is a major public health concern. Of an estimated 60,000 cases of illness due to *Giardia* (giardiasis) per year, it has been suggested that 60 percent of these are waterborne (Bennett et al. 1987). There has been an increase in the reported incidence of waterborne giardiasis since 1971 (Craun 1991). Because of animal reservoirs, the role sewage contamination has played in this increase has been impossible to determine. *Giardia* cysts have been detected in treated and untreated sewage at levels between 530 and 100,000/liter (Sykora et al. 1990).

Cryptosporidium was first recognized as a waterborne agent in 1985 (D'Antonio et al. 1985). Cryptosporidiosis is a serious and potentially fatal infection in the immunocompromised (infants less than six months of age, the elderly, and those with disease states that impair the immune system) and may be more prevalent in children under one year of age. In the United States, *Cryptosporidium* appears to account for between 0.1 and 1.9 percent of the incidences of acute diarrhea (CDC 1990). Yet sporadic outbreaks associated with drinking water have occurred in which 13,000 people became ill (Hayes et al. 1989) as have outbreaks from recreational exposure to water in lakes (Gallagher et al. 1989). Occurrence in treated wastewater effluents has been documented and the levels appear to be slightly less than *Giardia* (Rose et al. 1988).

Bacteria

Enteric bacterial pathogens remain an important cause of disease in the United States (Table B.3). Classical waterborne bacterial diseases such as dysentery, typhoid, and cholera, while still very important worldwide, have dramatically decreased in the United States since the 1920s (Craun 1991). However, *Campylobacter*, non-typhoid *Salmonella*, and pathogenic *Escherichia coli* have been estimated to cause 3 million waterborne illnesses per year (Bennett et al. 1987). Foodborne cases represent a much greater percentage. The specific role of polluted coastal waters in the acquisition of these infections has been difficult to determine. While the previous bacteria all have nonmarine animal reservoirs, the noncholera *Vibrio* sp. may be found naturally in the marine environment and contributes to a portion of the 50,000 cases of seafood-associated gastroenteritis annually reported in the United States.

No animal reservoirs have been identified for *Shigella*; therefore humans appear to be the only source. This agent was responsible for the majority (52 percent) of recreational waterborne outbreaks between 1981 and 1987 in lakes and rivers for a total of 428 cases of shigella gastroenteritis

TABLE B.3 Characterization of Enteric Bacterial Pathogens (Feachem et al. 1983, Bennett et al. 1987)

Bacteria	Reported Cases in the U.S.	Percent Waterborne	Mortality Rates (%)	Levels in Sewage/100ml
Campylobacter	8,400,000	15	0.1	NR[1]
Pathogenic *E. coli*	2,000,000	75	0.2	NR
Salmonella	10,000,000	3	0.1	2.3 - 8,000
S. typhi	600	10	6	NR
Shigella	666,667	10	0.2	1 - 1,000
Vibrio cholera	25	NR	1	NR
Vibro non-cholera	50,000	10	4	10 - 10,000
Yersinia	5,025	35	0.05	NR

[1]NR = Not reported.

(shigellosis). The source of the contamination was never fully described in these outbreaks. Like the enteric viruses, the overuse of recreational sites may lead to the contamination, degradation of water quality, and disease outbreaks (CDC 1990). Recreational outbreaks in marine waters have not been as well documented.

Intestinal bacteria have been used for more than 100 years as indicators of the presence of feces in water and overall microbial water quality. These indicator bacteria live in the intestinal tract of humans and other warm-blooded animals without causing disease. They are naturally excreted in feces in large numbers (10^9 to 10^{10} per gram of feces). Commonly measured are total coliforms and a subset of this group, the fecal coliforms, which are considered to be more predictive of fecal contamination. Generally greater than 90 percent of the coliforms found in feces of warm blooded animals are a specific fecal coliform *Escherichia coli* (*E. coli*). In addition to the coliform bacteria, fecal streptococci and enterococci have been used to monitor water quality and are also natural flora of the intestines of animals, including humans. The use of bacterial indicators is discussed in Chapter 4.

Animal and Wildlife Sources

Domesticated animals and wildlife may excrete pathogenic microorganisms that are infectious to humans. Agricultural runoff, stormwaters, and direct input from animals leads to the contamination of waterways, which eventually discharge to coastal areas. There is no evidence at this time that animal enteric viruses infect humans.

Two of the enteric protozoa carried by animals, *Giardia* and *Cryptosporidium*, may be pathogens of concern in nonpoint sources. *Giardia* is found in 90 percent to 100 percent of the muskrat populations and is prevalent in beavers (Erlandsen et al. 1988). *Cryptosporidium parvum* is found widely distributed in mammals, and zoonotic (animal to human) transmission has been well documented (Current 1987). Infections in cattle along with rainfalls, which washed the oocyst (the environmentally resistant and infectious form of the organism) into the water supply, were hypothesized as contributing to a large outbreak in the United Kingdom, resulting in 55,000 illnesses (Smith and Rose 1990). Cattle and sheep may represent a large reservoir for human infections. Both *Cryptosporidium* and *Giardia* can be found at prevalences of 68 percent and 29 percent, respectively, in polluted waters (waters receiving sewage and agricultural discharges) and 39 percent and 7 percent, respectively, in pristine waters (Rose et al. 1991b). In one watershed, animals were the major source of the contamination rather than sewage discharges (Rose et al. 1988). These studies suggest that domestic sewage discharges are a larger source of *Giardia*, while animals may be the major source of *Cryptosporidium* (Rose et al. 1991b).

Among the bacteria, *Salmonella*, *Yersinia*, and *Campylobacter* are associated with animal reservoirs. *Salmonellae* are common in poultry (chickens, turkeys, ducks) and in gulls, pigeons, and doves but have been identified in other wild birds much less frequently (Feachem et al. 1983). Between 15 and 50 percent of domestic animals and 10 percent of mice and rats may be infected. Wild mammals do not appear to be a major source for human infections. Both wild and domestic animals may serve as reservoirs for *Yersinia enterocolitica.* The organism has been identified in foxes and beavers as well as cattle, sheep, and pigs. *Campylobacter* has been found in a wide variety of animals. Domestic animals (cattle, sheep, and pigs) and birds (poultry and caged birds) have been documented as sources of infections in humans.

Animals may also contribute significant numbers of indicator bacteria (total coliforms, fecal streptococci, and enterococci) to waters (Crane et al. 1983). Gannon and Busse (1989) suggested that animals were the source of the elevated indicator bacterial levels in storm water. An epidemiological study of recreational waters has suggested that the indicator bacteria arising from agricultural inputs are not associated with human bacterial and viral infections (Calderon et al. 1991).

Toxins in Shellfish and Fish

Several illnesses are associated with the consumption of shellfish and fish as a result of toxic algal blooms (NRC 1991), including neurotoxic shellfish poisoning, paralytic shellfish poisoning, and scromboid poisoning (Table B.4).

TABLE B.4 Illnesses Associated with Consumption of Seafood and Associated with Toxic Algal Blooms (Adapted from NRC 1991)

Toxins	Source	Epidemiology	Symptoms
Neurotoxic Shellfish Poisoning	Red tide "gymnodinium" accumulation in shellfish	53 cases reported 1973-88	Numbness, gastrointestinal effects, dizziness, muscle aches
Paralytic Shellfish Poisoning	Dinoflagellates accumulate in shellfish	137 cases reported 1978-85	Neurologic symptoms, paralysis, death
Ciguatera Poisoning	Reef algae gambiordiscus toxins in tropical reef fish	791 cases reported 1978-87	Gastrointestinal symptoms, neurological symptoms
Scromboid Poisoning	Histive production by bacterial contaminants during storage	757 cases reported 1978-87	Vomiting, diarrhea, headaches, palpitation

The blooms of red tide, dinoflagellates, and reef algae are seasonal and in some cases geographically restricted. There has been some suggestion that nutrient additions to marine waters may affect size of blooms, frequency, and seasonality of occurrence (see Appendix A). Scromboid is believed to be due to improper handling of shellfish after harvest and no association with polluted marine waters has been suggested.

OCCURRENCE OF PATHOGENS IN COASTAL WATERS

In the United States, rarely are monitoring programs designed to determine level of pathogenic agents in marine waters. The bacteriological indicator system has been used primarily to determine the microbial quality of estuaries and recreational waters. No information is available on the occurrence of enteric protozoa in marine waters. Specialized studies have been directed at specific pathogenic bacteria, but the greatest amount of information on the occurrence of pathogens in marine waters has been reported for the enteric viruses. There may be several reasons for this. Viruses have long been recognized as a major cause of shellfish associated disease. It has also been determined that the indicator concept is inadequate for determining viral water quality. Methods were developed for the recovery and detection of viruses and studies implemented as a result of the European Economic Community recognizing a need for virological monitoring. Past and ongoing studies continue to evaluate virus contamination of the marine environment in the United States.

Bacteria

With a better appreciation of the limitations of the indicator system, new methods are being used to detect the presence of bacterial pathogens in coastal waters. In a study in Spain, *Salmonella* was detected in 32 percent of 256 samples collected from 21 bathing beaches along the north coast (Perales and Audicana 1989). Similarly, 16 sites in New York Harbor, the Hudson and East rivers offshore in the Hudson River plume, Chester River, and the upper Chesapeake Bay were sampled for the presence of *Salmonella* (Knight et al. 1990). Salmonellae were detected at 75 percent of the sites and in 50 percent of these samples, cultivation techniques failed to isolate the organism. Previous work has demonstrated that non-cultivatable organisms can remain infectious (Colwell et al. 1985, Grimes and Colwell 1986).

DePaola et al. (1990) investigated the occurrence of *Vibrio parahaemolyticus* in shellfish growing waters in Washington, California, Texas, Louisiana, Alabama, Florida, South Carolina, Virginia, and Rhode Island. They found no correlation of *V. parahaemolyticus* with fecal coliforms. Average densities were 3, 11, and $2.1 \times 10^3/100$ g of oyster in samples from the Atlantic, Gulf, and Pacific coasts, respectively. Concentrations were 100 times greater in oysters than in the water. Temperature appeared to be a significant factor in the seasonal and geographical distribution of this organism.

Newly recognized bacterial pathogens have also been studied in coastal estuarine waters. *Listeria monocytogenes* has been associated with foodborne gastroenteritis. This organism was detected in 62 percent of the samples in the Humboldt-Arcata Bay in California. The organism was found in 17 percent of the sediment samples and was not detected in oysters. It was suggested that domestic animals, such as horses and cattle, were responsible for the contamination (Colburn et al. 1990).

Enteric Viruses

During the 1960s, several studies were published reporting the occurrence of enteroviruses in marine waters and shellfish (Metcalf and Stiles 1968, Bendinelli and Ruschi 1969). The next two decades brought with them improvements in the methods for the recovery and detection of enteric viruses, which were applied to surveys primarily in the Gulf and Atlantic coastal areas. These advances made it possible to study the significance of viral contamination of marine waters, and currently the scientific consensus supports several conclusions.

- Enteric viruses persist significantly longer when compared with the bacterial indicators;
- There is no qualitative or statistical association between the enteric viruses and the bacterial indicators, and
- Enteric viruses have been isolated from both waters and shellfish within current bacterial standards for water quality.

Table B.5 gives a summary of some of the more recent studies on the occurrence of viruses in shellfish and their overlying waters in areas opened and closed for harvesting based on the bacteriological indicators. In areas open to harvesting, viruses were recovered from 4 percent to up to 50 percent of the samples and levels ranged from 0.2 to 31 plaque forming units (PFU) per 100 grams of shellfish and 2.9 to 46 PFU/100 liters of water.

Two locations were selected for study in Mississippi (Ellender et al. 1980). The waters of the Pass Christian reef were approved for shellfish harvesting. The Graveline Bayou was closed to harvesting due to influences by rainfall, tidal flushing, wastewater treatment plant discharges, and septic tanks. The year-long study demonstrated no significant correlations of the viruses with bacterial indicators, temperatures (4-32°C), or salinities (which would reflect fresh water inputs due to rain). Within the Texas Gulf coast, sewage outfalls may have been responsible for the contamination of the estuaries (Goyal et al. 1979). While rainfall was associated with increases in bacterial counts, it was not associated with viral contamination.

Along the Atlantic coast, Wait et al. (1983) investigated viral and bacterial pathogens as well as the coliform indicators. Viruses were isolated without any correlation with the indicator system. Researchers investigating the Oyster River system in New Hampshire (A.B. Margolian, University of New Hampshire, personal communication, 1992) recently speculated that wastewater treatment plant discharges and septic tank leachate to rivers, which then flow into estuaries, are responsible for the viral contamination of the water as opposed to a sewage outfall.

Although fewer studies have been directed toward recreational areas, there is evidence that sewage outfalls can impact bathing beaches. Along the shoreline of South Wales, viruses were isolated from 35 to 50 percent of the bathing beaches at levels averaging 21 PFU/100 liters. Wastewater treatment plant discharges from a short outfall were the most probable source of the contamination, as no other source of human enteric viruses could be

TABLE B.5 Enteric Virus Isolations from Shellfish Harvesting Areas

Area	State or Location	Open Shellfish Beds[1]		Closed Shellfish Beds[1]	
		Water (%)[2] PFU[3]	Shellfish	Water	Shellfish
Gulf	Michigan (Ellander et al. 1980)	ND[4]	(9%) 1.95	ND[4]	(34%) 1.02
	Texas (Goyal et al. 1979)	(50%) 2.9 - 18[5]	(20%) 19	(63%) 4.8 - 11	(40%) 47 - 94
Atlantic	North Carolina (Wait et al. 1983)				
	Central	ND	(25%) 0.2	ND	(37%) 1.85
	Southern	ND	(12%) 6.0	ND	(37%) 0.35
	New Port River Estuary System (Carrick et al. 1991)			1.27[6]	0.43[6]
	New Jersey	(4.3%) 46	(40%) 8.0	(43%) 36	(28%) 4.3
	New York (Vaughn et al. 1980)	(12%) 9.2	(25%) 31.0	(0%)	(37%) 9.5

[1]Based on bacteriological indicators standards.
[2]Percent of samples positive for viruses.
[3]PFU/100 liters water/100 grams shellfish.
[4]Not determined.
[5]Ranges.
[6]Two of 11 stations positive for viruses were within total and fecal coliform standards for microbial contamination.

identified (Tyler 1982). Enteroviruses have also been isolated in sediments along a Florida bathing beach (Schaiberger et al. 1982). Two of three stations, 3.6 kilometers from the outfall, were positive for viruses, averaging 2.2 PFU/100cc of sediment. Viruses were not isolated from the water column and a significant association was demonstrated between the concentrations of viruses and the distance from the outfall. Indicator bacteria and viruses could not be detected in water samples at distances greater than 200 meters from the outfall. Indicator bacteria could not be isolated from sediments at distances greater than 400 meters from the outfall. It has been suggested that viruses may be sequestered in sediments and then transported shoreward. The resuspension of viruses in sediments in deep coastal waters near the outfall pipe would be insignificant due to dilution of the viruses in the large volume of overlying waters. However, in shallow coastal waters, sediments could significantly impact public health by serving as reservoirs for viral contamination of the water column resuspended by currents, storms, boats, swimmers, dredging, etc.

SURVIVAL OF ENTERIC MICROORGANISMS IN MARINE WATERS

Several factors influence the survival of enteric microorganisms in the marine environment. These include salinity, type of microorganism, temperature, sediments, nutrients, antagonistic factors, light, and dissolved oxygen.

Research has demonstrated that inactivation or die-off rates for enteric microorganisms are greater in waters of greater salinity, such as estuarine waters and seawaters, than in fresh waters (Table B.6). Coliforms survive poorly in marine waters, and this is one of the major reasons that this group of bacteria are inadequate predictors of the presence of pathogens. *E. coli* survival rates are more reflective of the pathogens; but at warmer temperatures the die-off rates for many of the pathogens appear to be slower than *E. coli*. (Figure B.1)

The intrinsic nature of the organism will influence the longevity of the pathogen in the marine environment. Viruses and protozoa are unable to replicate in the environment, but many of the enteric bacteria can grow under appropriate conditions of temperature and nutrients. In tropical areas, coliforms may be a part of the natural freshwater microbial flora, so that fresh water flows in the Gulf and southern Atlantic states may be contributing to the coliform levels in marine waters in the absence of any association with pathogens.

Survival for the bacteria and viruses has historically been measured by cultivation techniques and may underestimate counts ten-fold (Garcia-Lara et al. 1991). In stressed environments such as marine waters, bacteria have been shown to remain viable even when noncultivatable (Colwell et al.

TABLE B.6 Survival of Enteric Pathogens and Indicator Bacteria in Fresh and Marine Waters (Source: Feachem et al. 1983)

	Marine Waters		Fresh Waters	
Microorganism	Temperature °C	T-90[1]	Temperature °C	T-90[1]
Coliforms	10 - 20	0.025 - 0.33 avg. 0.083	10 - 20	0.83 - 4.8 avg. 2.5
E. coli	0 30	1.6 0.58	15	3.7
Salmonella	4 37	0.96 0.7	10 - 20	0.83 - 83
Yersinia	4 - 37	0.6	5 - 8.5	7
Giardia	2 - 5	14 - 143	12 - 20	3.4 - 7.7
Enteric Viruses	20 18 - 20 4- 15	0.67 -1.0 6.0[2] 14.0[2]	4 - 30	1.7 - 5.8

[1]Time in days for 90 percent reduction in microbial levels.
[2]In sediments.

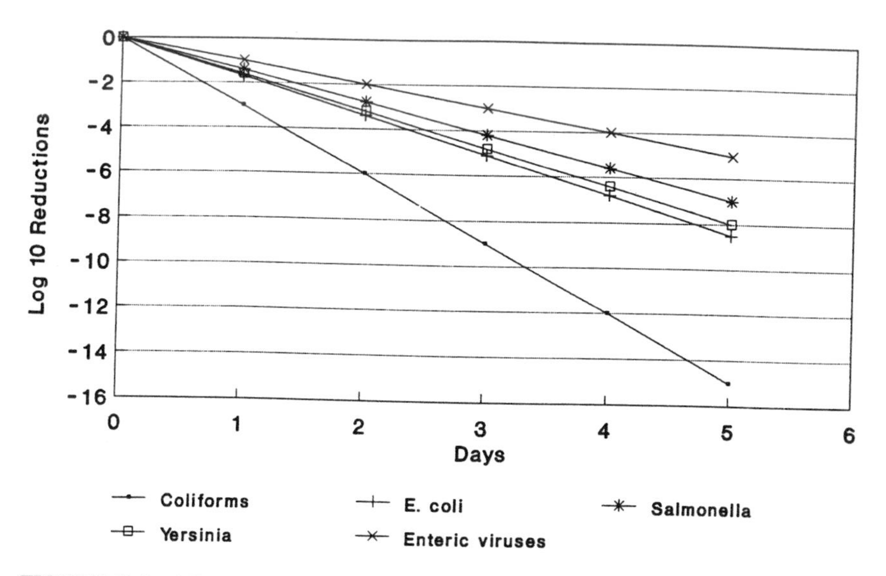

FIGURE B.1 Microbial die-off in marine waters (Feachem et al. 1983, Garcia-Lara et al. 1991).

1985, Grimes and Colwell 1986). These noncultivatable organisms could be infectious.

Very little information is available on the survival of the protozoa in marine waters. Investigators in the 1920s and 1930s reported that *Entamoeba* survival was unaffected by salt concentrations found in seawater. *Giardia* cysts maintained the ability to encyst at the same rate for up to 12 days in seawater, surviving for 26 days at 10 to 20°C and up to 28 days in fresh waters (DeRegnier et al. 1989).

Temperature is perhaps one of the most important factors influencing microbial survival and has been used as the primary parameter in developing predictive models. At cooler temperatures, below 10°C, the survival of enteric pathogens is enhanced. Enteric viruses may survive for months in marine waters at low temperatures. At temperatures above 25 to 30°C, however, bacteria may be able to proliferate. Surveys of viruses in the Gulf of Mexico have demonstrated no association with detection and temperature. This implies that other factors influence the occurrence of viruses, which may or may not affect survival (i.e., infection in the community or association with sediments). Figure B.2 shows the relationship between temperature and inactivation rates in marine waters. The study on which Figure B.2 is based found no significant correlation between virus inactivation rates and salinity.

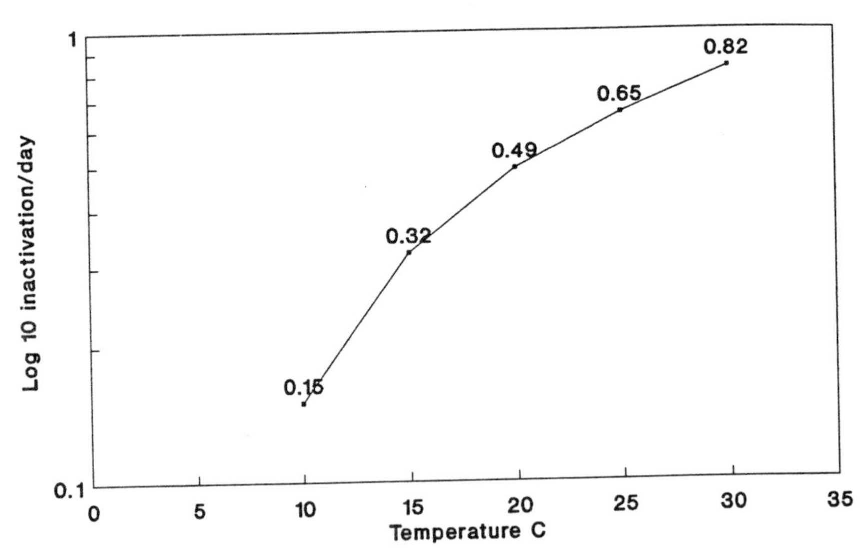

FIGURE B.2 Effects of temperature on virus inactivation rates in water (Source: Goyal 1981).

Enteric organisms may accumulate in sediments. Virus levels may be 100 times greater in the sediments than in the water column, and sediments have been found to contain 100 to 1,000 times greater levels of indicator and pathogenic bacteria than the overlying waters (Volterra et al. 1985, Van Donsel and Geldreich 1971). Greater than 99 percent of the enteric viruses were found to adsorb to marine sediments, and suspension in sewage effluents did not alter this pattern (LaBelle and Gerba 1979). It has been well documented that survival time is greatly enhanced for enteric organisms associated with sediments or with shellfish.

Nutrient addition to waters, nitrogen, phosphorus, and organics has been related to increased bacterial numbers generally affecting the indigenous microflora. This may indirectly influence the survivability of introduced microorganisms. Algal blooms may protect microorganisms at the surface from photoinactivation. Decreases in dissolved oxygen are also related to increases in survival, and naturally occurring organisms may show an antagonistic effect on pathogens. These effects are tied to the temperature of the water. Biological inactivating factors have been shown to be antiviral and are often associated with particulates and bacteria (Fujioka et al. 1980), but their significance under natural conditions remains unknown. Although natural solar light may affect bacteria and viral particles through direct inactivation, this would only occur at the surface in waters with low turbidity. This phenomenon would have little impact for submerged outfalls or in coastal waters with greater turbidity.

Much more research is needed in order to understand the fate of enteric pathogens introduced into the marine environment. The complexities of the interactions between the factors effecting survival and transport will ultimately determine the public health impact of pathogen-laden discharges to coastal waters.

ILLNESSES FROM BATHING

A number of epidemiological studies have documented the risks of acute gastroenteritis among those bathing in contaminated seawater (Cabelli et al. 1983, Cheung et al. 1990, Balarajan et al. 1991, Fleisher 1991, Alexander et al. 1992, Fewtrell et al. 1992). One study showed the risks to be three times greater for children under the age of two who immerse their heads in water than for adults (Cheung et al. 1990).

It has been estimated that for each swimming event, for those individuals who submerge their heads in the water, the exposure is on the average of 100 milliliters of seawater. This value may represent a child's potential dose rather than an adult's. Recreational exposure generally will come from the contaminants suspended in the water column and via oral ingestion. Although aerosols and inhalation of water may also be potential exposure routes, these are probably very minor. Pathogens that accumulate in the sediments may be protected from adverse environmental conditions, and sediments may act as a reservoir for enhanced survival of enteric pathogens, which may later be resuspended due to currents, wave action, and human activities.

In 1986 the United States moved toward the use of an enterococci standard to govern sanitary quality of marine waters for recreational uses. This was based on a series of epidemiological studies to determine the relationship between gastroenteritis among swimmers and the level of indicator densities (Cabelli et al. 1983). Several indicator bacteria were studied, including *E. coli*, total coliforms, fecal coliforms, and enterococci. The enterococci levels best predicted risk of illness and were chosen as the variable for measuring water quality and health impacts. Currently, the recommended allowable concentration of enterococci in bathing waters is 35/100 milliliters, which predicts a gastroenteritis rate of 19/1000 swimmers.

The use of the enterococci to govern marine water quality has been criticized due to methodological weaknesses in defining the curves. Fleisher (1991) has reported that different risks may be obtained depending on the site. He obtained estimated risks of 24, 82, and 36/1000 at New York City, Boston, and Lake Pontchartrain, respectively. One explanation for the differences between the three sites (which were merged in the Cabelli study) was the difference in the salinities at the three beaches. The waters of greater salinity were associated with the lower risks.

RISK ASSESSMENT APPROACH FOR MICROORGANISMS

The quantitative risk procedures most recently employed for infectious pathogens include development of a dose-response curve assuming no safe level of exposure. The dose-response method has been used to estimate infections after exposure to varying levels of enteric microorganisms in drinking water (Haas 1983; Rose and Gerba 1990, 1991, Rose et al. 1991c; Regli et al. 1992). This was used to evaluate low level exposure, to determine appropriate water treatment needed for reducing the risk of microbial infection, and to develop appropriate standards.

The risk of infection is a function of hazard (the probability of infectivity from a given unit dose) and exposure (Haas 1983). The two components of the model that aid in characterizing the risk are 1) the level of exposure and 2) the interaction of the particular pathogen and host (defined by the dose-response curve). Each pathogen or strain has an intrinsic ability to cause infection, morbidity, and mortality. Secondary and tertiary person-to-person transmission should also be accounted for. The host population tested would also influence the model, and a different model might be

TABLE B.7 Components of Waterborne or Foodborne Microbial Risk Characterization

Probability Model	Exposure
Defined by dose response (infectious dose)	Level of microorganism in water or food
Dependent on type of microorganism and/or strain variance	Level after any processing or condition which may decrease numbers (or potentially increase numbers[1])
Dependent on population tested (age, immune status)	Amount of food or water consumed
Possibly influenced by the type of food	

[1]Bacterial regrowth.

developed for each population with varying sensitivities to the pathogen. Host factors include general and specific immunity, genetic factors, age, sex, and other underlying diseases or conditions that might influence susceptibility.

Exposure depends on the initial concentration of the pathogen in the water or shellfish, processes that would decrease the numbers (i.e., wastewater treatment), and environmental conditions that would influence microbial survival. The final level of the pathogen in the food or water and the amount or volume consumed determine the exposure. See Table B.7.

Dose-Response Assessment: Probability of Infection, Morbidity, and Mortality

Dose-response experiments for microorganisms of concern have been conducted in human volunteers for some bacteria, protozoa, and viruses. In these experiments, several sets of volunteers were exposed to known doses of microorganisms. The resulting percentage of infected individuals was then determined. Generally, in laboratory studies, the distribution of microorganisms is found to follow the Poisson distribution. In waters, it has been found that in some cases variability of microbial counts is greater than that given by the Poisson distribution. In particular, a number of workers have indicated that the microbial counts may often be better described by the negative binomial distribution (Pipes et al. 1977, El-Shaarawi et al. 1981, Maul et al. 1990). While this phenomenon has been observed for a number of indicator groups, there is little definitive work on pathogen distributions.

Laboratory dose-response studies generally have been conducted under conditions where the counts of microorganisms in the administered dose approximates the Poisson distribution. Under these conditions, if one microorganism is sufficient to cause an infection, and if host-microorganism

TABLE B.8 Probability Models Used for Microorganisms (Haas 1983)

single-hit exponential model	beta-distributed "infectivity probability" model
$p = 1 - \exp(-rN)$	$p = 1 - [1 + (N/\beta)]^{-\alpha}$

p = probability of infection (risk)
N = exposure
α, β, r = parameters characterized by dose response curves

interactions are constant, then the probability of an infection resulting from ingestion of a single exposure containing an average number of organisms may be given by an exponential model. An alternative model that has a better fit with experimental data is the beta-distributed model (Table B.8) (Furumoto and Mickey 1967a and 1967b; Haas 1983).

Using these models, dose response curves were plotted for a number of pathogens for a number of studies using the maximum likelihood method (Regli et al. 1992). The Rotavirus and Poliovirus 3 were found to be more infective than the Echovirus 12 and Poliovirus 1 (Figure B.3). This may have been due to the use of nonvirulent strains. The probability of infection from exposure to one viral unit ranged from 2.8×10^{-1} to 7.2×10^{-5}.

The development of clinical illness (symptoms) depends on numerous

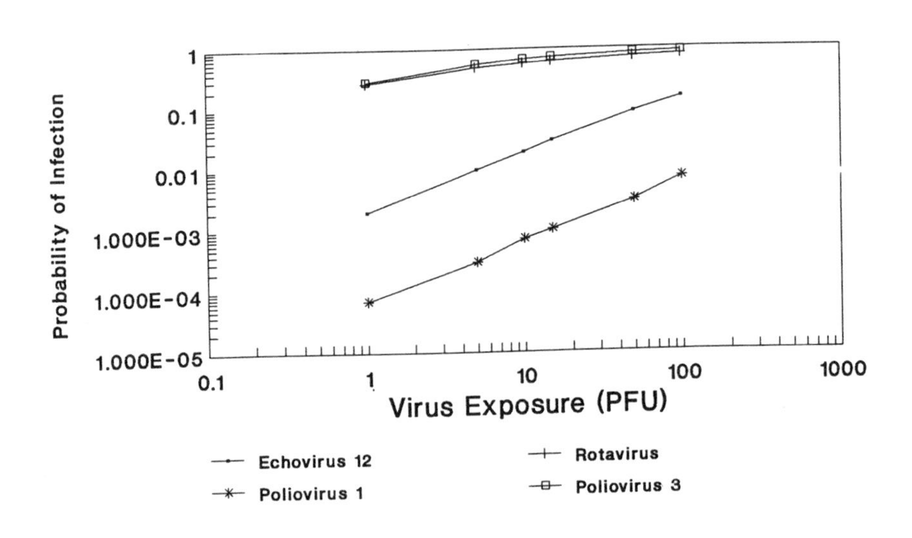

FIGURE B.3 Dose response relationships for various enteric viruses (Regli et al. 1992).

factors, including the immune status of the host, age of the host, virulence of the microorganisms and type, strain of microorganism, and route of infection. Clinical illness may also be influenced by the dose (Haas 1983, Graham et al. 1978). For hepatitis A virus, the percentage of individuals with clinically observed illness is low for children (usually < 5 percent) but increases greatly with age (Evans 1982). The frequency of clinical hepatitis A virus in adults is estimated at 75 percent. However, during waterborne outbreaks, it has been observed as high as 97 percent (Lednar et al. 1985). In contrast, the frequency of clinical symptoms for rotavirus is greatest in childhood (Gerba et al. 1985) and lowest in adulthood. The observed illness rates for various enteroviruses may range from 1 percent for poliovirus to more than 75 percent for some of the coxsackie B viruses (Cherry 1981) (Table B.9).

Case fatality rates are also affected by many of the same factors that determine the likelihood of the development of clinical illness. The risk of mortality for hepatitis A virus has been estimated at 0.6 percent (CDC 1985). Mortality from other enterovirus infections in North America and Europe has been reported to range from < 0.1 to 1.8 percent (Assaad and Borecka 1977). Case fatality rates for selected enteroviruses are summarized in Table B.9. The values for enteroviruses probably only represent hospitalized cases. For some pathogens, the risk of infection may be low but the consequences of infection may be more drastic. Therefore, infectivity and case fatality rates can be added to the model to further estimate disease and death.

The ultimate aim of developing standards, treatment approaches, and intervention strategies is to provide an acceptable degree of protection for the susceptible population. A risk assessment model targeting infection can be used to emphasize the initial step in the chain of events that leads to the mortality associated with waterborne or foodborne pathogens. Since infec-

TABLE B.9 Morbidity and Mortality Rates Associated with Various Viral Pathogens (Assaad and Borecka 1977, Cherry 1981, Evans 1982, CDC 1985, Gerba et al. 1985, Lednar et al. 1985.)

Microorganism	Morbidity Rates (%)	Mortality Rates (%)
Poliovirus 1	1	0.01
Rotavirus	56	0.01
Hepatitis A virus (in children)	5	not known
Hepatitis A virus (in adults)	75	0.6

tion is the primary event that leads to disease, by preventing infection one prevents any range of morbidities and mortalities associated with the pathogen. Secondary spread of enteric infections range from 30 to 70 percent. Thus three to seven of ten individuals who came into contact with the first infected person may also become infected. Prevention of initial infections will also prevent the rippling effect of secondary spread.

Exposure Assessment

One can use these models once exposure has been determined. For example, viral contamination of recreational waters at levels between 0.1 to 100 PFU/100 liters may be associated with risks of infection from 2×10^{-7} to 5.4×10^{-2} depending on the type of virus and assuming ingestion of 100 ml during each swimming event.

Limited studies have been undertaken to evaluate virus contamination in shellfish. Viruses were found in 9 to 40 percent of the shellfish in waters open to harvesting and in 13 to 40 percent of the shellfish in areas closed due to coliform levels in the water. The concentrations of enteroviruses ranged from 10 to 200 virus plaque-forming units per 100 grams of shellfish. Table B.10 shows the results of four studies on virus contamination of shellfish in waters open to harvesting, based on the bacterial indicator. Studies have determined that for an average meal, 6 to 12 shellfish ranging in weight between 10 and 20 grams each may be consumed (M.D. Sobsey, University of North Carolina, personal communication, 1992). These values were used to determine virus levels in a risk assessment model to evaluate potential health impacts of consuming raw shellfish.

Application of a Virus Risk Model to Characterize Risks from Consuming Shellfish

It is well known that infectious hepatitis and viral gastroenteritis are caused by consumption of raw or, in some cases, cooked clams and oysters. The number of documented viral outbreaks seems to be on the increase in the United States (DeLeon and Gerba 1990).

Using the data presented in Table B.10 and the Echo-12 virus probability model, the individual risk was determined for consumption of raw shellfish (Table B.10). The percentage of samples contaminated with viruses ranged from 9 percent (Mississippi oysters) to 40 percent (New York clams). The levels of viruses ranged from 0.3 to 200 viruses/100 grams. In the model, one exposure was used, representing a single serving of six shellfish (60 grams). Risks ranged from 3.5×10^{-2} to 2.2×10^{-4}, and on average there is a 1/100 chance of infection when consuming raw shellfish.

The risk calculations are shown below:

TABLE B.10 Risk of Infection for a Single Serving of Shellfish from Samples of Viral-Contaminated Shellfish Based on Infectivity of Echovirus (Goyal et al. 1979, Ellender et al. 1980, Vaughn et al. 1980, Wait et al. 1983.)

Study Site	Shellfish	Total Samples Collected	Total Samples Positive (levels)[1]	Average Viruses (PFU/100g)	Individual[2] Risk
Mississippi	Oysters	22	2(0.3) (3.6)	0.18	2.2×10^{-4}
New York	Clams	5	2(10) (30)	8.0	9.4×10^{-3}
New York	Oysters	8	2(48) (200)	31.0	3.5×10^{-2}
North Carolina	Clams	13	3(0.8) (48)	3.8	4.5×10^{-3}
Texas	Oysters	10	2(17) (59)	7.6	9.0×10^{-3}
Total/Averages		58	11	10.0	1.2×10^{-2}

[1]Levels are PFU/100 grams.
[2]Consuming 60 grams.

Echovirus Model $$P = 1 - \left(1 + \frac{N}{186.69}\right)^{-0.374}$$

Rotavirus Model $$P = 1 - \left(1 + \frac{N}{0.42}\right)^{-0.26}$$

where,

P = the probability of infection,

N = exposure as measured in PFU/60 grams.

Morbidity and mortality risks can be estimated from probabilities of infection and will vary depending on the virus. Figure B.4 compares infection, illness, and death risks for rotavirus and hepatitis A virus (HAV). (The Echo-12 virus model was used in the absence of a model for HAV.) The morbidity rates used were 56 percent for rotavirus and 75 percent for HAV as shown in Table B.9. The mortality rates used were those shown on Table B.1. Exposure was set for a small serving (60 g) of shellfish with contamination ranging from 0.1 to 100 viruses per serving. These exposures correspond to virus concentration levels ranging from 0.17 to 177 viruses per 100 grams and are within the ranges detected in surveys of shellfish from waters open to harvesting in the United States (see Table B.10). The risk of infection is more than 10 times greater for rotavirus than HAV if one assumes HAV infectivity is similar to Echo-12; however, mortality is much more significant for HAV infections. For even a single serving of shellfish that is greatly contaminated with viruses, the risk of death is very high at between 1.7×10^{-3} to 7.8×10^{-3}. The risk of becoming infected with the exposure to even one virus was estimated at between 10^{-1} to 10^{-2}.

SUMMARY OF SHELLFISH AND RECREATIONAL MICROBIOLOGIC RISKS

Acceptable recreational risks based on indicator bacterial levels and epidemiological studies have suggested an acceptable risk level of 8×10^{-3} (Cabelli et al. 1983). This risk would correspond to between 1 rotavirus and 100 echoviruses per 100 milliliters. However, these same concentrations in the water column can lead to concentrations 100 to 900 times as large in underlying sediments underlying and shellfish. This concentration would increase virus levels to between 0.1 and 100 viruses per 60 grams (representing a single meal of 6 oysters for example) accordingly (Figure B.4). Therefore, protective levels for bathing water may be inadequate to protect against food-borne infections.

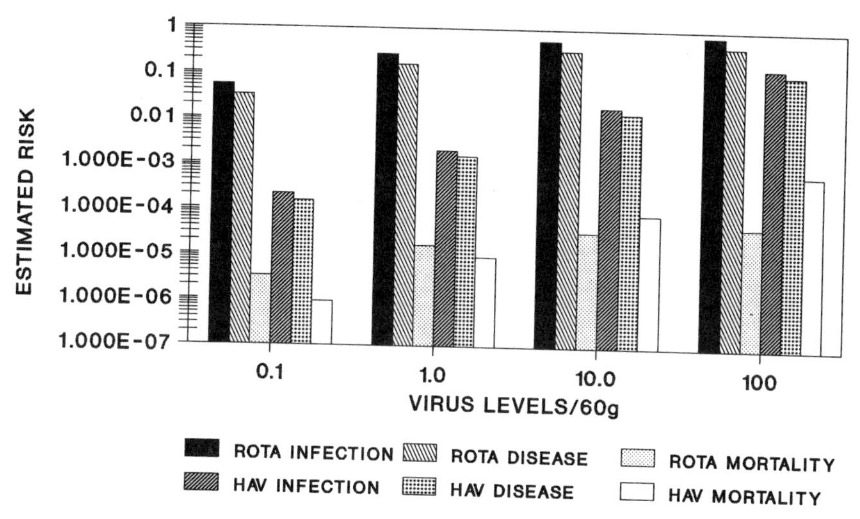

FIGURE B.4 Risks of infection, disease, and mortality for contaminated shellfish (Rose and Sobsey undated).

Risk assessment may be used to evaluate potential health impacts of food or water contaminated with pathogens. However, it is important to be cautious in interpreting monitoring data; most methods do not recover more than 40 percent of the organisms present; and therefore the exposure may be underestimated. Past surveys may not be useful for prediction of future contamination due to the variation in concentrations of microbial pathogens and the differences in die-off rates in the environment.

Several factors will influence the significance of wastewater inputs associated with risks from microbial pathogens. The level of infection in the human and animal populations producing the wastes will influence the initial concentrations. It is likely that large cosmopolitan cities with larger immigrant populations have a greater concentration and variety of pathogens present in their sewage. Cities such as Los Angeles, San Francisco, Miami, and New York may fall into this category. Coastal water temperatures will influence the survival of the pathogens and the potential for regrowth of the coliforms. The warmer waters in the Gulf and off the southern Atlantic states may enhance the inactivation rates for pathogens while enhancing the potential for inputs of coliforms that have grown in fresh water environments. There is no doubt that the cooler temperatures of the waters off the west coast and northeast coast will maintain pathogen viability for a longer period of time.

Of greatest concern are combined sewer overflows which carry untreated wastes, and short outfalls, which have the potential for contributing to microbial contaminants that can be transported back to shore. Wastewater treatment plants that achieve secondary treatment remain a threat to recreational areas and shellfish harvesting areas by direct discharge through short outfalls or through discharge to freshwaters, which then flow into the marine environment. Enteric viruses remain the primary concern for any wastewaters carrying human sewage (Asano et al. 1992). Secondary treatment and disinfection, operated and monitored for the 200 fecal coliform per 100 ml standard, do not guarantee the removal of such pathogens to levels that are safe for discharges close to shellfish beds or bathing areas. See Appendix D for discussion of wastewater treatment options, disinfection procedures, and combined sewer overflow controls.

Despite requirements that shellfish harvesting waters must meet established bacterial indicator standards and sanitary survey criteria, disease outbreaks due to consumption of contaminated shellfish continue to occur in the United States. For example, a recent outbreak of hepatitis A virus associated with shellfish consumption affected several southern states. The virus was detected in shellfish harvested from waters approved for harvesting (Desenclos et al. 1991). There is no doubt that such outbreaks should be prevented. Better detection methods and risk assessment methods are needed in order to provide adequate protection of consumers from disease transmitted through the nation's seafood.

Disease transmission through exposure to recreational waters has been demonstrated, however, the associated risks are not as well documented as those for seafood consumption. The development of better detection techniques, additional epidemiological information, and improved risk assessment methodologies will allow for more certain determinations of the risks associated with recreational exposure to contaminated waters and the development of better recreational water protection strategies.

REFERENCES

Alexander, L.M., A. Heaven, A. Tennant, and R. Morris. 1992. Symptomatology of children in contact with sea water contaminated with sewage. J. Epid. Commun. Hlth. 46:340-344.

Asano, T., L.Y.C. Leong, M.G. Rigby, and R.H. Sakaji. 1992. Evaluation of the California wastewater reclamation criteria using enteric virus monitoring data. Water Science and Technology 26(7-8):1513-1524.

Assaad, F., and I. Borecka. 1977. Nine-year study on WHO virus reports on fatal virus infections. Bulletin of the World Health Organization 55:445-453.

Balarajan, R., V.S. Raleigh, P. Yuen, D. Wheeler, D. Machin, and R. Cartwright. 1991. Health risks associated with bathing in sea water Brit. Med. J. 303:1444-1445.

Bendinelli, M. and A. Ruschi. 1969. Isolation of human enterovirus from mussels. Appl. Microbiol. 18:531-532.

Bennett, J.V., S.D. Homberg, M.F. Rogers and S.L. Solomon. 1987. Infectious and parasitic diseases. Am. J. Preventative Med. 55:102-114.

Cabelli, V.J., A.P. Dufour, L.J. McCabe, M.A. Levin. 1983. A marine recreational water quality criterion consistent with indicator concepts and risk analysis. J. Water Pollut. Control Fed. 55:1306-1314.

Calderon, R.L., E.W. Mood, and D.P. Dufour. 1991. Health effects of swimmers and nonpoint sources of contaminated water. Int. J. Env. Hlth. Res. 1:21-31.

CDC (Centers for Disease Control). 1985. Hepatitis Surveillance Report No. 40. Atlanta, Georgia: CDC.

CDC (Centers for Disease Control). 1990. Waterborne Disease Outbreaks. U.S. Department of Health and Human Services, Atlanta, Georgia. Morbidity and Morality Weekly Report 39(ss-1):1-57.

Cherry, J.D. 1981. Nonpolio Enteroviruses: Coxsackieviruses, Echoviruses, and Enteroviruses. Pp. 1316-1365 in Textbook of Pediatric Infectious Diseases, R.D. Geigin and J.D. Cherry, eds. Philadelphia, Pennsylvania: W.B. Saunders.

Cheung, W.H.S., K.C.K. Chang, and R.P.S. Hung. 1990. Health effects of beach water pollution in Hong Kong. Epidemiol. Infect. 105:139-162.

Colburn, K.G., C.A. Kaysner, C. Abeyta, Jr. and M.M. Wekell. 1990. *Listeria* species in a California coast estuarine environment. Appl. Environ. Microbiol. 56:2007-2011.

Colwell, R.R., P.R. Brayton, D.J. Grimes, D.B. Roszak, S.A. Huq and L.M. Palmer. 1985. Viable but non-culturable *Vibrio cholerae* and related pathogens in the environment: Implications for release of genetically engineered micro-organisms. Bio. Tech. 3:817-820.

Crane, S.R., J.A. Moore, M.E. Grismer, and J.R. Miner. 1983. Bacterial pollution from agricultural sources: A review. Trans. American Society of Agricultural Engineers 26(3):858-866.

Craun, G.F. 1986. Waterborne Diseases in the United States. Boca Raton, Florida: CRC Press.

Craun, G.F. 1991. Statistics of waterborne disease in the United States. Water, Science and Technology 24(2):10-15.

Current, W.L. 1987. *Cryptosporidium*: Its biology and potential for environmental transmission. CRC Crit. Rev. Environ. Control 17:21-51.

D'Antonio, R.G., R.E. Winn, J.P. Taylor, T.L. Gustafson, W.L. Current, M.M. Rhodes, G.W. Gary, and R.A. Zajac. 1985. A waterborne outbreak of cryptosporidiosis in normal hosts. Ann. Intern. Med. 103:886-888.

DeLeon, R. and C.P. Gerba. 1990. Viral disease transmission by seafood. Food Contam. Env. Sources 639-662.

DePaola, A., L.H. Hopkins, J.T. Peeler, B. Wentz and R.M. McPhearson. 1990. Incidence of *Vibrio parahaemolyticus* in U.S. coastal waters and oysters. Appl. Environ. Microbiol. 56:2299-2302.

DeRegnier, D.P., Cole, L., Schupp, D.G., and S. L. Erlandsen. 1989. Viability of *Giardia* cysts in lake, river, and tap water. Appl. Environ. Microbiol. 55(5):1123-1129.

Desenclos, J.C.A., K.C. Klontz, M.H. Wilder, O.V. Nainan, H.S. Margolis, and R.A. Gunn. 1991. A multistate outbreak of hepatitis A caused by the consumption of raw oysters. Amer. J. Pub. Hlth. 81(10):1268-1272.

El-Shaarawi, A.H., Esterby, S.R. and Dutka, B.J. 1981 Bacterial density in water determined by poisson or negative binomial distributions. Appl. Environ. Microbiol. 41:107.

Ellender, R.D., J.B. Map, B.L. Middlebrooks, D.W. Cook, and E.W. Cake. 1980. Natural enterovirus and fecal coliform contamination of Gulf coast oysters. J. Food Protec. 42(2):105-110.

Erlandsen, S.L., L.A. Sherlock, M. Januschka, D.G. Schupp, F.W. Schaefer, W. Jakubowski, and W.J. Bemrick. 1988. Cross-species transmission of *Giardia spp*: Inoculation of beavers and muskrats with cysts of human, beaver, mouse and muskrat origin. Appl. Environ. Microbiol. 54:2777-2785.

Evans, A.S. 1982. Epidemiological concept and methods. Pp. 1-32 in Viral Infection of Humans, A.S. Evans, ed. New York: Plenum.

Feachem, R.G., Bradley, D.H., Garelick, H., and Mara, D.D. 1983. Sanitation and Disease Health Aspects of Excreta and Wastewater Management. New York: John Wiley and Sons.

Fewtrell, L., A.F. Godfree, F. Jones, D. Kay, R.L. Salmon, and M.D. Wyer. 1992. Health effects of white-water canoeing. Lancet 339:1587-1589.

Fleisher, J.M. 1991. A reanalysis of data supporting U.S. federal bacteriological water quality criteria governing marine recreational waters. Research Journal of the WPCF 63:259-265.

Fujioka, R.S., P.C. Loh, and L.S. Lau. 1980. Survival of human enteroviruses in the Hawaiian ocean environment: Evidence for virus-inactivating microorganisms. Appl. Environ. Microbiol. 39:1105-1110.

Furumoto, W.A., and R. Mickey. 1967a. A mathematical model for the infectivity-dilution curve of tobacco mosaic virus: Theoretical considerations. Virology 32:216.

Furumoto, W.A., and R. Mickey. 1967b. A mathematical model for the infectivity-dilution curve of tobacco mosaic virus: Experimental tests. Virology 32:224.

Gallagher, M.M., J.L. Herndon, L.J. Nims, C.R. Sterling, D.J. Grabowski, and H.H. Hull. 1989. Cryptosporidiosis and surface water. American Journal of Public Health 79:39-42.

Gannon, J.J., and M.K. Busse. 1989. *E. coli* and enterococci levels in urban water and chlorinated treatment plant effluent. Water Res. Journal 23:1167-1176.

Garcia-Lara, J., P. Menon, P. Servais, and G. Billen. 1991. Mortality of fecal bacteria in seawater. Appl. Environ. Microbiol. 57:885-888.

Gerba, C.P., S.N. Singh, and J.B. Rose. 1985. Waterborne viral gastroenteritis and hepatitis. CRC Crit. Rev. Environ. Control 15:213-236.

Goyal, S.M., C.P. Gerba, and J.L. Melnick. 1979. Human enteroviruses in oysters and their overlying waters. Appl. Environ. Microbiol. 37:572-581.

Goyal, S.M. 1981. Development of Management Strategies for Assessment and Control of Viral Pollution of Coastal Waters, Final Report. National Oceanic and Atmospheric Administration Grant NA80RAD0056.

Graham, D.Y., G.R. Dufour, and M.K. Estes. 1987. Minimal infection dose of rotavirus. Arch. Virol. 92:261-271.

Grimes, D.J. and R.R. Colwell. 1986. Viability and virulence of *Escherichia coli* suspended by membrane chamber in semitropical ocean water. FEMS Micro. Lett. 34:161-165.

Haas, C.N. 1983. Estimation of risk due to low doses of microorganisms: A comparison of alternative methodologies. Am. J. Epidemiol. 118:573-582.

Hayes, E.B., T.D. Matte, T.R. O'Brien, T.W. McKinley, G.S. Logsdon, J.B. Rose, B.L.P. Ungar, D.M. Word, P.F. Pinsky, M.L. Cummings, M.A. Wilson, E.G. Long, E.S. Hurwitz, and D.D. Juranek. 1989. Contamination of a conventionally treated filtered public water supply by *Crystosporidium* associated with a large community outbreak of cryptosporidiosis. N. Engl. J. Med. 320:1372-1376.

Jarroll, E.L. 1988. Effects of disinfectants on *Giardia* cysts. CRC Crit. Rev. Environ. Control 18:1-28.

Knight, I.T., S. Shults, C.W. Kaspar and R.R. Colwell. 1990. Direct detection of *Salmonella* spp. in estuaries by using a DNA probe. Appl. Environ. Microbiol. 56:1059-1066.

Korick, D.G., J.R. Mead, M.S. Madore, N.A. Sinclair, and C.R. Sterling. 1990. Effects of ozone, chlorine dioxide, chlorine and monochloramine on *Cryptosporidium parvum* oocyst viability. Appl. Environ. Microbiol. 56:1423-1428.

LaBelle, R.L., and C.P. Gerba. 1979. Influence of pH, salinity and organic matter on the adsorption of enteric viruses to estuarine sediment. Appl. Environ. Microbiol. 38:93-101.

Lednar, W.M., S.M. Lemon, J.M. Kirkpatrick, R.R. Redfield, M.L. Fields, and P.W. Kelley. 1985. Frequency of illness associated with epidemic hepatitis A virus infections in adults. American Journal of Epidemiology 122:226-233.

Maul, A., A.H. El-Shaarawi, and J.C. Block. 1990. Bacterial distribution and sampling strategies for drinking water networks. Chapter 10 in Drinking Water Microbiology, G.A. McFeters, ed. New York: Springer-Verlag.

Melnick, J.L., and C.P. Gerba. 1980. The ecology of enteroviruses in natural waters. CRC Crit. Rev. Environ. Control 10:(1):65-93.

Metcalf, T.G., and W.C. Stiles. 1968. The accumulation of enteric viruses by the oyster *Crassostrea virginica*. Journal of Infectious Diseases 115:68-76.

NRC (National Research Council). 1991. Seafood Safety. Washington, D.C.: National Academy Press.

Perales, I., and A. Audicana. 1989. Semisolid media for isolation of *Salmonella* spp. from coastal waters. Appl. Environ. Microbiol. 55:3032-3033.

Pipes, W.O., P. Ward, S.H. Ahn. 1977. Frequency distributions for coliform bacteria in water. Journal of the American Water Works Association 69:12:664.

Regli, S., J.B. Rose, C.N. Haas, and C.P. Gerba. 1992. Modeling risk from *Giardia* and viruses in drinking water. Journal of the American Water Works Association 83:76-84.

Richards, G.P. 1985. Outbreaks of shellfish-associated enteric virus illness in the United States: Requisite for development of viral guidelines. Journal of Food Protection 48:815-823.

Roos, R. 1956. Hepatitis epidemic conveyed by oysters. Sven. Lakartidningen 53:989-1003.

Rose, J.B., and C.P. Gerba. 1990. Assessing potential health risks from viruses and parasites in reclaimed water in Arizona and Florida. Water Science and Technology 23:2091-2098.

Rose, J.B., and C.P. Gerba. 1991. Use of risk assessment for development of microbial standards. Water Science and Technology 24:29-34.

Rose, J.B., and M.D. Sobsey. undated. Quantitative risk assessment for viral contamination of shellfish and coastal waters. submitted to Journal of Food Protection.

Rose, J.B., H. Darbin, and C.P. Gerba. 1988. Correlations of the protozoa, *Cryptosporidium* and *Giardia* with water quality variables in a watershed. Water Science and Technology 20:271-276.

Rose, J.B., C.N. Haas, and S. Regli. 1991a. Risk assessment and control of waterborne giardiasis. American Journal of Public Health 81(6):709-713.

Rose, J.B., Gerba, C.P., and Jakubowski, W. 1991b. Survey of potable water supplies for *Cryptosporidium* and *Giardia*. Environ. Sci. Technology 25(8):1393-1400.

Rose, J.B., C.N. Haas and S. Regli. 1991c. Risk assessment and control of waterborne giardiasis. Am. J. Pub. Hlth. 81:709-713.

Rose, J.B., R.L. Mullinax, S.W. Singh, M.V. Yates, and C.P. Gerba. 1987. Occurrance of rota- and enteroviruses in recreational waters of Oak Creek, Arizona. Water Research 21:1375-1381.

Schaiberger, G.E., T.D. Edmond and C.P. Gerba. 1982. Distribution of enteroviruses in sediments contiguous with a deep marine sewage outfall. Water Research 16:1425-1428.

Smith, H.V., and J.B. Rose. 1990. Waterborne cryptosporidiosis. Parasitology Today 6:8-12.

Stevenson, A.H. 1953. Studies of bathing water quality and health. American Journal of Public Health 43:529.

Sykora, J.L., C.A. Sorber, W. Jakubowski, L.W. Casson, P.D. Gavaghan, M.A. Shapiro, and M.J. Schott. 1990. Distribution of *Giardia* cysts in wastewater. Water Science Technology 24:187-192.

Tyler, J.M. 1982. Viruses in fresh and saline waters. Pp. 42-63 in proceedings of the International Symposium of Viruses and Disinfection of Water and Wastewater. Guilford: University of Surrey.

Van Donsel, D.J., and E.E. Geldreich. 1971. Relationships of *Salmonellae* to fecal coliforms in bottom sediments. Water Research 5:1079-1087.

Vaughn, J.M., E.F. Landry, T.J. Vicale, and M.C. Dahl. 1980. Isolation of naturally occurring enteroviruses from a variety of shellfish species residing in Long Island and New Jersey marine embayments. Journal of Food Protection 43(2):95-98.

Volterra, L., E. Tosti, A. Vero, and G. Izzo. 1985. Microbiological pollution of marine sediments in the southern stretch of the Gulf of Naples. Water, Air and Soil Pollution 26:175-184.

Wait, D.A., C.R. Hackney, R.J. Carrick, G. Lovelace, and M.D. Sobsey. 1983. Enteric bacterial and viral pathogens and indicator bacteria in hard shell clams. Journal of Food Protection 46(6):493-496.

C

Transport and Fate of Pollutants in the Coastal Marine Environment

INTRODUCTION

This appendix presents an assessment of current knowledge of the various physical, chemical, and biological processes that determine the transport and fate of pollutants associated with wastewater and stormwater inputs to coastal waters, and how well the behavior of these inputs can be modeled and predicted for engineering purposes. Specifically, how do the quantity, quality, and method of discharge of the wastewater to the coastal ocean affect the ambient water-quality and the quality of the sediments? With increasing knowledge of environmental engineering and marine sciences, it is now possible to design a waste management system by the water-quality and sediment-quality driven approach, namely finding the most cost-effective combination of source control, wastewater treatment, and outfall configuration. This process is explained in a later section on Overall Design following the next two sections which address Mechanisms of Input and Transport and Fate.

Coastal areas include a continuum from poorly-flushed small estuaries to the well-flushed open coastlines. This study focuses on larger estuarine and coastal systems subject to major urban impacts and that have significant exchanges of marine water with also the possibility of internal recirculation and entrapment of pollutants in the sediments within these bodies.

Federal law classifies inputs into point and nonpoint sources, according to whether discharge permits are required or not. As the regulations have changed (e.g., storm drains for cities over 100,000 people now require permits), distinctions on the basis of physical characteristics have become blurred. Traditional point sources at the time of passage of the Clean Water Act of 1972 included only outfall discharges from defined municipal and industrial installations; these sources, the focus of most control efforts heretofore, are generally well characterized now by types and fluxes of pollutants, although that was not the case before the Clean Water Act was passed. Outfalls (with very few exceptions) are submarine pipelines or tunnels discharging from a few hundred meters up to 15 kilometers (10 miles) from shore depending on the volume and character of discharge and the nature of the receiving water body.

The term nonpoint sources is a poor descriptor because this term includes all inputs that are not point sources. Also, the definition of point sources changes with new laws and regulations. Here, the broad classification of *diffuse sources* is used to include all sources except the traditional point sources. This category includes (but is not limited to) streams, storm drains and flood control channels, combined sewer overflows (CSOs), discharges from boats, ground water seepage, and atmospheric deposition. These sources have three common features: 1) the original pollutant sources are widely distributed, 2) the rates of delivery to coastal waters are highly irregular depending primarily on the occurrence of rain, and 3) control measures other than at the original sources are limited. In some locations, the release of pollutants from existing contaminated sediments can be a significant diffuse source.

Inputs of storm runoff, CSOs, and streams occur in a very unsteady manner at, or close to, the shoreline. Storm drains and flood channels (separate from sewers) discharge significantly when it rains, bringing as pollutant loads whatever wastes have accumulated in the drainage basin since the last storm; but also, smaller dry-weather flows may be highly polluted by illegal or unregulated waste disposal practices. Combined sewer overflows occur when runoff combined with sewage flows exceeds the capacity of a system, which then discharges at numerous predesignated places into various bodies of water in an urban area, including into streams and estuaries as well as the open coast. Natural streams and rivers may bring other pollutants from upstream areas, such as agricultural chemicals, atmospheric deposits, and nutrients washed off the land.

Mathematical and conceptual models are used extensively to explain processes that disperse and modify pollutants in the ocean and to predict their effects on ecosystems. Various submodels may be combined to produce an overall model to relate pollutant inputs to water and sediment quality for single and multiple sources. These models are fundamental to management by the environmental-quality driven approach because the limits on emissions for any outfall discharge or diffuse source may be back calculated through the models. These models are analogous to the emissions-to-air-quality models used in developing air pollution control programs.

The main purpose of this appendix is to assess the knowledge of all the relevant processes and evaluate the modeling capability for management of the quality of coastal waters and sediments by environmental-quality driven approaches. To be successful there must be good predictive capability for the dominant factors that determine the engineering choices for satisfying the standards. These factors can be determined based on sensitivity analyses and the experience of the modeler. Thus, for engineering purposes it is not necessary to understand every process if more knowledge would have no effect on the choice of control strategy. For example, it is not necessary to understand the behavior of a certain pollutant at a location where the exposure is far below any possible threshold value of concern.

Since modeling for design of a management plan for pollution control always has some uncertainty covered by safety factors, it is cost-effective to implement a system (such as a waste treatment plant and an outfall) in a stepwise flexible way to allow for continuous feedback of the operating experience and the observed impacts on the coastal waters. In fact, there are very few situations where there is not already an existing discharge that serves as a prototype to study before and during upgrading the system. For example, the full effect of upgrading primary treatment on coastal water quality might well be observed before proceeding to secondary treatment levels if there is significant uncertainty about the need for secondary treatment. Or source control efforts for specific chemicals can be focused on those observed to be too high. This approach is always self-correcting as the discharger commits itself to take as many steps as necessary to solve any known problems. This incremental approach is one of the important features of integrated coastal management as proposed in the main body of this report.

MECHANISMS OF INPUT

Outfalls

An outfall is a pipeline that discharges liquid effluent into a body of water. In the last four decades, there have been great advances in technology for ocean outfalls to achieve high initial dilutions and submerged plumes that are trapped beneath the pycnocline (or by the density stratification of the ambient water). Outfalls have advanced from simple open-ended pipes not far from shore to long outfalls with large multiple-port diffusers discharging in deep water. Figure C.1 provides an example of a deep water ocean outfall with a long multiport diffuser. The characteristics of major

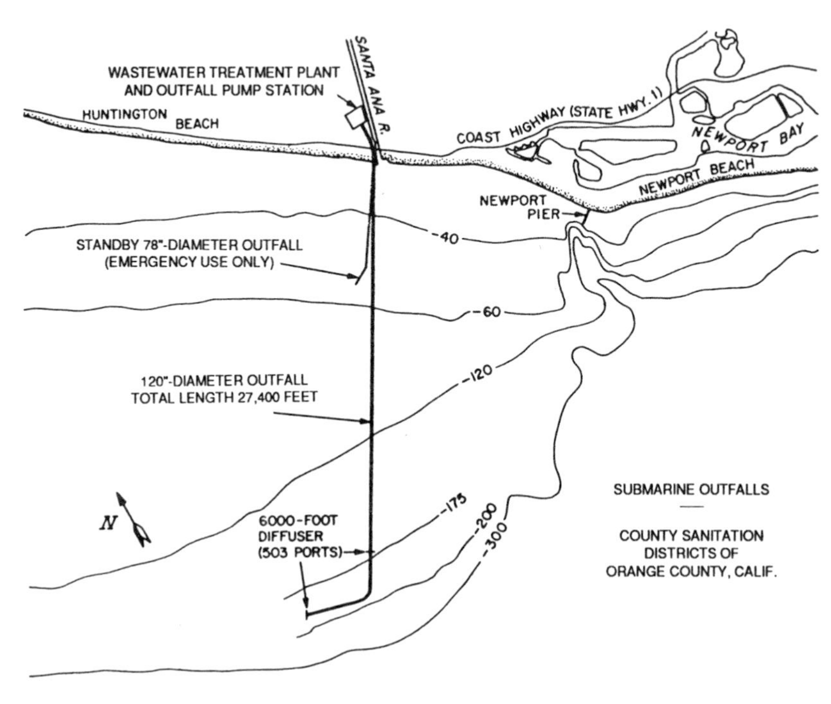

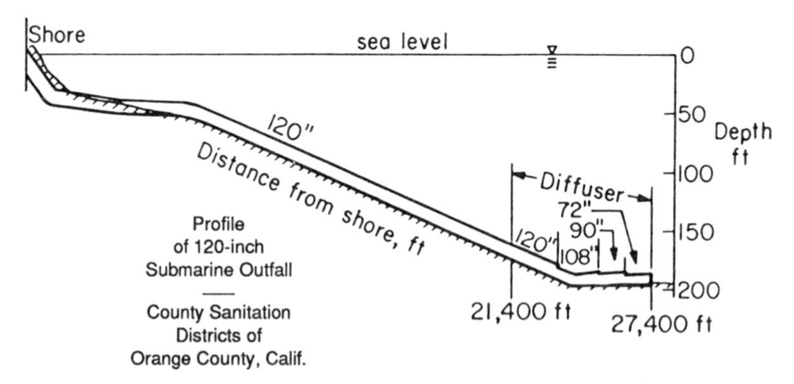

FIGURE C.1 Schematic plan and profile of the 120-inch outfall, County Sanitation Districts of Orange County, California. (In metric units, the overall length is 8.35 kilometers, the diffuser length is 1.83 kilometers, the diffuser depth is 53-60 meters, and the pipe diameter is 3.05 meters). (Source: Koh and Brooks 1975. Reproduced, with permission, from the Annual Review of Fluid Mechanics, Vol. 7, © 1975 by Annual Reviews Inc.)

outfalls on the Pacific coast of the United States constructed prior to 1978 are summarized in Fischer et al. 1979.

The construction of large outfalls in the marine environment has most commonly been accomplished with reinforced concrete pipe (RCP) with flexible joints. Recently, steel pipes have become more common because of improved manufacturing processes, better corrosion protection technology, proven constructibility (from technology transfer from the offshore oil industry), and construction costs for steel pipes, which can be significantly less than for RCP. One reason is that steel pipes are made into much longer lengths, requiring fewer junctions to be made in the marine environment. Two steel pipes of 64-inch diameter were used for the two outfalls of the recently built Renton outfall system in Puget Sound (Metro Seattle). They discharge through 500-foot long diffusers at a depth of about 185 meters (600 feet), which is probably beyond the capability of RCP construction. Tunnels have also become more competitive because of great advances in tunnel boring machines in the last 15 years. For example, the Boston outfall now under construction will be a 15 kilometer (9.4 mile) long tunnel, 7.39 meters (24.2 foot) in diameter, including a 2,000 meter (6,600 foot) long diffuser section with 55 vertical risers, each with 8 discharges ports. Three new successful outfalls in Sydney, Australia, are also tunneled.

The combination of source control, treatment plant, and outfall is an engineering system that has achieved often dramatic improvements in coastal water quality. Even today, however, while many major discharges have state-of-the-art systems, there are still others that discharge through short outfalls with poor initial dilution.

Figure C.2 shows schematically a typical multiport diffuser at the end of an ocean outfall discharging buoyant effluent into a density-stratified receiving water. Sewage effluent, being effectively fresh water, rises in the ocean, mixing intensely with the receiving water. The ocean is also usually density-stratified due to temperature and/or salinity gradients. Thus, the effluent mixing with the near-bottom denser ocean water can give rise to a mixture that is neutrally buoyant before the rising plume reaches the surface, leading to the formation of a submerged waste field, which is in turn advected by the prevailing currents (Figure C.2). This region of initial mixing is often called the *near-field*.

The mixing that occurs in the rising plume is affected by the buoyancy and momentum of the discharge and is referred to as initial dilution. It is typically completed within a matter of minutes. Dilution as used in the engineering community is defined to be the ratio of the volume of the mixture to that of the effluent (i.e., the reciprocal of the fraction of effluent in the mixture). This initial dilution phase of the mixing process is under some control by the design engineer since it depends on the diffuser details such as length of diffuser, jet diameter, jet spacings, and discharge depth.

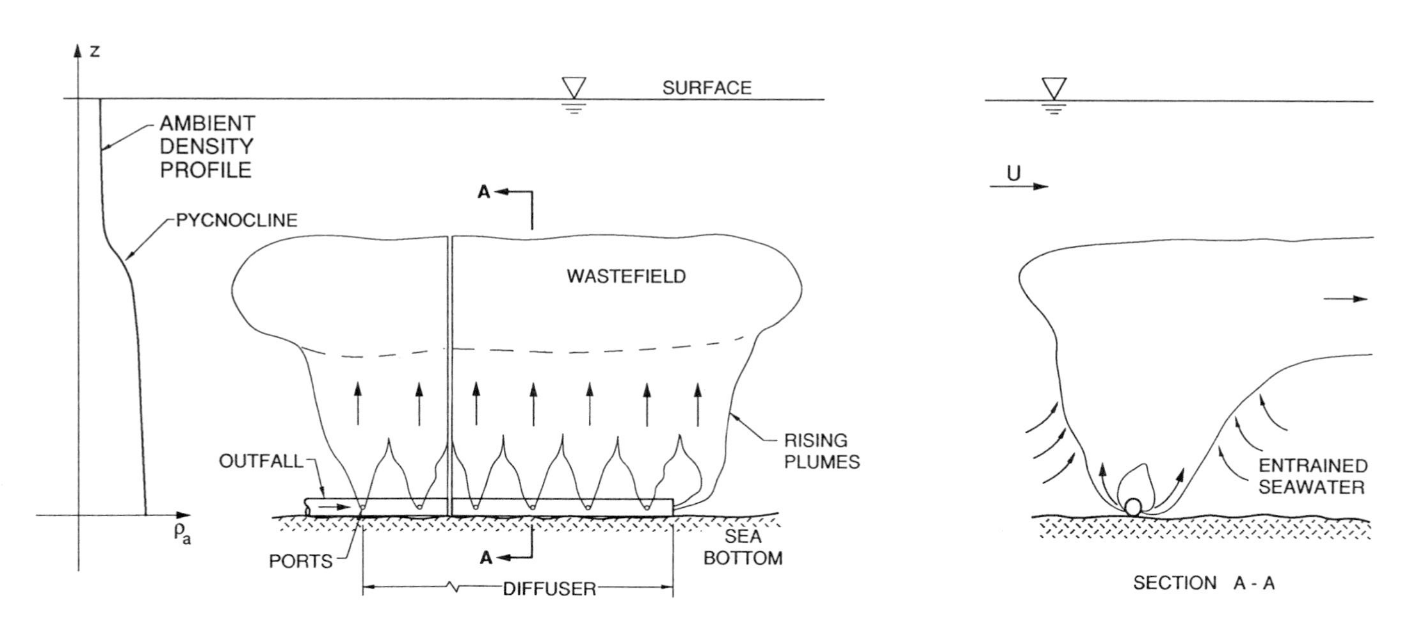

FIGURE C.2 Formation of a submerged effluent plume over a multiport diffuser in a stratified ocean with a current perpendicular to the diffuser. For clarity, only a few ports are shown, as typically there are hundreds for a large outfall.

The initial dilution is also controlled partially by nature since it depends on the density stratification and currents in the receiving water.

A typical large discharge diffuser (for a flow of 5 m^3/s) might be a kilometer in length and located in 60-meter water depth at a distance of 10 kilometers offshore. There might be several hundred discharge jets (typical diameter 10 centimeters) spaced along the 1-kilometer-long diffuser. The initial dilution obtainable for such a diffuser would be expected to be in the hundreds to a thousand depending on details (mainly flow rate and density stratification).

The initial dilution and waste field submergence can now be estimated with a fair degree of confidence as a result of three decades of engineering research on the mixing processes in buoyant jets and plumes (Koh and Brooks 1975, Fischer et al. 1979). A number of computer models can provide such estimates of sufficient reliability as to make decisions on design choices. The most commonly available ones in the United States are the ones published by the U.S. Environmental Protection Agency (EPA) (Muellenhoff et al. 1985; Baumgartner et al. 1992, based in part on laboratory tank experiments at the EPA by Roberts et al. 1989a, b, and c).

All the models for dilution and submergence calculations are based on analyses of buoyant jets discharged into a large receiving body of water. Usually the equations of conservation of mass, momentum, and buoyancy fluxes are integrated across the plume cross-section, having first assumed similarity of cross-plume profiles for the velocity and density deficiency (usually Gaussian). The resulting equations in this integral method are a system of nonlinear ordinary differential equations with the auxiliary conditions being in the form of initial conditions. Such systems are readily solved numerically. This is indeed the backbone of the available models.

For approximate start-up calculations, it often suffices to use the formula for a simple line plume in a linearly stratified environment based on assuming that the multiple-port diffuser is well approximated by a line plume—a source of buoyancy flux only. For this case, Brooks and Koh (1965) derived simple formulas as follows:

$$y_{max} = 2.84 \frac{q^{1/3}\left(\dfrac{g\Delta\rho}{\rho}\right)^{1/3}}{\left(\dfrac{-g}{\rho}\dfrac{\partial\rho_a}{\partial z}\right)^{1/2}}$$

$$S = \frac{0.31\left(\dfrac{g\Delta\rho}{\rho}\right)^{1/3} y_{max}}{q^{2/3}}$$

where:

q = discharge rate per unit length of diffuser
g = gravitational acceleration
ρ = density of discharge
$\Delta\rho/\rho$ = relative density difference between effluent and receiving water
y_{max} = maximum height of rise of the plume
S = initial dilution (centerline or minimum time-averaged dilution)
$\partial\rho_a/\partial z$ = average ambient density gradient

These formulas can be applied to the typical case described in the previous paragraphs. If we assume that the ambient density gradient is

$$\frac{-1}{\rho}\frac{\partial\rho_a}{\partial z} = 10^{-5} m^{-1}$$

(which can be due to a temperature difference of about 2°C over a depth of 50 meters), then y_{max} = 31 m (halfway from the discharge depth of 60 meters to the surface) and the initial dilution would be 200. Note that as the stratification increases, the plume rise and dilution are both reduced.

For actual design calculations with mathematical models, one also needs to examine many different ambient conditions such as density profiles and current speeds and directions. The effluent flow also varies. Thus the initial dilution for an outfall is not a constant value but fluctuates considerably depending on ocean conditions and the effluent flow rate.

It is important to point out here that dilution, being the ratio of the volume of the mixture to that of the effluent, can be converted to concentration c of a particular pollutant provided we know the concentration of that pollutant in both the effluent c_e and the receiving water c_b. Thus,

$$c = c_b + \frac{(c_e - c_b)}{S}$$

If c_b, the concentration of the pollutant in the receiving water were zero, then

$$c = \frac{c_e}{S}$$

The value of c_b includes the increase of the regional background concentration (background buildup) in the receiving water due not only to the continuous discharge from the outfall itself but also to all other sources.

Discharges from Barges and Ships

Ocean dumping from vessels has been practiced in the past by many coastal communities in various countries. It is still being practiced by some and is being planned by others. Nations are not in total agreement regarding ocean dumping, although the practice has seen a dramatic decline, particularly in the developed countries.

By far the largest amount of material involved in ocean dumping is dredged material (formerly known as dredge spoil). In the past, other materials dumped have included digested sewage sludge, various industrial wastes (including acids), oil well drilling mud and cuttings, coal ash, and mine tailings. Refuse has also been dumped in the past, but the practice has ceased (except for occasional illegal acts). Bilge water and ballast water are also discharged by ships in coastal waters.

Ocean dumping has been mandated by law to cease in the United States, with the exception of dredged material. Other developed countries have also largely agreed to stop dumping of sewage sludge. In less developed countries, the status of ocean dumping is unclear. Rules and regulations may not exist. It is unrealistic to expect ocean dumping of nonhazardous polluted materials to be eliminated worldwide any time soon or even in a few decades. Clandestine dumping and dumping where not allowed are difficult to police in most areas of the world's ocean, usually for the simple reason that the necessary infrastructure is inadequate or nonexistent.

Procedurally, the barge (or ship) is loaded with the waste material by placement into the ship's compartments. The vessel is moved to the designated dump site, which is generally a rectangular area with typical linear dimension of several kilometers. As long as the vessel is in the dump site, the material is allowed to be discharged into the ocean. Frequently a bottom-opening hopper barge is used. Here the barge bottom is equipped with doors, which can be opened to permit the material to fall out by gravity. Sometimes, the material is pump-discharged into the wake of the moving vessel to take advantage of the high turbulent energy that increases the initial dilution.

Modeling of the mixing, transport, and fates of materials after disposal from barges and ships is less well developed and much less well verified than the corresponding models for outfalls. While the physical processes involved in the two cases are similar, the situation for ocean dumping is less amenable to analyses because the discharge conditions (discharge rate, bulk density of the material, and characteristics of its contents) may be ill-defined. This has the most effect on near-field predictability but extends also to the intermediate and far-field because the near-field equilibrium vertical location of the discharged material depends on the discharge condition.

A detailed discussion of ocean disposal of digested sewage sludge has been presented with policy recommendations in a previous NRC report (NRC 1984).

Diffuse Sources

The term *diffuse sources* describes the inputs other than municipal and industrial wastewater to coastal water bodies. These pollutant sources include urban storm drains, combined sewer overflows, natural streams and rivers, ground water outflow (under the sea), discharges from recreational boats and commercial shipping, and atmospheric deposition. The input of pollutants into these delivery pathways is widely distributed, and more challenging to control at the points of origin. Furthermore, there is little opportunity to manage the hydraulics of the inputs to achieve high dilution far from shore as for wastewater from publicly owned treatment works (POTWs). Nonetheless, the same principles of transport and fate apply. For modeling the water and sediment quality, it is, of course, important to include all these diffuse sources along with the outfall discharges from publicly owned treatment works.

TRANSPORT AND FATE

Following plume rise and the attainment of initial dilution, the diluted effluent cloud (often submerged below the thermocline) is advected with the currents and undergoes a variety of physical, chemical, and biological processes, referred to as transport and fate of pollutants. These processes occur in the natural environment and are beyond the direct control of the engineers, other than the initial conditions determined by the characteristics of the outfall and the effluent. For example, if a plume is kept submerged below the surface mixed layer, the subsequent transport, fate, and effects may be greatly different from a surface plume in the near-term (on the order of days to weeks). This region, which is dominated by natural processes beyond the near-field, is called *far-field*. This section describes the major processes affecting the behavior of pollutants in the coastal ocean, that is, transport and fate.

Far-Field Transport and Dispersion of Contaminants

Scientific knowledge of far-field transport and dispersion of contaminants has advanced significantly in the last several decades. When this knowledge is coupled with modeling and site-specific programs to measure currents, density stratification, and dispersion, engineering designs for outfall diffusers can be made by the water-quality driven approach. Far-field transport and dispersion can be modeled for design purposes with reasonable factors of safety to cover uncertainties. This section addresses the current knowledge and gaps in science and modeling. While predictions now are adequate for project design, increased knowledge will lead to improved management techniques and predictions.

Beyond the near-field region where the delivery of the effluent has a dominant influence on its dispersal, the subsequent transport and dilution depends primarily on the currents in coastal waters. Persistent currents cause advection away from the outfall site, while currents that fluctuate over short time and space scales result in dispersion of the effluent. Dispersion results in dilution of the effluent, while advection carries it away from its point of entry. The exposure of the receiving waters to the environmental hazards introduced in the effluent depends sensitively on the advection and dispersion rates. Higher concentrations, and hence greater exposures, occur in regions of sluggish transport, and lower concentrations and exposure occurs in rapidly flushed environments.

There has been considerable effort mounted over the last 20 to 30 years to measure, model, and better understand transport and dispersion processes in coastal waters with application to the siting of outfalls and assessing the risks of oil spills and other toxic contamination, as well as developing an understanding of the interaction between the physics and the ecology of coastal waters. Because of the diversity of coastal water bodies and the complexity of the interactions between topography, density stratification, freshwater inflows, tidal motions, and the wind, it is not possible to predict *a priori* the magnitude of advective and dispersive transports at a given location. However, as will be discussed in more detail later in this section, it is possible to combine our general understanding of coastal processes with site-specific measurements to yield quantitative estimates of these processes that are accurate at least to an order of magnitude and often within a factor of two. This level of confidence is usually adequate to support water- and sediment-quality based analyses, with suitable safety factors to cover any errors of prediction.

A general discussion of the transport and dispersion in coastal waters must first acknowledge the great diversity in the physical characteristics of coastal environments, from lagoons to estuaries and bays of various sizes to continental shelves with widths that vary from several kilometers along the southern California coast to more than 100 kilometers on the east coast of the United States. The driving forces vary tremendously from place to place as well. For example, the currents on the west coast of the United States are driven primarily by the along-shelf winds, while in other areas, such as the Gulf of Alaska and the South Atlantic Bight (Georgia, South Carolina, and part of North Carolina), the currents are strongly influenced by the input of freshwater from rivers. Tidal motions, which are more important with respect to dispersion than net transport, are also highly variable in strength and relative importance, being tremendously important, for example, in Puget Sound and the Gulf of Maine. Finally, the currents in the ocean margins adjacent to the continental shelf often influence the transport on the shelf—the most famous example being the Gulf Stream, which intermittently spews warm-core rings (100-kilometer diameter rings of warm water originating from the subtropical Atlantic) against the continental shelf.

Transport also varies considerably depending on the location and timing of the delivery of effluent. Effluent being discharged through an outfall at the edge of the continental shelf will be directly exposed to oceanic currents, while stormwater, discharged at the shore, may be trapped in a nearshore region for a considerable period before it is exposed to the more energetic and dispersive motions further offshore. Discharge rates of effluent from POTWs are relatively constant, but nonpoint source discharges are highly intermittent, occurring during periods of high freshwater input. Thus, the fate of nonpoint source inputs is sensitive to the buoyancy-driven motions occurring during runoff events. Transport in coastal waters varies as a function of depth, so the fate of the effluent may depend sensitively on the vertical location of the discharge. Surface waters tend to be more energetic than deep waters, hence the advection and dispersion tend to be more rapid there. In regions with significant freshwater input and upwelling zones, there is a net offshore transport in the near-surface layer, favoring the dispersion of effluent. Because the introduction of nutrients in the euphotic zone may have the undesirable consequence of stimulating algal production, the recent tendency is to design submerged outfall plumes.

The fate of different waste substances varies considerably depending on whether they are dissolved or part of the particulate fraction of the effluent. Virtually all of the toxic metals and organic compounds in effluent are strongly particle-reactive. Hence their fates depend on sedimentary processes as well as fluid motion. The rapid coagulation or flocculation that occurs soon after the effluent is exposed to seawater causes settling of much of the solid fraction on time scales of 1 to 3 days. Subsequent transport of that material depends on resuspension from the seabed. Dissolved material, such as nutrients, is carried with the ambient water, but its distribution may change rapidly as a result of biogeochemical processes within the water column. Both in the case of the dissolved and the particulate constituents, residence times are generally short enough that the transport processes that occur within the first several days of their entry into the coastal environment are most important. Given typical transport rates, this represents a region extending 10 to 20 kilometers from the outfall. Transports of larger spatial and temporal scales are important with respect to the ultimate fate of substances and for basin-scale or regional ecological impacts.

Dispersion

Dispersion refers to the tendency of a parcel of water to increase in spatial dimensions, and hence be diluted, with time. In small-scale fluid dynamics, this tendency is referred to as diffusion (either molecular or turbulent). Dispersion is distinguished from diffusion in that it includes motions at various scales that may not be formally defined as turbulence but that have the effect of spreading out the fluid in a manner analogous to diffusion. Examples of processes leading to dispersion include tidal motions, eddies shed from coastal currents, and vertical or horizontal shears in the mean or low-frequency flow. In the context of far-field dispersion, vertical spreading is generally much less significant than horizontal spreading due to the smallness of vertical scales (10 to 100 meters) as compared with the horizontal scales (0.1 to 100 kilometers).

Dispersion has been notoriously difficult to predict, whether in estuaries, coastal waters, or the deep ocean, due to the complexity and wide range of scales of motion that may contribute to the mixing. The dispersion rate can be estimated by the time rate of increase in the size of a patch of fluid, with the dispersion coefficient defined as

$$K = \frac{1}{2}\frac{\partial \sigma^2}{\partial t}$$

where σ^2 is the spatial variance of the patch. Stronger flows as well as flows of large spatial scales tend to disperse more rapidly. Dispersion coefficients have been estimated directly in a variety of coastal environments by measuring the spreading of dye patches. Okubo (1971) combines the results of a number of studies, obtaining a consistent relationship between the diffusion coefficient (K) and the scale σ of the patch, from scales of 100 meters up to 100 kilometers, over which the dispersion coefficient increases from 10^{-1} m^2/sec to 1000 m^2/sec.

Alternative formulations of dispersion in coastal waters express the rate of patch growth as $\partial\sigma/\partial t = P$ where P is a dispersion velocity that has been found to be in the range 10^{-3} to 10^{-2} m/sec (Kullenberg 1982). Empirical correlations have also been developed (usually in terms of a dispersion coefficient) for shear-induced dispersion in rivers and longitudinal dispersion in estuaries (Fischer et al. 1979). These studies have succeeded in demonstrating consistency among different experiments in different environments, but the scatter in the data generally reflects a five to ten fold uncertainty in the magnitude of dispersive transport. Consequently, many studies of potential environmental impacts have been based on site-specific dispersion measurements using suitable tracers or drifters.

How critical is this uncertainty for predicting the reduction in concentration and increase in horizontal extent of a pollutant distribution over a time scale of several days? In relatively open coastal waters, the tracer data indicate that a pollutant patch grows at a rate of about 1 kilometer per day. So a continuous plume that is initially 100 meters in size, typical of a relatively small discharge, will experience a ten fold increase in width and a ten fold decrease in average and peak concentration that may be a significant addition to the initial dilution. On the other hand, a larger discharge, say from a submerged diffuser on the order of a kilometer in length, will only about double in width over a day or so. The consequent dilution factor of two increases dilutions from initial values of the order of 50-200 to 100-400. In this case, the uncertainty in the estimated rate of dispersion is less critical to the analysis.

Where net advection is absent or weak, the more rapid dispersion that occurs over longer times and larger spatial scales, due to fluctuating tidal and wind-driven currents, may determine the residual background concentration of effluent (Csanady 1983, Koh 1988). In open water, the background concentration may be a negligibly small quantity. However, in relatively enclosed coastal regions, including estuaries, significant accumulation of pollutant mass may be controlled by large-scale dispersive processes that should be quantified by tracer studies. As discussed in the previous section, an increase in background concentration reduces the effective initial dilution.

Net Advective Processes

Currents occur at a broad range of time scales, from seconds (e.g., surface waves and turbulence) to hours (e.g., tidal motions) to days (e.g., wind-driven motions) to seasonally varying flows and finally to steady flows. Generally the spatial scale of the motion increases as the temporal scale increases simply because water of a given velocity will be carried farther in a longer time period. Motions of short time scales do not carry water large distances, hence they do not contribute to net advection except at small scales (although as mentioned above, they may be important with respect to dispersion). In moderate sized embayments and the continental shelf, where spatial scales are at least tens of kilometers, the motions responsible for net transport tend to have time scales longer than 24 hours; these include wind-driven motions, buoyancy-driven flows, and flows forced by oceanic motions.

Buoyancy-Driven Flows. Buoyancy-driven flows predominate in estuarine environments, where the input of fresh water whose density contrasts with seawater produces a pressure gradient, which drives the less dense water seaward in the upper water column and pulls more dense seawater landward underneath. This so-called estuarine circulation is generally the most important flushing mechanism in estuarine systems. The well-documented variations in water quality in San Francisco Bay as a function of freshwater input clearly indicate the important role of estuarine circulation to the flushing and hence the water quality of estuarine systems.

The transport of suspended solids is highly coupled to density-induced vertical variations in advective transport. In many estuaries, the locations where fine sediments accumulate are known to be determined by the estuarine circulation pattern.

In larger estuarine systems and in coastal environments with large freshwater inputs, such as the Gulf of Maine and the Gulf of Alaska, buoyancy effects are still important for net advection, but rather than developing a two-layer, estuarine circulation, the buoyancy-driven flow is manifested as a coastal current, which flows along the coast to the right (in the northern hemisphere) of the offshore direction. Thus the coastal flow in the Gulf of Maine is southwestward, while in the Gulf of Alaska, it is northwestward. While these flows provide a substantial along-shelf component of flow, they do not by themselves contribute significantly to cross-shelf exchange. Effluent that is introduced in one portion of the coast may be carried away from its source region only to impact coastal areas downstream (after further dilution). A natural example of this type of downstream influence is the transport of toxic red-tide organisms from the estuary of the Kennebec River to the coasts of New Hampshire and Massachusetts by the buoyancy-driven coastal current (Franks and Anderson 1992).

Instabilities in buoyancy-driven coastal currents (e.g., Chao 1990) provide a mechanism of cross-shelf transport. The instabilities start out as undulations in the front, growing into eddies and extrusions into the adjacent oceanic waters. In addition, buoyant plumes are very sensitive to wind forcing due to their shallow vertical expression. Significant cross-shelf transport can occur when winds act on buoyant flows.

Wind-Driven Motions. The wind is often the most important driving force of net transport on the continental shelf, and it is often a major contributor to exchange between embayments and coastal waters. In the upper few centimeters of the water surface, the flow tends to proceed in the same direction as the wind at approximately 3 percent of the wind speed (Wu 1983). This rule of thumb is useful for the prediction of the trajectories of oil spills and floatable wastes. Below this very thin surface layer, the influence of the earth's rotation tends to turn the wind-driven transport to the right. In the absence of other forces, a steady wind will result in transport exactly 90 degrees to the right of the wind direction (or left in the southern hemisphere). This is known as Ekman transport, and the portion of the upper ocean in which it occurs, typically the upper 30 meters, is called the surface Ekman layer. The strongest wind-driven currents on the continental shelf occur as a result of along-shelf winds, which result in cross-shelf Ekman transport, but the pressure gradient induced by the coast causes the dominant flow to be in the along-shelf direction, in the same direction as the wind. Along-shelf current speeds tend to be 1 to 2 percent of the wind speed, their magnitudes varying with the shelf geometry, stratification, and other factors.

The cross-shelf flow based on Ekman theory should be a few centimeters per second (kilometers/day) for a moderate along-shelf wind stress. This is weak enough relative to the along-shelf flow that it has not been well resolved in measurements. It is also weak enough that it takes on the order of a week of persistent winds to transport material across a 20-kilometer-wide shelf and a full month to cross a 100-kilometer-wide shelf. A week of persistent winds is not uncommon, particularly along the west coast of the United States, hence the cross-shore Ekman transport often provides a relatively rapid offshore transport on the west coast. The wider shelf on much of the east coast and the absence of persistent winds renders cross-shelf Ekman transport less effective there.

A potentially important cross-shelf transport process is the convergence caused by a change in the along-shelf wind forcing, such as a change in wind direction. Relaxation from upwelling on the west coast causes a sudden shift from southward to northward currents. Where the oppositely-directed currents collide, a strong cross-shelf flow results. Such cross-shelf flows tend to be short lived, but they are strong enough to transport material all the way across the shelf, and they may be volumetrically as important as the Ekman transport itself in contributing to cross-shelf exchange. Similar convergences can also be caused by changes in coastline geometry such as headlands.

The vertical shear associated with wind-driven currents has consequences for effluent plumes that may be located near the surface, below the pycnocline, or near the bottom. In addition, the near-bottom currents may regulate the long-term transport of sediments (see later section on Sediment Processes).

Oceanic Currents Impinging on the Coast. Ocean margins are often the sites of major current systems, the most notable one bordering North America being the Gulf Stream. A strong steering influence of bathymetry generally keeps these currents from riding up onto the continental shelf, but there are often instabilities in these current systems that cause eddies to be shed, which impinge on the coast and may influence the cross-shelf transport. On the east coast, instabilities of the Gulf Stream generate warm core rings and smaller eddies called shingles, which impinge on the shelf along the east coast. They have been found to have a strong influence on the flow on the outer shelf and are likely important agents in fluid exchange between the outer shelf and the adjoining ocean. On the west coast, instabilities in the southward flowing California Current result in a complex field of eddies adjacent to the continental shelf. These eddies result in strong offshore flows between the outer shelf and the ocean, which carry cold, upwelled water from the shelf into the ocean interior.

Modeling and Measurements

Numerical models are well suited to investigating the nature of physical processes in complex coastal environments in which the mathematical problem does not lend itself to an analytical solution. Numerical models are particularly useful for describing flows in regions of complex geometry (Signell and Butman 1992) and for performing simulations of effluent motions (Baptista et al. 1984). There are many examples of accurate numerical model predictions of sea level variations, including tidal and wind-forced regimes and recently the problem of coastal trapped waves. However, sea level is of no relevance to effluent transport. Tidal motions have been modeled reasonably well (Tee 1987, Sheng 1990), and wind-driven, along-shelf motions can be predicted accurately under certain circumstances (Allen 1980). Nontidal currents in general are more difficult to model than sea level, and there are few examples in which models have accurately predicted the variations in currents in a particular environment, based on prescribed forcing variables such as winds and freshwater input. The quantities most relevant to the fate of effluent, such as dispersion and cross-shelf motions, are considerably more difficult to model since they depend on spatial gradients of the dominant currents.

It is clear from the above discussion that the scientific understanding of transport processes has not advanced enough to be able to model, *a priori*, the rates of the various processes at a given site. However, with an appropriate mix of measurements, theoretical calculations, and numerical modeling, the advection and dispersion rates can be determined with acceptable levels of uncertainty for use in environmental risk assessment.

In order to proceed effectively with a site assessment, it is important to start with a solid understanding of the regional transport processes. Once a region is well enough understood to focus an investigation on candidate outfall sites, a combination of field measurements, theoretical calculations, and modeling allows the transport to be quantified and the distribution of effluent components to be predicted. The measurements typically include moored measurements of currents and water properties (temperature, salinity, and in recent studies suspended sediment and dissolved oxygen), drifter releases, and shipboard survey measurements of relevant water column and sediment parameters. The measurements should (but generally do not) extend over a complete annual cycle. The outcome of such studies is usually a very good characterization of the dominant current regime, most often the tidal currents, and a somewhat less certain picture of the other transport processes. Even without the aid of a numerical model, such a measurement program can provide the basis for predictions of effluent transport (e.g., Koh 1988). By combining current measurement results with analyses of water column and sediment distributions of various measured constituents, the more subtle problems such as sediment transport and horizontal dispersion may be quantified.

This combination of tools for assessing physical transport processes will yield predictions with large uncertainty factors, but the errors in predicting advection and dilution will typically be well inside an order of magnitude. Considered in context with the overall risk management matrix, the assessment of water-column transport processes has a fairly narrow range of uncertainty. The fate of sediment is less easily predicted, but it is still amenable to prediction in a probabilistic sense, albeit with order-of-magnitude uncertainty in transport rates. Further discussion of sediments is provided in a later section, Sediment Processes.

Conclusions

1) Scientific knowledge of far-field transport and dispersion of contaminants has advanced significantly in the last several decades. When this knowledge is coupled with modeling and site-specific programs to measure currents, density stratification, and dispersion, the water- and sediment-quality driven approach incorporating transport and dispersion is feasible with reasonable safety factors. This approach is the only one that can be improved and adjusted in response to new research information.

2) Ongoing programs of research will provide continual refinements in the predictive capability of the transport and fate of effluent constituents, which will be useful for making future engineering decisions.

Recommendations

1) Coastal physical oceanographers should be encouraged to address questions of particular relevance to water quality, particularly dispersion and cohesive sediment transport.

2) The investigation of the fate and transport should not stop upon selection of an outfall site, but it should continue after a new or modified outfall is put into service, particularly to learn how good the preconstruction modeling and predictions were.

Behavior of Particles from Wastewater and Sludge: Flocculation and Sedimentation

Wastewater effluents and sludges contain particulate matter and particle-reactive pollutants. These particles vary in size over a broad range, from submicron dimensions to several tens of microns. These particles and the particle-reactive pollutants that accompany them can be deposited near the point of discharge or transported long distances in coastal zones. After deposition, substantial recycling and release of contaminants from bottom sediments to overlying waters can occur. Cycling of contaminants between suspended solid phases and seawater may also occur in the water column. While a substantial amount is known about these processes, much remains to be understood.

Knowledge of the transport and fate of particles and particle-reactive pollutants in marine environments is limited. This limitation constrains, but does not negate, the ability to design waste disposal systems that protect the environment. Conservative approaches can be used to meet appropriate water-quality and sediment-quality criteria. For example, an approach to addressing sediment quality criteria could assume that all of the particles and particle-reactive pollutants in a waste discharge are deposited in the region around a discharge site. This assumption would provide conservative estimates of the concentrations of contaminants to be expected in sediments and pore waters. If the resultant estimates did not meet sediment quality criteria, the contaminant discharge could be reduced to provide the needed environmental protection. Additional and more detailed modeling could be conducted to improve the estimates as necessary. A similar approach, in which all of the particles and contaminants are assumed to remain in the water column, could be used in addressing water quality criteria.

Particles in Marine Environments

Particulate matter in the ocean is comprised largely of aggregates of algae, bacteria, organic detritus, and inorganic particles (including natural sediments). These aggregates vary in size from submicron dimensions to many centimeters. Settling rates can be high. Hill (1992) has summarized published observations indicating that settling velocities of aggregates following phytoplankton blooms are in excess of 100 meters per day. These aggregates are formed by two major pathways: 1) biological aggregation by animal grazing and 2) physicochemical aggregation involving interparticle collisions and attachment. Biological activity can affect interparticle collisions and attachment as, for example, in the formation of the large aggregates known as marine snow. The relative importance of physicochemical and biological processes in aggregate formation and destruction has been debated for some time and remains unresolved at present. There are models for these processes, but their validity and accuracy require testing.

The production of these aggregates is important in the transport and cycling of carbon, energy, pollutants, and nutrients in marine environments. Large aggregates such as marine snow (Silver et al. 1978, Smetacek 1985, Alldredge and Gotschalk 1989) are important in particle transport and in marine chemistry (Fowler and Knauer 1986). Recently Hill and Nowell (1990) have assessed the role of rapidly settling particles in clearing nepheloid layers by physicochemical coagulation. To these can be added the scavenging of metals, including U and Th by rapidly settling particles in the ocean (Honeyman et al. 1988) and the importance of physicochemical coagulation in this process (Honeyman and Santschi 1991). Jackson (1990) has combined theory for the kinetics of coagulation with expressions for gravity sedimentation and the kinetics of algal growth to describe algal production, aggregation, and sedimentation during a bloom. The results indicate that physicochemical coagulation can place an upper limit on the accumulation of biomass in such blooms, leading to the formation of algal flocs commonly observed by divers.

Particles and Particle-Reactive Pollutants in Wastewaters and Sludges

The particles in wastewater effluents from primary or secondary treatment plants can be expected to contain particles that have settling velocities in fresh water that are less than about 40 m/day, corresponding to the hydraulic loadings used in the design of the settling basins in these treatment systems. (Hydraulic loadings for settling basins range from 300 to 2,400 gal ft^{-2} d^{-2} as discussed in Appendix D. A representative value used for comparison purposes here is 1,000 gal ft^{-2} d^{-1}.) This agrees fairly well with laboratory measurements of the settling velocities of these particles in seawater. For example, Faisst (1980) reviewed reported experimental results for sludge and wastewater effluent particles in seawater and found settling velocities to range from about 10^{-2} m/day to 25 m/day. Wang (1988) reported that 90 percent of the particulate mass in a secondary wastewater effluent had settling velocities of ≤ 1 m/day; for a digested sludge the corresponding figure was about 3 m/day. As with particles in the open ocean, particles in wastewater effluents and sludges range in size over several orders of magnitude, from the submicron range to several tens of microns.

Many pollutants in wastewater effluents and sludges are associated with particles in these suspensions. Karickhoff (1984) has reviewed the thermodynamics and kinetics of the sorption of organic pollutants in aquatic systems. The sorption of uncharged organic chemicals to particles is dominated by hydrophobic interactions and depends primarily on a chemical's affinity for water, typically described by an octanol-water partition coefficient, and on the organic carbon content of the solid sorbent phase. An estimate of the sorption of an uncharged organic pollutant in a wastewater or sludge can be made on the basis of the chemical's octanol-water partition coefficient, the organic carbon content of the solid phase, and the concentration of solids in the water. For hydrophilic organic pollutants, nonhydrophobic contributions to sorption can be important and inorganic surfaces, such as clays and metal oxides, can be significant adsorbents. The sorption of organic pollutants to suspended particles can generally be viewed as rapid, although true equilibrium may take weeks or more to achieve, and the process is often not completely reversible.

Several investigators have studied the transport and fate of nonpolar organic compounds in natural waters. Representative studies are given by Schwarzenbach and coworkers (Schwarzenbach and Westall 1981, Imboden and Schwarzenbach 1985). These studies and others show that the sorption of many organic compounds is proportional to the organic carbon content of the solids when this is greater than 0.1 percent and indicate that sorption can be predicted from the octanol-water partition coefficient of the nonpolar organic solute and the organic carbon content of the natural solid sorbent.

The adsorption of inorganic pollutants in aquatic systems has been reviewed by Dzombak and Morel (1987). In contrast to the nonspecific adsorption observed for hydrophobic organic pollutants, the adsorption of inorganic solutes is viewed as a site-specific process in which ions bind chemically at functional groups on solid surfaces. Surface complex formation models of varying complexity are available to model pollutant adsorption. The reaction is dependent upon pollutant type and concentration, pH, ionic strength, and solid concentration.

Particle-reactive pollutants from anthropogenic sources accumulate in marine sediments. An example of the effects of the organic particulate matter in a wastewater effluent on the transport and fate of synthetic organic compounds and of the deposition of metals in the discharge is given by Olmez et al. (1991). The site considered is the San Pedro Shelf off Southern California in the area of the outfalls of the Joint Water Pollution Control plant of Los Angeles County Sanitation Districts. Sediment distributions of organic carbon, DDT, and hydrocarbons as functions of depth or time correlate well with inputs of effluent particulate matter. Upper sediments are also enriched in light rare earth elements (La, Ce, Nd, Sm) and reflect use of fluid-cracking catalysts in the petroleum industry, and some release into the sewer system.

Transport and Fate of Wastewater Particles

When a wastewater effluent or sludge is discharged to the ocean, light, fresh water is introduced into dense, salty water. Three spatial regions are summarized here: 1) a zone of initial mixing or entrainment, often accomplished with a diffuser and characterized by a time scale of minutes; 2) subsequent transport and further dilution by tidal and wind-driven currents with time scales in the order of several hours to days; and 3) far-field transport driven by large scale circulation with time scales of several days to a few months. Brief descriptions of each follow.

The Outfall: Initial Mixing. In a point source discharge of wastewater to the ocean, fresh water containing particles at high concentration and that are moderately stable with respect to aggregation are mixed with seawater to produce a dilute suspension of particles that are probably chemically unstable with respect to their ability to attach to each other and form aggregates. When a diffuser is used to provide rapid effective initial dilution of the wastewater with seawater, the energy dissipated in the mixing process can provide contact opportunities between particles and contribute to particle aggregation. Extensive aggregation, if it occurs, will enhance the deposition of particles and particle-reactive pollutants in the sediments at the discharge site.

Extensive theoretical and laboratory studies of particle aggregation in wastewater plumes have been made by Wang (1988) and by Holman and Hunt (1986). These investigators disagree about the occurrence of coagulation during initial mixing in an outfall plume. Holman and Hunt concluded that there is a subregion of the zone of initial mixing within which coagulation occurs. It occurs because dilution is significant enough to provide salt for particle destabilization, while at the same time mixing and energy dissipation are fast enough and particle concentrations are high enough to provide particle contacts for aggregation. Further dilution with sea water slows coagulation appreciably because particle destabilization is not increased while particle concentrations continue to be lowered and energy dissipation continues to be reduced. Wang (1988) considered the changes in particle stability in the plume (from stable particles in the wastewater to unstable particles in the mixed discharge) and also the reductions in particle concentration and in energy dissipation that occur as dilution of the effluent proceeds; she concluded that coagulation is insignificant in the area of initial mixing or entrainment of a typical wastewater or sludge discharge because favorable conditions last for such a short time. Both of these studies are based on conceptual models and laboratory experiments; the occurrence and significance of coagulation in the zone of initial mixing of a wastewater plume have not been determined with accuracy.

The Outfall Region. The sedimentation of particles from sewage sludge discharged from ocean outfalls to coastal waters off southern California was modeled by Koh (1982), who concluded that sludge particles would be widely dispersed. Aggregation to enhance sedimentation was not considered. Coagulation of particles in wastewater and sludges has been studied conceptually and in the laboratory; results indicate that the process may affect particle transport and enhance particle deposition after discharge into coastal waters (Hunt 1980, 1982; Hunt and Pandya 1984). Farley (1990) has considered particle decomposition, aggregation, and settling together with advective and dispersive transport in modeling the deposition of particles and the accumulation of organic matter in the regions of the Orange County and Los Angeles County outfalls in southern California. Model predictions of both particle deposition rates and organic accumulations compared well with observations at these two locations. It was concluded that coagulation, sedimentation, and tidal motion affect particle deposition and sediment accumulation around outfalls. The model contains empirical coefficients that must be determined from laboratory or field data.

Far-Field Transport. Modeling of the transport of particles in wastewaters at the scale of tens or hundreds of kilometers has been sparse. O'Connor et al. (1983a) considered the fate of wastewater sludge dumped at a deep water site (Dumpsite 106) in the New York Bight and predicted little effect on the bottom sediments. These authors used information about the settling properties of Los Angeles sludge in arriving at this prediction. Some observations indicate, however, that sediments below the dump site have been affected by the sludge inputs (Van Dover et al. 1992).

The Sediment-Water Interface

Wastewater and sludge particles that reach the sediment-water interface are then subject to a variety of physical, chemical, and biological processes that can enhance deposition, lead to resuspension, produce chemical dissolution or biological mineralization, and lead to the release or burial of particle-reactive pollutants and nutrients. For example, Stolzenbach et al. (1992) report that particles, including fine submicron particles, are removed from suspension by aggregation in a porous and mobile layer at the sediment-water interface driven by near bottom flow and also by the activities of benthic organisms. The authors suggest that this process may be particularly important in shallow waters.

A study of benthic recycling in Lake Superior by Baker et al. (1991) provides a paradigm for pollutant cycling at the sediment-water interface in coastal waters. As in the ocean, degradation of organic particles in Lake Superior is efficient so that only a small fraction of the primary production in the epilimnion is incorporated permanently into the sediments. Inputs of hydrophobic organic contaminants to the lake are dominated by atmospheric deposition, while burial in the sediments and volatilization are the major removal mechanisms. Contaminants include polyaromatic hydrocarbons (PAHs) and polychlorobiphenyls (PCBs). Most of the hydrophobic organic compounds introduced into the lake are incorporated into rapidly settling particles produced in the epilimnion. On reaching the sediments, many of these contaminants are released and mixed back into the water column by biological processes in the benthic food web. Higher molecular weight PAHs, the most particle-reactive of the contaminants, are retained in the sediments and efficiently buried.

Discussion and Conclusions

A hypothetical description of the behavior of particles and particle-reactive pollutants in wastewater effluents and sludges prior to and after discharge to coastal waters is as follows.

- Particles in both wastewater effluents and sludges can be enriched in hydrophobic organic contaminants such as PAHs and PCBs. In aerobic discharges, many metals will partition so that substantial fractions occur in both the solution and the particulate phases; in anoxic discharges, such as sludges and some primary effluents, most metals of concern will reside in solid phases such as particulate organic matter and metal sulfides.
- Upon discharge, the particles will be chemically destabilized, thus favoring aggregation, but the extent of aggregation and enhanced settling that may occur during the initial dilution in the plume is not known. Chemical mineralization, oxidation, and sorption reactions are probably not important in this step, nor are biodegradation processes.
- Some particles, perhaps most of them, will be deposited in a region up to several tens of square kilometers in size in the vicinity of the discharge. Aggregation in the water column among the wastewater or sludge particles and with marine particles may and probably does occur, but the process cannot be predicted accurately. Horizontal transport of particles by currents is dominated by tides, wind, and density gradients; vertical transport is driven by gravity, perhaps enhanced by aggregation and by processes at the sediment-water interface. The time scale in the water column is from several hours to a few days. Chemical dissolution, oxidation, and desorption reactions may occur, releasing some contaminants to the water column, but the results of these reactions cannot be forecast with accuracy. Biological degradation of particulate material in the water column is probably not great on this time scale.
- The sediments in the region of the discharge will receive inputs of particles and particle-reactive pollutants. Rates of deposition and sediment accumulation cannot be predicted accurately. Biological degradation of organic matter and other processes can lead to a release of particle-reactive pollutants (metals, nutrients, and hydrophobic organic contaminants) to the overlying waters. Some fraction of many contaminants will be buried permanently in these sediments. The rates and effects of these reactions cannot be predicted quantitatively.
- Some particles and particle-reactive pollutants in the wastewater or

sludge discharge will be transported horizontally for long distances. Predictions are few, and their accuracy has not been demonstrated. To these materials in a discharge will be added those pollutants previously deposited in the discharge region and then released from bottom sediments by biological and chemical processes.

This summary of the behavior in marine environments of particles from wastewater and sludges indicates that, while a substantial amount is known, much remains to be understood. The science is challenging. The summary also indicates that our ability to make accurate and precise predictions of the fate of these particles and of particle-reactive contaminants in the coastal zone has not been tested sufficiently. The design of wastewater or sludge discharges to coastal waters is constrained by our limited knowledge of these environments. This does not mean that discharges cannot be designed to protect the marine environment; it does mean that these designs must be conservative, incorporating factors of safety into the design process.

Two different approaches can be taken in evaluating the effects of particles and particle-reactive pollutants in the coastal zone. One is directed at sediment quality criteria and the other addresses water quality criteria. Considering sediment quality, all of the particles in the wastewater or sludge discharge can be assumed to settle to the bottom in a zone around the outfall, with the area of this zone determined by depth and tidal motion. The rate of deposition will depend on the mass rate of input of particulate matter by the discharge. Estimates of the pollutant concentrations in the sediments and the pore waters can be made and compared with appropriate sediment-quality criteria. Where a sediment quality problem is anticipated, pollutant emission can be reduced to meet the standard. Alternatively, additional and more detailed modeling and testing can be performed to improve the estimates. This approach to the problem is conservative because dilution of contaminated particles with ambient particles is neglected, transport of particles away from the discharge area is ignored, and release of contaminants from the bottom sediments is not considered. An example of this type of calculation is provided in the subsequent section on Sediment Quality Modeling.

With respect to water quality criteria, all of the particles and particle-reactive contaminants can be assumed to remain in the water column and be transported by wind, tides, and large-scale circulation. Initial concentrations in the water column would be determined by the initial dilution of the plume, corrected for background buildup. Transport and fate would be modeled assuming that the contaminants are conservative with respect to chemical or biological degradation and are not removed from the water by sedimentation. Where a water quality problem is anticipated, site-specific modeling and testing can be performed to achieve more reliable predictions, and pollutant emission can be reduced to meet the standard.

Chemical and Biological Conversions of Toxics *in situ*; Biological Availability/Bioaccumulation

Chemical speciation serves as the common conceptual foundation on which geochemical, biochemical, and modeling studies have built our present understanding of the environmental chemistry of both organic contaminants and trace elements. Advances in the design of wastewater management systems will necessarily utilize this foundation. The term *speciation* includes important physical and chemical distinctions in the form in which an organic compound or metal ion is found. The most commonly measured type of chemical speciation of a toxicant involves separation of particulate and dissolved forms of a substance. Characterization of the chemical nature of particle-associated substances is currently the subject of intensive research. A second type of speciation involves identifying the chemical nature of solutes by distinguishing species engaged in reversible equilibrium reactions, i.e., acidic or basic forms of organic acids and inorganic or organic complexes of metal ions. Although these reactions are readily described by classical solution thermodynamics, the speciation of metals in the environment has only begun to be elucidated in recent years. Finally, speciation can involve distinguishing forms of trace elements that are not readily interconvertible by equilibrium reactions, e.g., alkylated metals and different redox states of metals. All of these kinds of speciation come into play when modeling the environmental fate of organic compounds and trace elements.

Speciation-based models of the environmental fates of potentially toxic trace metals and organic compounds couple equilibrium representations of the reversible reactions with rate laws for the transport and transformation processes. The importance of speciation lies in the fact that different species behave differently, e.g., particulate species settle and those associated with dissolved or colloidal organic matter are generally unavailable to biota. Consequently, the rates of the transport and transformation processes making up the environmental cycle of a toxicant are determined by the availability of its various species to the governing biogeochemical processes and the magnitude of those processes. Speciation-based models have been developed for organic solutes (Imboden and Schwarzenbach 1985) and trace metals such as manganese (Johnson et al. 1991) and mercury (Hudson et al. 1992) in lakes. The major limitation that these models currently face, however, is that in many cases our understanding of the principles governing metal cycling is ahead of our knowledge of the mechanisms and rate dependencies governing the processes, particularly transformation processes.

Transformation Processes

Waste materials entering the marine environment can undergo a wide variety of transformations such as photolysis, biodegradation, and hydrolysis. Transformation processes can be biotic or abiotic and are influenced by a variety of physical and chemical conditions. Biodegradation processes, for example, depend upon the population of microorganisms present as well as the structure of the particular organic compound being degraded.

The transformation of wastewater constituents in the marine environment can influence the bioavailability, transport, and fate of the materials. Degradation products can be more or less hazardous than the original compound. With metals, for example, methylated mercury is far more hazardous than inorganic mercury species. The debutylization of tin, however, results in a less toxic form of the metal.

There has been significant progress in understanding and unravelling transformations in the marine environment over the past decade. For example, understanding has evolved from the concept of partitioning between the particulate and dissolved phases to a recognition that there is a continuum of phases from truly dissolved through the colloidal state to true particulate. In this case, improved recognition of the importance of colloids in natural waters reveals that concepts commonly accepted ten years ago were simplistic and somewhat arbitrary, having been based on separation techniques using filters with pore sizes ranging from 0.2 to 0.5 μm.

Organics in Sediments

A considerable amount of research has recently been directed toward understanding the distribution of organic contaminants in sediments and the biological effects of the contaminants in the various sediment phases. The observation that bulk chemical concentrations of hazardous chemicals in sediments do not necessarily reflect the biological availability of those substances was a major impetus for the effort. The result is the equilibrium partitioning theory to determine concentrations of a substance in question among the various phases that exist in sediments. These levels are then correlated with concentrations known or thought to be toxic (EPA 1991). The equilibrium partitioning (EqP) concept assumes that, for non-ionic organic substances, most of the chemical in the sediment solid phase will be sorbed to organic carbon. It also assumes that pore water concentrations of the chemical correlate best to biological effects. The partitioning coefficient between the organic carbon and pore water (K_{oc}) has been shown to be approximated by the partitioning coefficient between n-octanol and water (K_{ow}) (EPA 1991). The relationship is given by:

$$\text{Log } K_{oc} = 0.00028 + 0.983 \text{ Log } K_{ow}$$

Therefore, if the bulk sediment concentration, the organic carbon content of the sediment, and the K_{ow} for a particular non-ionic chemical are known, one can calculate the pore water concentration of that substance at equilibrium. In a retrospective mode, one can compare this number to some *acceptable* concentration to determine if the sediment is likely to be harming the environment. In a prospective mode, knowing the acceptable pore water concentration and the K_{ow} for a non-ionic organic, one can calculate a sediment quality criterion on a sediment organic carbon basis. An example of this type of calculation is provided in the subsequent section on Sediment Quality.

Trace Elements and the Importance of Speciation

Although knowledge of the concentrations and distributions of trace elements in coastal waters has advanced dramatically, it has become increasingly clear that information on just total concentrations is insufficient for providing an adequate understanding of a trace element's biological and geochemical interactions. Much of the uncertainty about the relationship between total metal concentrations and their toxicity to aquatic organisms results from a lack of definitive knowledge of the chemical forms of these metals in natural waters. Trace elements dissolved in seawater can exist in different oxidation states and chemical forms (species) including free solvated ions, organometallic compounds, complexes with inorganic ligands (e.g., with Cl^-, OH^-, CO_3^{2-}, SO_4^{2-}, etc.), and complexes with organic ligands (e.g., with phytoplankton metabolites or humic substances). Particulate forms include metals adsorbed onto or incorporated into a range of particles from small colloids on the order of 10 nm to large particles resuspended from the bottom sediments by episodic events such as storm activity or tidal flushing. Advances made over the past 15 years in the understanding of metal speciation and its relevance to toxicity in marine ecosystems have important implications for the development of wastewater management strategies for coastal urban areas.

Organometallic Compounds. Organometallic forms of trace elements are those in which the trace element is covalently bound to carbon (e.g., methyl forms of As, Ge, Hg, Sb, Se, Sn, and Te; ethyl-Pb forms; butyl-Sn forms). The existence of naturally-occurring methylated forms of As (Andreae 1977, 1979), Ge (Hambrick et al. 1984, Lewis et al. 1985), Hg (Mason and Fitzgerald 1990), Sb (Andreae et al. 1981), and Sn (Byrd and Andreae 1982, Andreae and Byrd 1984) has been demonstrated in seawater. For example, a most astonishing discovery is that 90 percent of the oceanic Ge exists as methylated forms ($CH_3Ge(OH)$ and $(CH_3)_2Ge(OH)$) that are so stable to degradation that they have been called the "Teflon of the sea" (Bruland 1988).

Examples of important highly toxic organo-metal species include methyl-mercury and butyl-tin compounds. Recent studies have indicated that methyl-mercury species, rather than inorganic mercury species, accumulate in fish and pose a potential problem in human diets (Clarkson 1990, Grieb et al. 1990, Weiner et al. 1990). Methyl-mercury, which can be formed in aquatic environments by microbial activity, is far more hazardous than inorganic mercury species. Butyl-tin and phenyl-tin compounds are used as algicides and fungicides in paints and in agriculture, respectively. These highly toxic organo-tin forms degrade into inorganic tin species, which are less toxic.

Examples of important organ-metalloid species include organo-selenium compounds and methylated arsenic acids. Organo-selenium species can be bioaccumulated much more effectively than inorganic forms of selenium such as selenate or selenite (Fisher and Reinfelder 1991, Luoma et al. 1992). Water quality criteria need to take into account the chemical form or species of the various elements in both source waters and receiving waters, together with a knowledge of the potential transformations that can occur in natural water systems. Better understanding must be gained about the kinetics and mechanisms of production, degradation, and transformation of these organometallic compounds to improve modeling and standard setting.

Trace Metal/Organic Ligand Complexation. Potentially toxic metal cations include Cu, Zn, Pb, Ag, Hg, Cd, Ni, etc. These metals can occur in the marine environments as free cations, as relatively labile inorganic complexes, or as relatively inert coordination complexes (or chelates) with various organic ligands. Recent research has demonstrated that the biological response of planktonic organisms to these metals is related to the free metal cation concentration (Sunda 1988-1989). The toxicity of Cu and Cd to phytoplankton (Sunda and Guillard 1976, Brand et al. 1986); cadmium to grass shrimp (Sunda et al. 1978); and even the availability of the nutrient metals, Fe, Nm, and Zn, to phytoplankton (Brand et al. 1983) are believed to be related to the concentrations of the respective free metal ions, rather than by their total concentrations or the concentrations of specific organic complexes.

It is the lowest trophic level organisms that are most sensitive to these hazardous trace metals. For example, ciliate protozoans isolated from estuarine waters are among the most sensitive organisms to free copper (Stoecker et al. 1986); certain phytoplankton species (Brand et al. 1983) and zooplankton such as copepods (Sunda and Hanson 1987) have exhibited toxicity at very low levels of free copper.

Recently, marine chemists have advanced their ability to characterize the forms (i.e., chemical species) in which some of these trace metals exist in seawater (Bruland et al. 1991). For example, recent studies demonstrate that greater than 90 percent (generally closer to 99 percent) of dissolved copper is organically complexed in coastal and estuarine waters (Sunda and Ferguson 1983, van den Berg 1984, Moffett and Zika 1987, Sunda and Hanson 1987, Coale and Bruland 1988, 1990). Complexation with these organic ligands appears to render copper essentially unavailable biologically and therefore, nontoxic. Consequently, the degree to which the metals in source waters from sewage effluents and urban runoff are complexed with organic ligands can determine their biological effect in the receiving waters. Measures of total copper concentrations, while easier to obtain, may have little relevance in predicting biological effects.

Bioavailability of Trace Metals. The factors controlling the biological availability of trace metals influence both their uptake by the food chain and their involvement in the biologically mediated reactions of their biogeochemical cycles. Because most biologically mediated processes require uptake of the element, either into cells or by surface sites of cells, biological availability can often be reduced to the relative rates of uptake of different metal species. A variety of mechanisms, each exhibiting different dependencies on metal speciation, can control the rates of biological uptake. Consequently, a complete understanding of the environmental fate of trace metals awaits the determination of the mechanisms by which metals are assimilated.

For example, the central mechanistic issue for mercury assimilation is whether it enters the cells in question by passive diffusion or by facilitated transport. Transport of neutral forms of some metals, e.g. elemental mercury and neutral complexes of methyl and divalent mercury, by passive means is rapid in lipid bilayer membranes (Gutknecht 1981, Boudou et al. 1983). Metal complexes with naturally-occurring organic chelators are generally not membrane permeable. Passive diffusion provides a baseline uptake rate that facilitated transport may supplement. In aquatic organisms, facilitated transport of toxic metals likely involves the uptake systems for nutrient metals, i.e., copper may enter cells through the uptake systems whose normal physiological function is to acquire zinc, manganese, or other essential metals. Such interactions have been previously observed in marine phytoplankton (Sunda et al. 1981, Harrison and Morel 1983). Facilitated transport generally involves complexation of the metals by specific ligands or binding sites. When this mechanism is significant for a toxic metal, the rate of its uptake would be influenced by both competition with the nutrient metal for transport and feedback between transport system kinetics and the nutritional status of the organisms. Both of these factors would cause the rates of toxicant uptake to be enhanced under conditions of low essential metal availability.

The best known relationship of facilitated transport rates to chemical speciation involves control by *free metal ion* concentration or activity. This observation reflects the control of a variety of biological effects by the equilibrium binding to specific cellular sites. In some cases, equilibrium between the site and solution are not attained, and the rate of the process reflects the complexation rates of labile species, not just that of the free metal ion (Hudson and Morel 1990). In both equilibrium- and rate-controlled cases, strong chelators reduce the metal's availability by reducing the concentration of free metal ions and labile species.

Finally, whenever the rate of cellular uptake is high enough, slow diffusion through the extracellular medium can cause the transported species to become depleted at the cell surface. When species interconvert rapidly within the boundary layer, the uptake rate under diffusion-limitation will depend on the total concentration of labile species rather than the species that controls the rate of the transport step (Jackson and Morgan 1978). Generally, inorganic and weak organic complexes of a metal interconvert rapidly enough that they remain in equilibrium throughout the diffusive boundary layer.

Nutrient Cycling and Biostimulation

Eutrophication is perhaps the greatest and most obvious impact of wastewater disposal in estuaries and some shallow coastal areas (see Appendix A). Eutrophication results from a damaging excess production of phytoplankton or aquatic plants due to abnormally high inputs of nutrients. Nutrients are elements required for plant growth and include nitrogen, phosphorus, silicon, and sulfur as well as trace metals such as iron and molybdenum. Damage caused by excess growth may simply include reduced transparency or, at the extreme, hypoxia or anoxia caused by respiration and decay of dying plants and animals. Intermediate effects can include damage to coral reefs and blooms of noxious algae.

Phytoplankton production in most coastal areas is nutrient limited; i.e., increased inputs lead to increased net production. Phosphorus is the limiting nutrient in some temperate estuaries and in most tropical lagoons and estuaries. However, nitrogen is the limiting nutrient in most of the United States coastal zone, including most estuaries.

Because nitrogen is quickly assimilated by phytoplankton and plants in the coastal zone, it is nearly impossible to define a safe concentration of dissolved nitrogen. Instead, researchers have had to develop criteria and guidelines based on relationships between rates of nitrogen input and phytoplankton production or standing crop. Assessment of actual conditions in various estuaries of the United States, coupled with controlled experiments in large tanks (mesocosms), have lead to what this Committee believes to be a convergence of opinion on rates of nitrogen inputs that may be of concern. This body of information is summarized in Appendix A, Figure A.6a, which suggests several ranges in inputs of interest. First, estuaries may maintain low phytoplankton standing crops (i.e., below 5 mg/m^3 chlorophyll a, the upper end of the oligiotrophic range) noneutrophic when the rate of dissolved inorganic nitrogen (DIN) inputs is below about 100 mmol/m^3/y. Estuaries and coastal areas with DIN loadings between about 100 and 500 mmol/m^3/y are mesotrophic, with phytoplankton standing crops on the order of 5 to 15 mg/m^3 of chlorophyll a. Above about 500 mmol/m^3/y of DIN (to 8,000) phytoplankton standing crop may range from about 15 to over 60 mg/m^3 (or 3 to 120 times the production in oligiotrophic estuaries). Interestingly, Jaworski (1981) proposed that inputs to estuaries be kept below 380 mmol N/m^2/y (a surface area criterion); this corresponds to a loading of 40 to 95 mmol/m^3/y for shallow estuaries, 4 to 9 meters deep, a range of noneutrophication rates not inconsistent with the first category (i.e., less than 100 mmol/m^3/y).

How do these rates relate to other coastal areas not included in the estuary studies, and how do they translate into actual emissions from real waste streams? An example is Santa Monica Bay. At present, no sewage DIN is discharged directly to the upper mixed layer, which on average is 20 meters deep (twice Jaworski's shallow estuary depth). In 1989, the Hyperion Treatment Plant (city of Los Angeles) discharged 9900 Mt DIN/yr, mostly below the mixed layer (data from SCCWRP 1991). If all the N went into the mixed layer (only a fraction may, due to accumulation by motile phytoplankton, Hendricks and Harding 1974), and if exchange and transport were in the same range as for shallow estuaries (i.e., Nixon 1988, Appendix A), the bay would still be in an oligotrophic range (DIN input = $70 < 100$ mml/m^3/yr) insofar as predicted chlorophyll a is concerned (assuming no other nutrient sources). This hypothesis is not inconsistent with observation (Eppley 1986) that suggests that the bay is, at most, mesotrophic. However, transport in this bay is in fact greater than in the estuaries reviewed in Appendix A. Only a fraction of the nitrogen input from the Hyperion Treatment plant is expected to enter the mixed layer in Santa Monica Bay. Again, there has been no evidence of damaging excess phytoplankton blooms in Santa Monica Bay over the past two decades (no dissolved oxygen depression, no green epicenter, no depression of secchi disc depth readings (light penetrations)), consistent with the expected lack of nutrient biostimulation.

The alternatives for controlling nutrient inputs from wastewaters include 1) no action, 2) relocation of the discharge to increase dilution and plume submergence and far-field dispersion, 3) source control, 4) treatment to remove limiting nutrients, and/or 5) changing the ratio of nutrients (provided that the limiting nutrient is at least partly removed). As noted above for Santa Monica Bay, the no action alternative for sewage wastewater is a reasonable one; at current sewage DIN input rates and location, there is no damaging excess production and, therefore, there is no action that would make an improvement. However, there may be a need to evaluate shallow nearshore conditions to make sure that this is the correct conclusion when all possible sources to the mixed layer are taken into account. These would include non-point source urban runoff and other diffuse sources (including upwelling), as well as other point sources.

The second alternative is discharge relocation. Officer and Ryther (1977) argued that eutrophic conditions in some east coast estuaries would not be relieved by secondary treatment (which removes little DIN) but could be greatly relieved by diverting primary-treated effluents to the ocean. During the 1950s, there was biostimulation of plankton (whole plankton volume) nearshore in Santa Monica Bay associated with a shallow-water (20 meters) nearshore (1.6 kilometers) discharge; the area of excess phytoplankton disappeared when effluent was diverted to a deepwater (60 meters) diffuser (2,440 meters [8,000 feet] long) located 8 kilometers (5 miles) offshore (SCCWRP 1973), with greatly increased initial dilution and plume submergence below the thermocline (except in winter).

Such opportunities for diversion exist along much of the United States west coast, Alaska, the Pacific islands, and Florida but not within east coast embayments such as the Long Island Sound or the Chesapeake Bay. In these areas, source control (of agricultural fertilizer applications, farm practices) and treatment (beyond secondary) are required. These actions are now recognized in many area plans that include specific goals for reducing nutrient input rates.

To control biostimulation due to excessive nutrients in areas where it is or may be a problem, it is necessary to use an integrated management approach using the water- and sediment-quality driven approach to devise appropriate control strategies.

Sediment Processes

Many of the chemicals that have caused environmental impacts in estuarine or marine systems readily partition to particles, either suspended in the water column or on the bottom in sedimentary deposits. As such, there are numerous cases of sediment contamination that exist even though the original source of the substance has been eliminated. These deposits have become persistent sources of the contaminants to the overlying waters and the biota that reside in them. Therefore, not only must this source term be considered in modeling efforts for some areas but also the factors that govern the sorption and desorption of contaminants from particles must be understood to allow accurate fate and transport models to be constructed.

In general, fine-grained sediments, such as silts and clays, exert a greater influence on the quality of overlying waters than coarse-grained sediments such as sand and gravel. This is due to a number of factors. The surface area per unit mass of the sediments is greater for finer grained sediments, which facilitates adsorption of some cationic species. In addition, the organic carbon content of sediments generally increases with decreasing sediment particle size, which enhances sorption of hydrophobic organic compounds. Therefore factors that enhance either transport, deposition, or resuspension must also be carefully considered in the formulation of fate and transport models.

Sediment Deposition, Resuspension, and Transport

Sediment deposition by gravitational settling is regulated by the size and density distribution of suspended particles. Particle coagulation by physical or biogenic processes may alter the size distribution and thus affect the rate of deposition by increasing the effective settling of fine particles that would otherwise remain suspended (Weilenmann et al. 1989). Because of the difficulty of making quantitative measurements of particle size, density, and coagulation efficiency, estimates of deposition are often based on observations of mass accumulation in sediment traps or the decrease of suspended mass in laboratory settling columns. Although neither of these methods is free from methodological problems (Wang 1988, U.S. GOFS 1989), they can provide valuable information on spatial and temporal variations in potential particle deposition.

Resuspension of deposited sediments occurs whenever the shear stress exerted on the bottom exceeds a critical value. Although the critical shear stress for larger, non-cohesive sediment particles (i.e., sand) can be fairly well predicted from an empirical relationship known as the Shields diagram, this method has been of little use for cohesive fine sediments, particularly when they are colonized by benthic organisms (Cacchione and Drake 1990). The rate of erosion once the critical stress has been exceeded has also been observed to vary from site to site for sediments with similar physical characteristics (Lavelle et al. 1984). Site-specific measurements of the critical stress have been obtained using inverted flumes in situ to produce controlled flow velocities over the bed surface for scour observations (Gust and Morris 1989).

Reasonably well-accepted methods for estimating the bottom stress from oceanographic measurements are now available and have been used in attempts to predict sediment resuspension in relatively open coastal waters (Grant and Madsen 1986). To date, these efforts seem to have been limited in predicting the long-term sediment accumulation by the variation in the critical shear stress as a function of grain size and depth in the sediment as well as by interactions between different grain sizes (armoring of fine sediment by larger grains) (e.g., Lyne et al. 1990). However, these methods are useful in predicting the frequency of episodic erosion by large storms (Drake et al. 1985).

Once eroded, particles may be transported as bed load (rolling, hopping, or sliding along the bottom) or as suspended load (moving with the water) with the latter being the most likely mode for contaminant-laden fine sediments. Predictions of suspended load transport are often based on the assumption of local equilibrium, i.e., that the upward flux of particles is balanced by the downward settling flux. However, this model has been difficult to apply in practice because of the uncertainties in particle settling speeds and in the suspended concentration at the bed-water interface (Hill et al. 1988). Direct measurement of the suspended flux is feasible using a combination of current meters and transmissometer deployments, but the transmissometer must be carefully calibrated to account for particle size effects (Lyne et al. 1990).

Finally, episodic events such as spills or hurricanes can have a major influence on the ultimate fate of pollutants in sediments. Materials that have been stored in unconsolidated sedimentary deposits for decades can be dispersed in a matter of hours by abnormal waves or currents. Therefore, these phenomena must be considered in treatment and disposal system designs. The effects of episodes depend not only on the intensity of the episode but also on the water depth, basin geometry, and other factors.

As a result of the difficulties noted above in quantifying the deposition, erosion, and transport of fine sediments, it is not surprising that virtually all estimates of net sediment accumulation or loss are made on the basis of observations of the deposited sediment. Regions of relative sediment deposition or erosion may be inferred from an examination of sediment characteristics such as grain size and organic content (Nichols 1990). Net accumulation is commonly determined from the measured profiles of radioactive elements (Thorbjarnarson et al. 1985) or back-figured from models of diagenetic processes in the sediments (Berner 1980).

Sediment Mixing, Contaminant Reactions, and Release to the Water Column

The subsurface distribution of contaminants associated with deposited sediments may be altered by biogenic or physical mixing. In the marine environment, sediment mixing by organisms, where present, usually exceeds that resulting from waves and currents. It may transport sediment particles and porewater from the bed-water interface to a depth of 10 cm or more on a time scale of weeks to years depending upon the intensity of organism activity (Berner 1980). This mixing has been modeled as both a diffusive (Officer and Lynch 1989) and an advective (Fisher et al. 1980) process in the sediments, and, although experience has established some bounds on the likely magnitude of mixing rates, the most reliable estimates are based on site-specific observations. Biogenic mixing may be particularly important in transporting contaminants that would otherwise be buried by net sediment accumulation to the bed-water interface where exchange with the water column and exposure of pelagic and benthic organisms may occur.

The subsurface distribution of a contaminant may also be affected by chemical and biochemical transformations, including, but not limited to, sorption and desorption of organic and inorganic compounds and elements, oxidation-reduction reactions, and microbial degradation of organic matter (Berner 1980). These diagenetic processes may result in remobilization of heavy metals and the detoxification of organic contaminants such as hydrocarbons and pesticides. Changes in the contaminant profile in the sediments and release of contaminants to the water column may also involve the processes of molecular diffusion and colloidal advection (Gschwend and Wu 1985). A variety of models of varying complexity have been developed to describe such reactions in conjunction with bioturbation and sediment accumulation (see Berner [1980] for a summary). Because of uncertainty in key parameters, these models have been used mostly to infer rates of sediment processes from measurements rather than to make quantitative predictions. Predicted rates of contaminant release from the sediments have been compared with observations obtained from flux chambers deployed on the bottom (Sayles and Dickinson 1991).

The United States Experience

Experience with contaminated sediments in the United States has been gained as a result of the Dredged Material Research Program of the Corps of Engineers (Palermo et al. 1989) and from studies at sites where PCB (Ikalainen and Allen 1989, Sanders 1989), DDT (Logan et al. 1989), or Kepone (Huggett 1989) are found at high concentrations in subsurface sediments. Future conditions at these sites have been predicted using models that parameterize advective and dispersive transport, sediment deposition and erosion, sediment mixing and reaction, and even contaminant uptake in the food chain (O'Connor et al. 1983b, Connolly and Tonelli 1985, Connolly 1991, Thomann et al. 1991). In most cases, both models and measurements indicate that, after source reduction has occurred, the contaminant will continue to be buried by subsequent accumulation of *cleaner* material and that the maximum concentration will be found some distance below the surface and may be reduced by *dilution* with the cleaner material (Logan et al. 1989, O'Connor 1990) (Figure C.3). However, even where the contaminant peak is buried, the flux of contaminant mass from the sediments to the water column may still be sufficient to maintain undesirable levels in aquatic organisms (Stull et al. 1986, Ikalainen and Allen 1989). Also, the effect of future source controls can depend on whether the contaminant source contributes significantly to the balance of sediment accumulation (Stull et al. 1986).

Conclusions and Recommendations

1. Because of the difficulty of measuring the separate components of the net sediment balance, accumulation (or erosion) rates should be estimated from site-specific observations of sediment properties and profiles of appropriate tracers.
2. Improved methods of predicting the flux of contaminants through the sediment-water interface will depend upon continued collaboration between specialists in sediment transport, contaminant transformation, and benthic biology.
3. Sites of historical contamination may be self-mediating in that natural accumulations may result in continuing burial of the highest levels of contamination. However, such sites should be carefully monitored to assess the potential for changes in sediment dynamics.
4. Quantitative modeling of future conditions is feasible but requires an extensive data set documenting historical changes for the purpose of calibrating the models.

Sediment Quality

Definition and Criteria

Until recently, environmental agencies in this country relied on either Best Available Technology (BAT) or aqueous chemical concentrations in receiving water to regulate hazardous chemicals in marine environments. In the latter case, concentrations that result in acute or chronic toxicities to aquatic organisms (called Water Quality Criteria) are determined through laboratory exposures, and these are used to establish acceptable levels, called Water Quality Standards. Permits for discharges into coastal waters are often based on these regulatory levels or treatment technologies.

Many of the chemicals that have the potential to adversely affect aquatic organisms are hydrophobic and sorb to sediments. Contaminated sediments can impact not only the organisms that live in direct contact with the solids but also those that reside in the overlying water since the sediments themselves can act as a source of the toxic substances. Water quality or BAT strategies do not directly address hazardous substances in sediments. This and the fact that there are numerous coastal areas that already have contaminated sediments (NRC 1989) have led to the development of methodologies to establish protection criteria for toxics in bottom materials. These levels, sometimes called Sediment Quality Criteria or Sediment Quality Values, can be used in engineering calculations to determine whether a

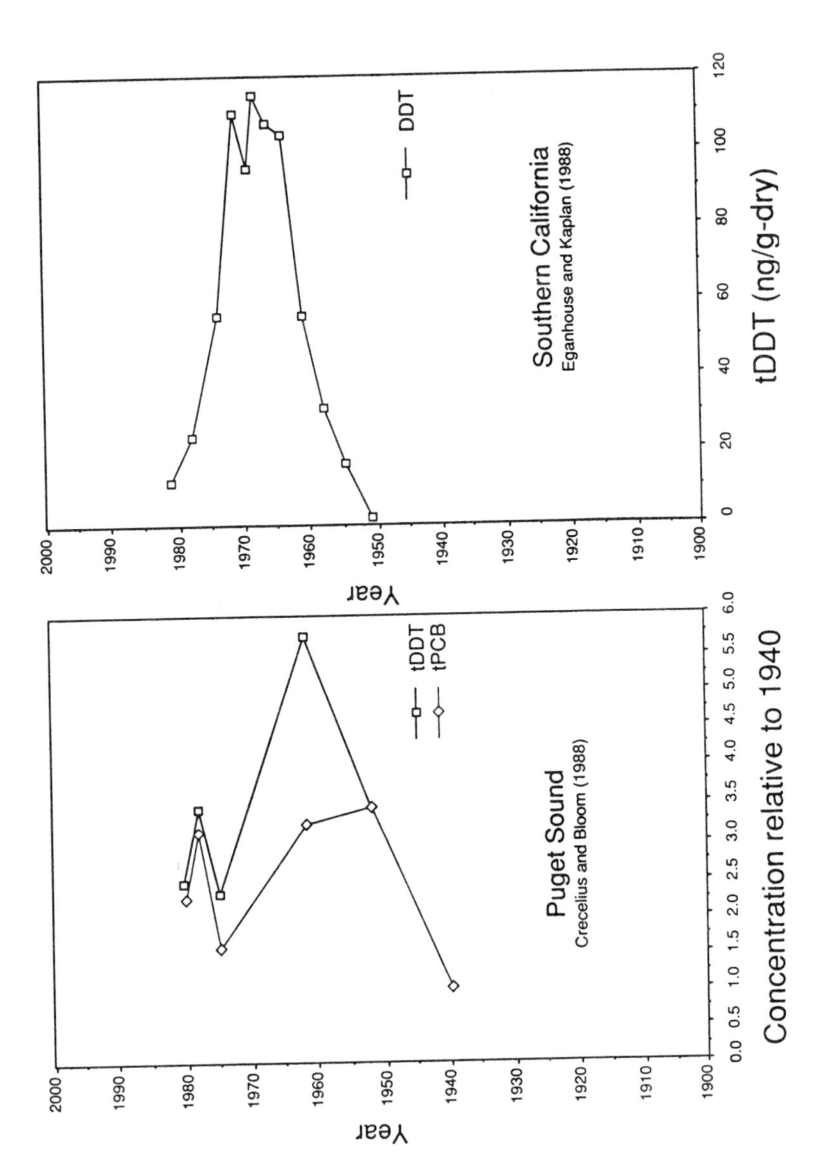
Southern California
Eganhouse and Kaplan (1988)
DDT
tDDT (ng/g-dry)
Year
0
20
40
60
80
100
120
1900
1910
1920
1930
1940
1950
1960
1970
1980
1990
2000
Puget Sound
Crecelius and Bloom (1988)
tDDT
tPCB
Concentration relative to 1940
Year
0.0
0.5
1.0
1.5
2.0
2.5
3.0
3.5
4.0
4.5
5.0
5.5
6.0

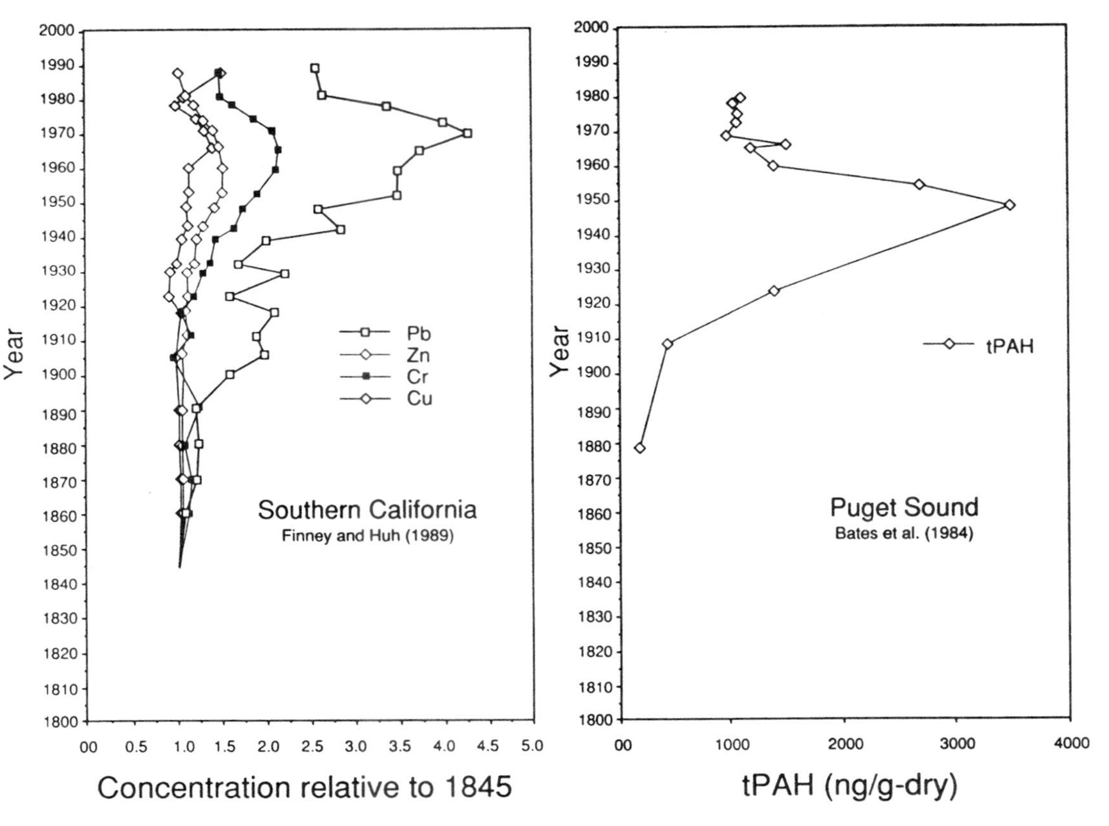

FIGURE C.3 Chronological profiles of chemical concentrations in sediment cores (Source: O'Connor 1990).

given mass loading of chemicals in an effluent will likely result in toxic sediments or what measures are needed to prevent toxic conditions. They can also be used to gauge the existing or potential adverse biological impacts of in-place contaminated sediments.

The EPA has compiled ten methodologies that have the potential to assess sediment quality relative to chemical contaminants (EPA 1989). Some of the methods involve chemical analyses that allow for the establishment of chemical specific criteria, e.g., an acceptable level for phenanthrene in sediments. Others involve only biological observations that limit the results to assessing whether or not a sediment is toxic. Others combine chemical and biological measurements. A brief description of the ten methods is given in Table 4.4 (EPA 1989).

The Equilibrium Partitioning (EqP) approach is the only method that relies, in part, on a fundamental thermodynamic parameter, i.e. fugacity, to derive a numerical estimate of sediment quality. Fugacity is the tendency of a chemical to flow from one phase of a system to another until the free energy of that particular system is at its lowest and equilibrium is attained. Under such conditions, one can calculate the equilibrium concentration of a substance in any one phase of the system. The concentrations in the remaining phases can be calculated if the distribution of partition coefficients among the phases is known.

The EqP method assumes that only chemicals dissolved in the pore water phase are biologically available and therefore potentially toxic. It further assumes that for hydrophobic organic chemicals, the organic carbon fraction of bottom sediments will contain most of the sorbed material. With these assumptions, one can calculate acceptable sediment levels if one knows: 1) the pore water concentration above which toxicity is exhibited, 2) the organic carbon content of the sediments, and 3) the partition coefficient between organic carbon and pore water. Experimental evidence indicates that for many hydrophobic organic compounds, the partition coefficient of the substance between n-octanol and water approximates the partitioning between sediment organic carbon and water. Therefore the sediment quality value (SQV) on an organic carbon basis is: $SQV = K_{oc}ATV$, where K_{oc} is the partitioning coefficient of the chemical between sediment organic carbon and water (≈ partitioning coefficient between n-octanol and water) and ATV is a chosen Acceptable Toxicity Value. The EPA is in the process of establishing sediment quality criteria for a number of organic compounds based on the EqP methodology (for example, for acenaphthene see EPA 1991).

Sediment Quality Modeling

A basic question is whether it is possible to predict sediment concentrations of hazardous chemicals due to a particular discharge in order to protect the coastal environment as well as is presently done by BAT. The answer is that it is possible for a given discharge to calculate quite simply an upper bound for sediment quality on an organic carbon basis or, conversely, for a specified SQV standard to back-calculate a safe limit for the effluent quality (the sediment-quality driven approach).

The calculation of an effluent limit is illustrated for a POTW discharge by considering acenaphthene, a polynuclear aromatic hydrocarbon (PAH), as an example. In EPA 1991, the proposed criterion for "acceptably" protecting benthic organisms in saltwater sediments is a concentration less than SQV = 240 μg acenaphthene/g organic carbon (g_{oc}), with confidence limits 110-520 μg/g_{oc}. To be safe, SQV = 110 μg/g_{oc} is used as the sediment quality standard to be met. Calculating backward to the source (for this example only one source is considered), the next question is: what fraction of the organic carbon in the sediment is derived from a particular POTW discharge; again to be conservative, assume 100 percent. If it is now conservatively assumed that there are no losses of the organic pollutant adsorbed on the suspended particles from point of discharge to sediment deposition, then the concentration of acenaphthene on the suspended solids in the effluent should be limited to 110 μg/g_{oc}. The concentration of particulate organic carbon c_{oc} is estimated to be ≈ 0.4 c_{ss}, where c_{ss} is the concentration of suspended solids. Therefore, on the basis of suspended solids, the limit should be reduced to 44 μg/g_{ss}. For the next step, it is necessary to know the suspended solids concentration of the POTW effluent (well known at every plant). For this example, assume that c_{ss} = 75 mg/l, which is the effluent standard (average value) of Table A in the California Ocean Plan (CWRCB 1990). The safe effluent concentration limit for acenaphthene (c_{as}) adsorbed to solids can now be calculated as follows:

$$c_{as} = c_{ss}\left(\frac{\text{mg}}{L}\right)44\,\frac{\mu\text{g}}{g_{ss}}\,\frac{\text{g}}{1000\text{ mg}} = (0.044)\,75\,\frac{\mu\text{g}}{L} = 3.3\,\frac{\mu\text{g}}{L}$$

To complete the calculation, one must estimate the dissolved concentration in the effluent, which would be in equilibrium with an adsorbed concentration of 110 μg/g_{oc}. Based on an organic carbon partition coefficient for acenaphthene of 6030 L/kg (log = 3.78) given by the EPA (1991), the calculated equilibrium dissolved concentration (c_{aw}) is

$$c_{aw} = \frac{SQV}{K_{oc}} = \frac{110 \times 1000}{6030}\,\frac{\mu\text{g}}{L} = 18.3\,\frac{\mu\text{g}}{L}$$

The total effluent safe limit is then

$$c_{at} = c_{as} + c_{aw} = 3.3 + 18.3 = 21.6 \approx 22\,\frac{\mu\text{g}}{L}$$

It is of interest to note that only 15 percent of the pollutant is attached to effluent particles in spite of its high K_{oc} value because of the low particle concentration in the effluent.

Note that most changes in the assumptions used could only increase the calculated effluent limit. For example, if the deposited organic carbon from the POTW is diluted with equal amounts of clean natural organic carbon, then the calculated answer would be c_{at} = 43 µg/*L*. Furthermore, if it was found that 10 percent of the particle-bound pollutant in the effluents is released to the ocean water before depositing to the sediments, the result would be multiplied by 1/0.9. The point is that with simple assumptions a safe effluent limit can be calculated, but if more were known, the required effluent limit would be larger—i.e., less stringent. Only if there were other sources—point of diffuse—would a lower effluent limit for the POTW be calculated to allow for the other sources or pre-existing contamination of sediments.

It is now interesting to compare the above result with the effluent limit derived from the California Ocean Plan (CWRCB 1990) for all PAHs (including acenaphthene). Table B gives a value of 8.8 ng/*L* as the standard for receiving water (specified for protection of human health [carcinogen]). For a typical initial outfall dilution of 100:1, this corresponds to an effluent limit of 0.88 µg/*L*. This value for *all* PAHs is about 25 times smaller than the limit for acenaphthene alone calculated above to meet the conservative sediment-quality standard. Thus, it would not have made any difference in the setting of the effluent limit to improve the conservative assumptions above because the health standard is much more stringent in this case. It is expected, however, that there will be other cases where the opposite is true, i.e., where the effluent standard needed to protect benthic organisms is smaller than that needed to protect human health. The sediment-quality driven approach provides a rational way to work this out in conjunction with water-quality driven analyses.

The reader is reminded that the above example is based on the Equilibrium Partitioning (EqP) approach, which is only one possibility. Long and Morgan (1991), in a National Oceanic and Atmospheric Administration study, give an extensive analysis and comparison of different methods of assessing toxicity using total sediment concentrations (mass of pollutant per unit mass of sediment) for a variety of trace metals, petroleum hydrocarbons, and synthetic organic compounds. The state of Washington has recently adopted sediment standards based on total sediment concentrations (see Becker et al. 1989 for discussion of criteria and approaches). The calculations to derive a corresponding effluent limit from a total sediment concentration standard would be different from the example above because the rate of deposition of organic carbon of sewage origin must be predicted, along with the losses from the sediments back into the water column. The EqP approach is simpler to model because it deals only with concentrations of contaminant per unit mass of organic carbon, thereby allowing an easy link to wastewater quality.

Ultimate Sinks for Pollutants: Distribution of Pollutants in Water and Sediments in the Ultra-Far-Field

The question of what constitutes the ultimate fate of material discharged into the marine environment has no simple answer, in part because most of the constituents also occur naturally, if not necessarily at the high concentrations present in municipal effluent. This is particularly true of the elements associated with the enrichment of biological productivity in receiving waters, such as carbon, nitrogen, and phosphorus, but is also true of such potential toxicants as trace metals. Furthermore, some substances of concern, such as chlorinated hydrocarbons, can be broken down chemically into compounds of no concern and thereby have an ultimate fate, while the carbon atoms of which they are composed continue to be recycled. Even material that is buried in the sediments, usually considered to be the ultimate fate of any material, can return to the ecosystem under the right circumstances.

Part of the problem in answering the question of "What is determining the ultimate fate of material?" is that it represents the poorly formulated question about what are the long term, subtle effects associated with a discharge—the ultimate impacts—for which we do not, and cannot, have enough information to answer for any anthropogenic interaction with the environment. Despite the impossibility of ever completely understanding all environmental effects of a given discharge, the major effects can usually be quantified and studied in order to learn about potentially important subtle effects. The fact that it is not possible to effectively understand all the cascading environmental effects associated with a discharge, be it on land, into the atmosphere, or into the ocean, should keep us from getting complacent about such a discharge.

Most studies of environmental effects associated with marine discharge have focused on changes in the sediment composition and associated ecosystem effects. The benthos is a particularly good place to study because particle sedimentation concentrates pollutants, providing larger signals from a discharge. Furthermore, the relative immobility of benthic organisms makes it easier to relate ecosystem responses to a documented environmental change. Such is not the case for planktonic systems, where it is difficult to separate those organisms that have been exposed to a discharge from those that have not and for which the exposure need not continue for lifetimes of organisms. This does not mean that planktonic effects cannot be important, as shown by the food chain transmission of DDT to pelicans off the coast of California until the 1970s, presumably mediated by the plankton.

From the narrow perspective, what constitutes the ultimate fate of a substance? For an undesirable substance with no natural sources, it would be its chemical degradation to an innocuous form or its permanent immobilization in the sediments at a location unlikely to be disturbed. The ultimate fate of a naturally occurring substance would be its dilution and dispersion into the environment to the point where its concentration is indistinguishable from the naturally occurring concentration. Here, again, there is ambiguity in the question of what is indistinguishable. On the one hand, it may be possible to measure even small differences between natural and altered systems. On the other hand, the natural spatial and temporal variability in natural systems can be so large that some anthropogenically induced differences may be meaningless when compared to the larger natural variability in total concentrations and ecosystem processes.

Dumpsite 106

Dumpsite 106 is a region offshore New York that has been designated for the surface disposal of sewage sludge. The surface is an average of 2,200 meters off the bottom (O'Connor et al. 1983a). The amount of material that reaches the bottom depends largely on the particle's sedimentation rates as well as the current patterns. O'Connor et al. (1983a) used simple settling and horizontal dispersion models to argue that only 20 percent of the sludge dumped there would reach the bottom in 200 days. The other 80 percent would continue to drift with the water. They argued that in the region underneath the actual discharge site, the anticipated impact of municipal dumping would be an increase in the benthic sedimentation rate of about 10 percent and in organic carbon sedimentation rate of about 20 percent.

Fry and Butman (1991) used more extensive information about sludge settling rates and about regional currents to predict that about 23 percent of the sludge would settle within 350 kilometers of the discharge site. This represented maximum increases in sedimentation rates of about 40 to 60 percent. Because most of this sedimentation is by faster settling particles, it would contain disproportionate amounts of denser grit and other inorganic particles.

Recent measurements of deposition rates and sedimentary composition in the region suggest that there are measurable increases (Van Dover et al. 1992). Preliminary results suggest that sedimentation rates of organic material are considerably larger than estimated above and represent a substantial fraction of the natural sedimentation rates. There is no evidence of ecological impacts of this enhanced flux.

It appears that models for the fate of sludge discharged at Dumpsite 106 do not yet adequately predict the fate of material discharged there. Measured values for sedimentation appear to be higher than predicted. Efforts are under way to increase the accuracy of these model predictions and to compare them against field data. This is a joint effort of the EPA, the National Oceanic and Atmospheric Administration, and the United States Coast Guard (EPA 1990).

Understanding from Seattle Puget Sound

The southern end of Puget Sound, Washington is dominated by the anthropogenic inputs from the Seattle and Tacoma metropolitan regions. Paulson et al. (1988, 1989) used the narrow, fjord nature of the sound to make budgets of the fate of trace metals discharged into Puget Sound. About 70 percent of the lead, and 40 percent of the copper and zinc discharged in the region is deposited in the sediments of the 70-kilometer-long central basin of Puget Sound. The rest moves out toward the Strait of Juan de Fuca and Pacific Ocean.

Understanding from Southern California

The chemical oceanography of the Southern California Bight has been the subject of a recent review by Eganhouse and Venkatesan (1991).

Trace metals are mostly advected from sites of municipal waste discharge. There have been several estimates that more than 85 percent of the trace metals discharged at the Whites Point outfall are not deposited in the sediments of the surrounding Palos Verdes Shelf. Evidence of enhanced trace metal deposition has been found in several basins offshore. Bruland et al. (1974) estimated the fraction of different trace metals associated with anthropogenic sources that were settling within three inner basins of the Southern California Bight: Santa Barbara, Santa Monica, and San Pedro. The fraction of released metals accounted for in the sediments varied from 15 percent for cadmium to 69 percent for silver.

Fallout of particulate organic carbon is locally dominant but represents a smaller fraction of deposition in local basins. This is due, in part, to the greater sedimentation of natural organic matter that could dilute out the anthropogenic component. There is evidence of the deposition patterns from different outfalls overlapping each other in the changes in benthic organism composition found along the 60 m bottom contour (Mearns and Young 1983). This suggests that there can be small but detectable effects of discharge over large areas.

Conclusion

Researchers have been able to establish the ultimate fate of only a fraction of material discharged into the ocean.

Afterword

The ability and effort put into the prediction of environmental impacts associated with marine discharges before a discharge is permitted are quite sophisticated in comparison to the limited data collection and analysis used to test predictive models after discharge commences. From a regulatory point of view, that of designing discharge systems that meet regulatory requirements, current models can predict environmental impacts. However, at the scientific level, the ability does not exist to predict detailed distributions of pollutants and their movement through the ecosystem. This disparity stems from the fact that there is no systematic program to test predictive models against actual discharges. Note that testing a model is distinct from running a monitoring program because the intensive data collection and analysis needed to test a model is not required to determine whether a system is complying with environmental regulations.

The main consequence of poor model predictions is likely to be overdesign of facilities. When models have uncertainties, safety factors are applied in the design of a facility to ensure that environmental objectives will be met. The application of excessive safety factors results in overbuilt facilities and excessive expenditures. Model comparison with real prototypes provides valuable information for the improvement of models. In the event a project was underdesigned, model prototype comparison will indicate what needs to be done to correct the problem.

Recommendation

Waste dischargers should institute programs to verify the models used to predict environmental impacts of discharges. These efforts should be distinct from monitoring requirements and of limited duration. Their costs might be considered as part of design and construction costs rather than as monitoring costs.

OVERALL DESIGN OF DISPOSAL SYSTEMS, CONTROL OF DIFFUSE SOURCES, AND USE OF MODELS

For a coastal community planning an ocean discharge, the task of choosing an outfall location, depth, design, and allowable mass loadings is basically an iterative one subject to definite but somewhat elastic boundary conditions. Some of these boundary conditions pertain to water and sediment quality, others to factors associated with such issues as foundation stability; earthquake loadings; forces due to waves, currents, and possible tsunamis; construction technology; and cost of construction. Additional factors that enter the overall design task of the outfall are related to system components associated with the outfall within the overall wastewater infrastructure system including interceptors, treatment plants, and pumping stations. This section addresses only issues related to water quality and not those associated with the other factors.

Steps in the Design of a Disposal System (new systems and upgrading existing systems)

The design of a system for wastewater disposal into the coastal wastes depends on three basic inputs:

- the water and sediment quality objectives,
- the characteristics of the receiving water, and
- the volume and composition of the wastewater to be discharged from an outfall.

Each of these items is a large array of information with temporal variations, or frequency distributions. Designs should not be for a single condition; they should be robust and work for all conceivable conditions with perhaps some intentional exceptions.

Furthermore, these inputs are not fixed and often change during the design in a feedback process. Consider the following examples. First, the regulatory agency or the discharger may set very high goals, but when faced with the engineers' estimate of the costs, decide to back off. This happened in San Francisco to the program to control combined sewer overflows to an average of only one spill per year; then, because of cost, it was relaxed in several steps to an expected average of 8 overflows per year! But if the cost was considered reasonable for the target of one per year, they would have stuck to it.

Second, consider the environmental characteristics of the receiving waters. These are not fixed either before the design starts because the outfall designer has some choice of location of the discharge point, i.e., depth and location, to optimize the system.

Third, the quality of the wastewater depends on the treatment processes chosen and the stringency of the measures for source control of toxics. From the overall systems point of view, the more treatment and source control implemented, the less that is required of the outfall to achieve the same water quality; conversely, a really good outfall reduces the need for treatment and source control measures. Thus, for a water-quality/sediment-quality driven approach, neither the optimum design of the outfall nor the appropriate selection of treatment levels and source controls can be made without analyzing the whole system for various combinations. Even the flow capacity of the system may be uncertain due to plans for storm water detention for later treatment or future wastewater reclamation for reuse.

The Water-Quality Driven Approach

The next step in risk management is to design engineering control systems that may be used to achieve compliance with various water-quality objectives established to manage risks to human health and ecosystems. In the use of water quality objectives, sediment quality objectives are also included. Sediments can also be significant routes of exposure to ecosystems and humans (through shellfish and benthic fish).

The design of engineering systems based on water quality objectives is called the water-quality driven approach, shown schematically in Figure C.4. The figure and the following discussion refer to a municipal wastewater disposal system in order to illustrate the concepts; however, the same concepts and procedures can be extended to water bodies with multiple inputs from point and diffuse sources.

The first step (top of Figure C.4) is the expression of the environmental objectives, in terms of water quality objectives or standards from which the engineering design may proceed. The chart then displays what appears as a *backwards* calculation: from these standards is derived the combination of engineering measures that is best suited to reach the prescribed water quality. There are three basic types of engineering components for a municipal wastewater system:

1. the *outfall(s)*, including location and characteristics of multiport diffusers used for high dilution.
2. the *treatment works* (POTWs), with various possible components and levels of treatment (see Appendix D).
3. *source control* (or source reduction) to limit the amount of toxic substances or other pollutants entering the sewer system and the treatment plant (see Appendix D).

These three parts constitute a system in which changes in one part will change the need for the others. For instance, a long outfall with high initial dilution generally reduces the need for secondary treatment; or better source control reduces the need for toxics removal at the POTW and simplifies the sludge disposal problem. Only by considering all three components and their associated environmental and financial costs and benefits at once can

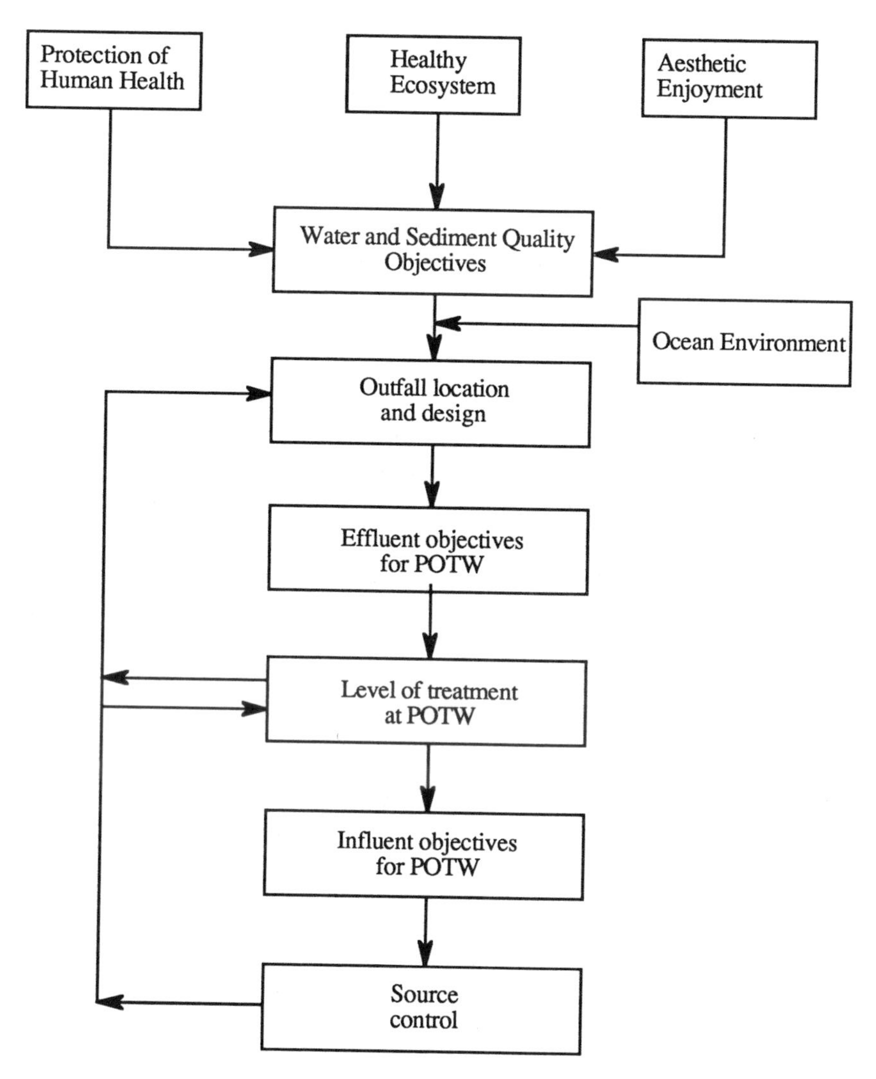

FIGURE C.4 Overview of water- and sediment-quality driven approach for design of municipal wastewater disposal system. The water-quality/sediment-quality driven approach involves working backward (compared to direction of flow) to establish what components are needed.

the optimum combination be found. The conceptual design then proceeds with a trial choice of outfall and related treatment levels and source control programs deemed necessary to meet the ambient water-quality standards. The process involves complex modeling of transport and fates of contaminants in the ocean after initial dilution. Given a certain outfall configuration, it can be determined in this way the effluent limits needed at the POTW. To meet these, appropriate treatment components and upstream source control measures are then selected.

As a practical matter, such modeling must be done in the *forward* sense (sources→treatment→outfall→transport and fate→effects), but with iterations it is conceptually the same as Figure C.4 with the reversed order of conceptual steps. However, with experience it is not difficult to work backwards from water quality and sediment quality standards to get approximate solutions for the three system components, which are then used as the initial iteration for the detailed forward water-quality modeling (Figure C.5).

Transport and Fates Modeling: Predicting Ambient Water and Sediment Quality

Mathematical and conceptual models are used extensively to explain observed processes that disperse and modify pollutants in the ocean. Various submodels may be combined to produce an overall model to relate pollutant inputs to water and sediment quality for single and multiple sources. These models are fundamental to management by the water-quality driven approach because the limits on emission for any discharge or nonpoint source may be back-calculated through the models. Because of the various length and time scales associated with different pollution problems, a variety of models is needed—one mathematical model cannot provide answers to all questions.

The following discussion relates to the wastewater disposal system for a municipality including source control, a treatment plant (POTW), and outfall. However, the three kinds of required information apply equally well to all other types of pollutant sources and the approach to devising an engineering system. For example, in the case of combined sewer overflows, there is a set of water quality objectives; some knowledge of the environment; and information on the amount, quality, frequency, and distribution of existing combined sewer overflows. If there are multiple contributing sources to the water quality, then the environmental modeling must integrate the effect of all sources and develop scenarios for different degrees of control and handling of different sources. Sources of the same kinds can be combined into classes to simplify the modeling, such as one for a large number of small POTWs all affecting the water quality of a large body of water such as Long Island Sound.

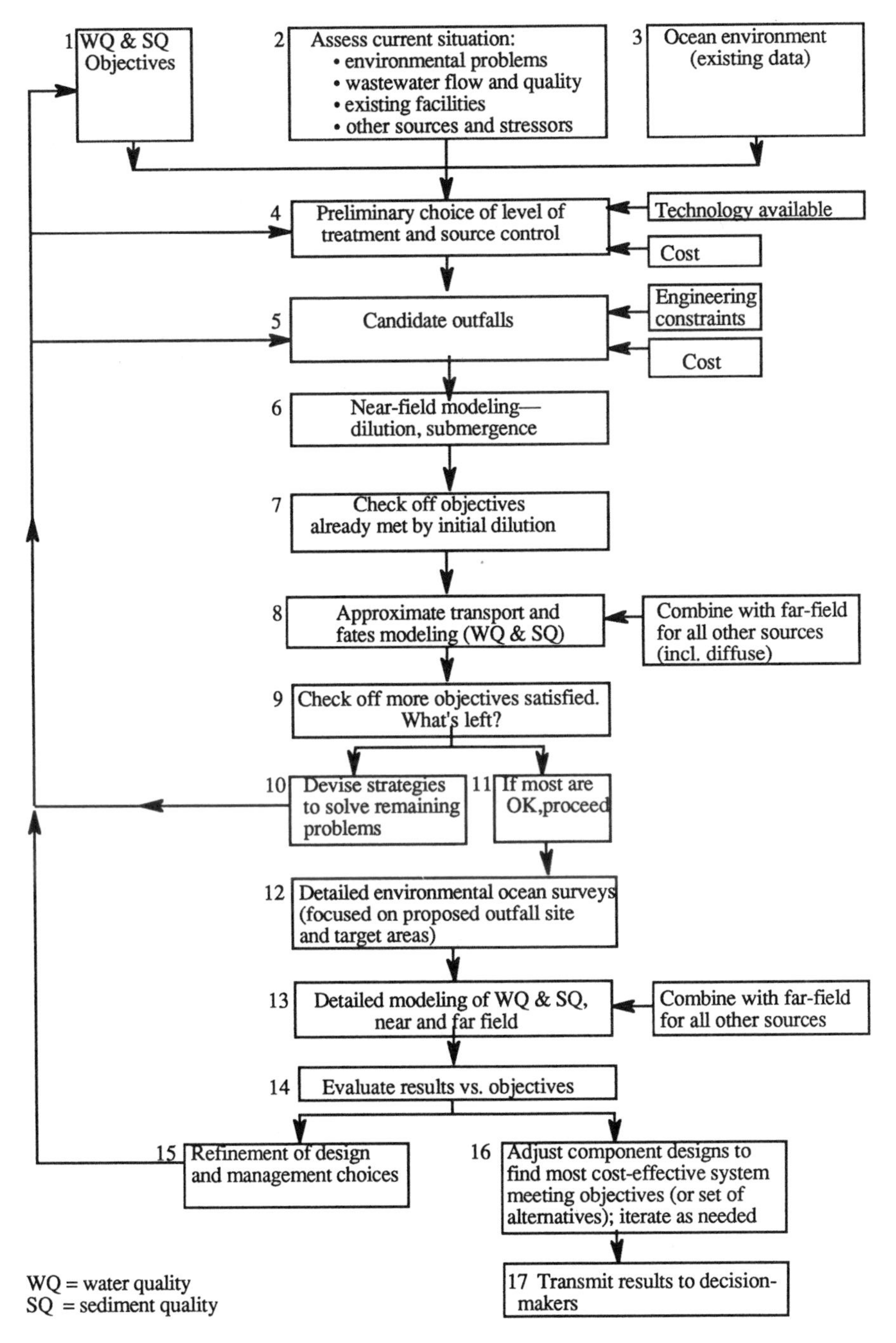

FIGURE C.5 Overall design process for ocean disposal system for POTW.

Predictive models have a number of uncertainties and need improvement, but nonetheless appropriate engineering systems for wastewater disposal and diffuse source control can be designed to meet prescribed water-quality and sediment-quality objectives.

Choice of System Components: Alternative Systems

The actual design process for an ocean disposal system for a POTW is illustrated in Figure C.5. It is an iterative process based on the water-quality driven approach. The following steps correspond to the numbered boxes in the figure.

Box 1. Water quality and sediment quality objectives are established. These objectives may either be regulatory standards for the receiving water and sediments or in some cases the effluent or they may be qualitative mandates related to health of ecosystems.

Box 2. The assessment of the current situation is the starting point for improvement, including identifying the problems that need to be addressed, such as those enumerated in Table 4.8 on page 122, with a full knowledge of the wastewater characteristics, including the whole range of toxics, the evaluation of existing facilities (whether they are to be used or replaced or upgraded), and a full accounting of all the sources and stressors for the location of the particular project. The latter is necessary in order to do the multiple source effects evaluation as required by the right-hand boxes adjacent to Boxes 8 and 13 in the diagram.

Box 3. It is important for the designer to make a full evaluation of the existing ocean data on temperature, salinity, and density profiles with seasonal variations; ocean currents; chemical constituents, including nutrients; dissolved oxygen profiles; present state of contamination of sediments; and a description of the biota and living resources to be protected and where they are. There is no use in acquiring new oceanographic data before evaluating what is already known and where the gaps are and are not.

Box 4. Based on experience, the designer can make a preliminary choice based on the logic of Figure C.4 of what treatment components and source control measures would most likely be required. For example, if a body of ocean water is poorly flushed and not very deep, then secondary treatment is very likely to be necessary; on the other hand, if the discharge is in deep water relatively well flushed (as Congress envisioned in establishing the 301(h) waiver provision), then enhanced primary treatment would be a likely choice rather than full secondary because the biochemical oxygen demand does not need to be removed for such a case.

Box 5. Different outfall choices are usually considered on a parametric basis, for example, maybe two outfall lengths (and different depths) and three diffuser lengths. Engineering constraints include, for example, the bottom bathymetry and geology, which may preclude siting in certain places. The type of outfall, such as tunneled versus steel pipe versus concrete pipe, may be dependent on various engineering considerations and costs. The longer outfalls also cost more so that different candidates will have different costs as well as different benefits associated with them.

Box 6. The near-field modeling consists of initial dilution calculations by various buoyant plume formulas of which there are a number of acceptable models (e.g., the EPA's group of computer models: Muellenhoff et al. 1985, Baumgartner et al. 1992). Also calculated is the height of rise of the wastewater plume, which may be kept submerged below the pycnoline due to the density stratification of the ambient water. Submergence is highly advantageous in many situations to control or completely prevent impacts to the surface-mixed layer or the shoreline. Initial dilution and sewage field submergence can be predicted reasonably well (~±25 percent for no current, worse for currents). Further research can improve the predictions but probably will have little impact on design choices.

Box 7. At this stage, it is useful to do a quick scan to see which water quality objectives and sediment objectives are already met. For example, if there are requirements for 50 constituents, it is no use to carry forward the modeling for those that are obviously going to meet the water and/or sediment quality objectives. Rather, attention should be focused on those that look as though they are going to be the key drivers of the design. In the past, coliform concentrations were often the critical factors, whereas now it is likely to be nutrients or toxics in sediments.

Box 8. The next step is approximate far-field modeling considering further mixing, advection, biological and chemical conversions, sedimentation, and final fate of particles. This may be done initially in an approximate way in order to screen out outfall locations that are going to turn out to be unsatisfactory so that no further time and effort is wasted in data collection at such sites in the ocean or doing analyses that will lead to unacceptable results. This is an extremely important step in order to focus resources on engineering alternatives that are going to be successful. One cannot wait until after one gets the full-blown oceanographic study because one needs this preliminary judgment in order to plan and focus the oceanographic study (Box 12). The overlap with far-field effects of other point or diffuse sources is also considered here in a preliminary way.

Boxes 9 & 10. The requirements can be further screened and it is possible to see which remain to be satisfied and what to do about them. For example, if only coliforms (pathogens) are still excessive at certain target areas, one needs to decide what to do next. One choice would be to disinfect the effluent full time or intermittently—that represents the arrow that goes from Box 10 back to Box 4. Or alternatively, one may decide to make the outfall longer and deeper, which is the arrow from Box 10 back to Box 5. Also, if the chosen candidate outfalls and treatment seem to be excessive, i.e., run up more cost than is necessary to achieve the desired objectives, then revisions in the opposite direction might be made. Finally, if it appears that the objectives are impossible to satisfy at any reasonable cost, then the arrow from Box 10 back to Box 1 indicates a possible change of objectives. For example, a small kelp bed or shellfish bed near the proposed discharge might be closed to beneficial use at a small environmental cost compared to the possibly large monetary cost to do otherwise.

Boxes 11 & 12. If the planned project appears viable and cost effective, then a major discharger would proceed typically with a year-long detailed environmental survey of the physical, chemical, and biological characteristics in the ocean in the area of the proposed outfall site and the target areas where water quality is to be protected. Included would be continuously recorded currents at different depths; profiles and transects of temperature, salinity, density, dissolved oxygen, and nutrients; measurement of toxicants in organisms and sediments; and biological assessments.

Box 13. With this information then, a full-blown mathematical modeling is undertaken with the near-field and far-field behavior of the projected waste discharge with full consideration of the combination with other sources. The predicted results will include spatial variations and frequency distributions of various water-quality parameters in the water column and rates of accumulation of pollutants in sediments.

However, the ability to do this is not perfect, but it is growing and it would be advantageous for the profession to have more post-construction evaluations of designs of systems that have been put into operation. The whole water-quality driven approach can easily assimilate new scientific information and oceanographic data as it becomes available for future adjustments or corrections of management plans.

Box 14. By this time, a completely satisfactory design has been developed with perhaps only minor refinements needed.

Box 15. These refinements are then carried in an iterative process by the arrows shown back to Boxes 1, 4, and 5 from Box 15.

Box 16. The next question is whether the system proposed is the most cost-effective system to meet the objectives, and, if not, adjustments can be made as needed.

Box 17. The proposed design and some viable alternatives can be presented to decisionmakers. Sometimes systems are developed that will go well beyond existing requirements in anticipation of future upgrading or tightening of requirements. Decisionmakers may often prefer such an alternative because the life of outfalls may be 50 to 100 years, far longer than the lifetime of many regulations.

Many coastal outfalls have been designed and built in the past two to three decades. Those that were designed generally using procedures outlined above have performed well and largely in accordance with predictions or better; some have probably been over designed because of large safety factors to cover uncertainties. High dilutions and plume submergence have been obtained when predicted.

Present day upgradings are being driven mostly by more stringent water-quality requirements and increased loadings, rather than incorrect predictions of performance at the time of initial design. A new factor is sediment quality, which was usually not directly included in design considerations before about 1980.

Quality driven approaches are not as well developed for nonpoint sources, but the principles are basically the same. Multiple point and diffuse sources can all be logically integrated into environmental-quality driven calculations as part of integrated coastal management.

Discussion

The Quality-Driven Approach

The preceding sections address the range of scientific knowledge and engineering techniques related to the processes by which wastewater treatment plant effluents can be discharged to coastal waters safely. Much of this knowledge has evolved within the past three decades. Since scientists and engineers now have a good basic understanding of these processes, the management of coastal water and sediment quality can be addressed through a logical scientific framework. Predictions can be made of the benefits and costs of various control actions on a case-by-case basis or by classes. Future research will contribute to this improved understanding of coastal waters, and allow for a shift from mandated technologies to the water-quality and sediment-quality driven approach. It is only through the latter approach that the most technically and cost effective control measures can be identified.

Because many uncertainties have been described, it may seem that the integration of all the necessary scientific information and engineering techniques for the purpose of making a decision is a hopeless task. But scientists, engineers, and other professionals working together with the public can sort out the key factors and focus on solving the most important problems in a cost effective manner through the integrated coastal management process discussed in Chapters 3, 4, and 5.

It is clearly possible to design pollution control systems to achieve water and sediment quality sufficient to meet specified standards with appropriate safety margins. The outlines of this approach are provided in this appendix with additional information presented in Appendix A on nutrients, Appendix B on pathogens, and Appendix D on wastewater treatment and stormwater management. The water-quality and sediment-quality driven approach is the only one that can be applied logically to multiple point and diffuse sources and rapidly assimilate new research and monitoring results.

While water quality modeling is well developed, sediment quality modeling is new within the past decade and is not fully developed yet. While currently there is limited agreement on the way in which sediment quality standards should be specified, it can be anticipated that a consensus will develop over the next few years.

Toxicants

Toxicants have received great attention during the last two decades (but little before that), and strong control measures have been implemented in many areas. Source control and source reduction have proved to be effective measures for many POTWs. As effective source control programs are implemented the toxics problems will evolve toward one primarily associated with either sediment beds contaminated with past deposits or diffuse sources that are still unregulated or uncontrolled. Further work on the chemical speciation of metals in relation to toxicity will help to focus and refine requirements. In the meantime, toxicity limits for metals are established without regard to speciation. Many POTWs have found source control programs to be easier to implement than expected. Still illegal discharges continue to pour into many storm drains and are polluting the shorelines.

Particles

Residual particles (or suspended solids) in the effluent may be a concern for several reasons: they may be carriers of adsorbed pollutants; they may reduce light levels; they may contribute to nutrient enrichment; and they may, by settling, decrease the dissolved oxygen in the water column and sediments. But whether these problems actually exist depends on the circumstances. For example, if toxics are well controlled by source control in the sewer system, then toxics transport by particles may be below the levels where any standards would be violated. Similarly if the wastewater plume is confined below the thermocline where light levels are already low, then there is no effect on the euphotic zone above the thermocline. Also, when a region is well-flushed, then nutrient buildup will not likely be a problem. Finally, for outfalls producing high dilutions, all of the effects are reduced.

Thus, acceptable limits for suspended solids concentrations are site-specific, but they can be worked out by the water quality/sediment quality approach explained in this appendix.

Nutrients

The need to limit nutrient inputs to coastal waters is also site or region specific. Coastal waters that are impacted by excessive nutrients (usually nitrogen) usually receive these inputs from a variety of sources, including some natural inputs (for example Long Island Sound). In such cases, it is absolutely essential to follow an integrated coastal management plan (as explained in the main report) in order to achieve any results. Tightening up on minor sources may be a real waste of effort if major sources are left uncontrolled. An overall water-quality modeling including all sources is necessary to first understand the system then to devise the most effective control measures.

In the long run, the nutrient enrichment problems in some areas may be the most difficult and expensive problems to solve; by comparison, toxics appear to be coming under control, with the residual in sediments being the remaining issue.

Better Integration of Field with Laboratory and Computer

An existing outfall discharge (or multiple discharges) is a full-scale prototype that can be studied and compared with mathematical and laboratory models. If models can reproduce what occurs now, then they reduce the uncertainties in planning the next level of improvement of environmental quality. Furthermore, post-construction field investigations are valuable to compare predictions with the actual performance. Such information provides a valuable feedback for planning future control measures.

SUMMARY

1. Integrated coastal management requires the use of the water-quality and sediment-quality driven approach to model and manage the effects of single or multiple discharges and diffuse pollution sources and to make effective regional control strategies.
2. Predictive models have a number of uncertainties and need improvement, but nonetheless appropriate engineering systems for wastewater disposal and diffuse source control can be designed to meet prescribed water- and sediment-quality objectives.
3. There is much to be learned from existing problem discharge situa-

tions that is useful to support modeling and engineering efforts for designing new or upgraded facilities.

4. Our ability and effort to develop and use mathematical and conceptual models is ahead of our field confirmation of the accuracy of models. More effort is needed to study prototype systems after construction to evaluate the pre-construction modeling and analysis.

5. A continuous, responsive approach is needed for future management of major discharge areas, including on-going ocean studies and flexibility of management to modify the discharge system as needed in response to new research findings, new problems, or new environment objectives.

6. Coastal water-quality management must be site (or region) specific because of widely varying conditions along the coastline of the United States.

7. Because of the wide range of length and time scales of various ocean processes, and the various time scales of various water-quality problems, different modeling approaches are required for different pollutants.

REFERENCES

Alldredge, A.L., and C.C. Gotschalk. 1989. Direct observations of the mass flocculation of diatom blooms: Characteristics, settling velocities, and the formation of diatom aggregates. Deep-Sea Research 36:159-171.

Allen, J.S. 1980. Models of wind-driven currents on the continental shelf. Annual Review of Fluid Mechanics 12:389-433.

Andreae, M.O. 1977. Determination of arsenic species in natural waters. Anal. Chem. 49:820.

Andreae, M.O. 1979. Arsenic speciation in seawater and interstitial waters: The influence of biological-chemical interactions on the chemistry of a trace element. Limnol. Oceanogr. 24:440.

Andreae, M.O., and J.T. Byrd. 1984. Determination of tin and methyltin species by hydride generation and detection with graphite-furnace atomic absorption or flame emission spectrometry. Anal. Chim. Acta 156:147.

Andreae, M.O., J.F. Asmod, P. Foster, and L. Van't dack. 1981. Determination of antimony (III), antimony (V), and methylantimony species in natural waters by graphite furnace atomic absorption spectrometry with hydride generation. Anal. Chem. 53:287.

Baker, J., S.J. Eisenreich, and B.J. Eadie. 1991. Sediment trap fluxes and benthic recycling of organic carbon, polycyclic aromatic hydrocarbons, and polychlorophenyl congeners in Lake Superior. Environmental Science and Technology 25:500-508.

Baptista, A.M., E.E. Adams, and K.D. Stolzenbach. 1984. The solution of the 2-d unsteady convection—diffusion equation by the combined use of the DE method and the method of characteristics. Proc. 5th International Conf. on Finite Elements in Water Resources. Burlington, Vermont: University of Vermont.

Baumgartner, D.J., W.E. Frich, P.J.W. Roberts, and C.A. Bodine. 1992. Dilution Models for Effluent Discharges. EPA Draft Report.

Becker, D.S., R.A. Pastorok, R.C. Barrick, P.N. Booth, and L.A. Jacobs. 1989. Contaminated Sediments Criteria Report. Olympia, Washington: PTI Environmental Services, for the Washington Department of Ecology.

Berner, R.A. 1980. Early Diagenesis. Princeton, New Jersey: Princeton University Press.

Boudou, A., D. Georgescauld, and J.P. Desmazes. 1983. Ecotoxicological role of the membrane barriers in transport and bioaccumulation of mercury compounds. Pp. 117-136 in Aquatic Toxicology, J. Nriagu, ed. New York: Wiley-Interscience.

Brand, L.E., W.G. Sunda, and R.R.L. Guillard. 1983. Limitation of marine phytoplankton reproductive rates by zinc, manganese, and iron. Limnol. Oceanogr. 28:1182-1195.

Brand, L.E., W.G. Sunda, and R.R.L. Guilland. 1986. Reduction of marine phytoplankton reproduction rates by copper and cadmium. J. Exp. Mar. Biol. Ecol. 96:225-250.

Brooks, N.H., and R.C.Y. Koh. 1965. Discharge of Sewage Effluent from a Line Source into a Stratified Ocean. Paper 2.19. Leningrad: XIth Congress of Intern. Assoc. for Hydr. Research.

Bruland, K.W. 1988. Trace element speciation: Organometallic compounds and metal-organic ligand complexes. Applied Geochemistry 3:75.

Bruland, K.W., J.R. Donat, and D.A. Hutchins. 1991. Interactive influences of bioactive trace metals on biological production in oceanic waters. Limnol. Oceanogr. 36:1555-1577.

Bruland, K.W., K. Bertine, M. Koide, and E.D. Goldberg. 1974. History of metal pollution in southern California coastal zone. Env. Sci. Technology 8:425-432.

Byrd, J.T., and M.O. Andreae. 1982. Tin and methyltin species in seawater: Concentrations and fluxes. Science 218:565.

Cacchione, D.A., and D.E. Drake. 1990. Shelf sediment transport: An overview with applications to the northern California continental shelf. In The Sea, Vol. 9, Part B, B. Le Mehaute and D.M. Hanes, eds. Wiley Interscience.

Chao, S. 1990. Instabilities of fronts over a continental margin. Journal of Geophysical Research 95:3199-3211.

Clarkson, T.W. 1990. Human health risks from methylmercury in fish. Environmental Toxicology and Chemistry 9:957-961.

Coale, K.H., and K.W. Bruland. 1988. Copper complexation in the northeast Pacific. Limnol. Oceanogr. 33:1084-1101.

Coale, K.H., and K.W. Bruland. 1990. Spatial and temporal variability in copper complexation in the north Pacific. Deep-Sea Res. A 37:317-336.

Connolly, J.P. 1991. Application of a food chain model to polychlorinated biphenyl contamination of the lobster and winter flounder food chains in New Bedford Harbor. Environmental Science and Technology 25:760-770.

Connolly, J.P., and R. Tonelli. 1985. Modelling kepone in the striped bass food chain of the James River Estuary. Estuarine, Coastal and Shelf Science 20:349-366.

Csanady, G.T. 1983. Dispersal by random varying currents. Journal of Fluid Mechanics 132:375-394.

CWRCB (California Water Resources Control Board). 1990. California Ocean Plan, Water Quality Control Plan, Ocean Waters of California. Sacramento, California: CWRCB.

Drake, D.E., D.A. Cacchione, and H.A. Karl. 1985. Bottom currents and sediment transport on San Pedro Shelf, California. Journal of Sedimentary Petrology 55(1):15-28.

Dzombak, D.A., and F.M.M. Morel. 1987. Adsorption of inorganic pollutants in aquatic systems. Journal of Hydraulic Engineering 113:430-475.

Eganhouse, R.P., and M.I. Venkatesan. 1991. Chemical oceanography and geochemistry. In Ecology of the Southern California Bight.

EPA (U.S. Environmental Protection Agency). 1989. Sediment Classification Methods Compendium. Draft Final Report.

EPA (U.S. Environmental Protection Agency). 1990. Monitoring, Research, and Surveillance Plan for the 106-Mile Deepwater Municipal Sludge Dump Site and Environs. Tech Rept. EPA-503/4-91/001. Washington, D.C.: Office of Water, U.S. Environmental Protection Agency.

EPA (U.S. Environmental Protection Agency). 1991. Proposed Sediment Quality Criteria for the Protection of Benthic Organisms: Acenaphthene. Washington, DC: U.S. Environmental Protection Agency.

Eppley, R.W., ed. 1986. Plankton Dynamics of the Southern California Bight. Lecture Notes on Coastal and Estuarine Studies. New York: Springer-Verlag.

Faisst, W.K. 1980. Characterization of particles in digested sewage sludge. Pp. 259-282 in Particulates in Water: Characterization, Fate, Effects, and Removal, M. Kavanaugh and J. O. Leckie, eds. ACS Advances in Chemistry Series No. 189. Washington, D.C.: American Chemical Society.

Farley, K.J. 1990. Predicting organic accumulation in sediments near marine outfalls. Journal of Environmental Engineering 116(1):144-165.

Fischer, H.B., E.J. List, R.C.Y. Koh, J. Imberger, and N.H. Brooks. 1979. Mixing in Inland and Coastal Waters. New York: Academic Press.

Fisher, N.S., and J.R. Reinfelder. 1991. Assimilation of selenium in the marine copepod Acartia Tonsa studied with a radiotracer ratio method. Marine Ecology Progress Series 70:157-164.

Fisher, J.B., W.J. Lick, P.L. McCall, and J.A. Robbins. 1980. Vertical mixing of lake sediments by tubificid oligochaetes. Journal of Geophysical Research 85(C7):3997-4006.

Fowler, S.W., and G.A. Knauer. 1986. Role of large particles in the transport of elements and organic compounds through the oceanic water column. Progress in Oceanography 16:147-194.

Franks, P.J.S., and D.M. Anderson. 1992. Alongshore transport of a toxic phytoplankton bloom in a buoyancy current, Alexandrium Tamarense in the Gulf of Maine. Marine Biology 112(1):153-164.

Fry, V.A., and B. Butman. 1991. Estimates of the seafloor area impacted by sewage sludge dumped at the 106-mile site in the Mid-Atlantic Bight. Mar. Env. Res. 31:145-160

Grant, W.D., and O.S. Madsen. 1986. The continental-shelf bottom boundary layer. Annual Review of Fluid Mechanics 18:265-30.

Grieb, T.M., C.T. Driscoll, S.P. Gloss, C.L. Schofield, G.L. Bowie, and D.B. Porcella. 1990. Factors affecting mercury accumulation in fish in the upper Michigan peninsula. Environmental Toxicology and Chemistry 9:919-930.

Gschwend, P.M., and S.C. Wu. 1985. On the constancy of sediment-water partition coefficients of hydrophobic organic pollutants. Environmental Science and Technology 19(1):90-96.

Gust, G., and M.J. Morris. 1989. Erosion thresholds and entrainment rates of undisturbed *in situ* sediments. Journal of Coastal Research 5:87-100.

Gutknecht, J. 1981. Inorganic mercury (Hg^{2+}) transport through lipid bilayer membranes. J. Membrane Biology 61:61-66.

Hambrick, G.A. III, P.N. Forelich Jr., M.O. Andreae, and B.L. Lewis. 1984. Determination of methyl-germanium species in natural waters by graphite furnace atomic adsorption spectrometry with hydride generation. Anal. Chem. 56:421.

Harrison, G.I., and F.M.M. Morel. 1983. Antagonism between cadmium and iron in the marine diatom *Thalassiosira weisflogii*. J. Phycol. 19:495-507.

Hendricks, T.J., and J.M. Harding. 1974. The Dispersion and Possible Uptake of Ammonia in a Wastefield, Point Loma. Technical Memorandum 210. El Segundo, California: Southern California Coastal Water Research Project.

Hill, P.S. 1992. Reconciling aggregation theory with observed vertical fluxes following phytoplankton blooms. Journal of Geophysical Research (Oceans) 97(C2):2295-2308.

Hill, P.S., and A.R.M. Nowell. 1990. The potential role of large, fast-sinking particles in clearing nepheloid layers. Philosophical Transactions of the Royal Society London, Series A. 331:103-117.

Hill, P.S., A.R.M. Nowell, and P.A. Jumars. 1988. Flume evaluation of the relationship between suspended sediment concentration and excess boundary shear stress. Journal of Geophysical Research 93(C10):12,499-12,509.

Holman, H.Y.N., and J.R. Hunt. 1986. Particle Coagulation in a Turbulent Plume, Report No. 86-6. Berkeley, California: Sanitary Engineering and Environmental Health Laboratory, University of California, Berkeley.

Honeyman, B.D., and P.H. Santschi. 1991. Coupling adsorption and particle aggregation—laboratory studies of colloidal pumping using FE-59 labelled hematite. Environmental Science and Technology 25(10):1739-1747.

Honeyman, B.D., L.S. Balistrieri, and J.W. Murray. 1988. Oceanic trace metal scavenging: The importance of particle concentration. Deep-Sea Research 35:227-246.

Hudson, R.J.M., and F.M.M. Morel. 1990. Iron transport in marine phytoplankton: Kinetics of cellular and medium concentration reactions. Limnol. Oceanogr. 35:1002-1020.

Hudson, R.J.M., S.A. Gherini, C.J. Watras, and D.P. Porcella. 1992. Modelling the biogeochemical cycle of mercury in lakes: The mercury cycling model and its application to the MTL lakes. In Mercury as a Global Pollutant: Toward Integration and Synthesis, R.C.J. Watras, and J.W. Huckabee, eds. Chelsea, Michigan: Lewis (in press).

Huggett, R.J. 1989. Kepone and the James River Estuary. Pp. 417-424 in Contaminated Marine Sediments—Assessment and Remediation. Washington, D.C.: National Academy Press.

Hunt, J.R. 1980. Predictions of oceanic particle size distributions from coagulation and sedimentation mechanisms. Pp. 243-257 in Particulates in Water: Characterization, Fate, Effects, and Removal, M. Kavanaugh and J. O. Leckie, eds. ACS Advances in Chemistry Series No. 189. Washington, D.C.: American Chemical Society.

Hunt, J.R. 1982. Particle dynamics in seawater: Implications for predicting the fate of discharged particles. Environmental Science and Technology 16:303-309.

Hunt, J.R., and J.D. Pandya. 1984. Sewage sludge coagulation and settling in seawater. Environmental Science and Technology 18:119-121.

Ikalainen, A.J., and D.C. Allen. 1989. New Bedford Harbor Superfund Project. Pp. 312-350 in Contaminated Marine Sediments—Assessment and Remediation. Washington, D.C.: National Academy Press.

Imboden, D.M., and R.P. Schwarzenbach. 1985. Spatial and temporal distribution of chemical substances in lakes: Modeling Concepts. Pp. 1-30 in Chemical Processes in Lakes, W. Stumm, ed. New York: Wiley-Interscience.

Jackson, G.A. 1990. A model of the formation of marine algal flocs by physical coagulation processes. Deep-Sea Research 37:1197-1211.

Jackson, G.A., and J.J. Morgan. 1978. Trace metal-chelator interactions and phytoplankton growth in seawater media: Theoretical analysis and comparison with reported observations. Limnol. Oceanogr. 23:268-282.

Jaworski, N.B. 1981. Sources of nutrients and the scale of eutrophication problems in estuaries. In Estuaries and Nutrients, B.J. Neilson and L.E. Cronin, eds. New York: Humana.

Johnson, C.A., M. Ulrich, L. Sigg, and D.M. Imboden. 1991. A mathematical model of the manganese cycle in a seasonally anoxic lake. Limnol. & Oceanogr. 36:1415-1426.

Karickhoff, S.M. 1984. Organic pollutant sorption in aquatic systems. Journal of Hydraulic Engineering 10:707-735.

Koh, R.C.Y. 1982. Initial sedimentation of waste particulates discharged from ocean outfalls. Environmental Science and Technology 16:757-762.

Koh, R.C.Y. 1988. Shoreline impact from ocean waste discharges. Journal of Hydraulic Engineering, American Society of Civil Engineers 114(4):361-376.

Koh, R.C.Y., and N.H. Brooks. 1975. Fluid mechanics of waste-water disposal in the ocean. Ann. Rev. of Fluid Mechanics 7:187-211.

Kullenberg, G., ed. 1982. Pollutant Transfer and Transport in the Sea. Vol. 1. Boca Raton, Florida: CRC Press.

Lavelle, J.W., H.O. Mofield, and E.T. Baker. 1984. An *in situ* erosion rate for a fine grained marine sediment. Journal of Geophysical Research 89(C4):6543-6552.

Lewis, B.L., P.N. Froelich, and M.O. Andreae. 1985. Methylgermanium in natural waters. Nature 313:303.

Logan, B.E., R.G. Arnold, and A. Steele. 1989. Computer simulation of DDT distribution in Palos Verdes sediments. Pp. 178-198 in Contaminated Marine Sediments—Assessment and Remediation. Washington, D.C.: National Academy Press.

Long, E.R., and L.G. Morgan. 1991. The Potential for Biological Effects of Sediment-Sorbed Contaminants Tested in the National Status and Trends Program. NOAA Technical Memorandum NOS OMA 52. Seattle, Washington: National Oceanic and Atmospheric Administration.

Luoma, S.N., C. Johns, N.S. Fisher, N.A. Steinberg, R.S. Oremland, and J.R. Reinfelder. 1992. Determination of selenium bioavailavility to a benthic bivalve from particulate and solute pathways. Environmental Science & Technology 26(3):485-491.

Lyne, V.D., B. Butman, and W.D. Grant. 1990. Sediment movement along the U.S. east coast continental shelf-II. Modelling suspended concentration and transport rate during storms. Continental Shelf Research 10(5):429-460.

Mason, R.P., and W.F. Fitzgerald. 1990. Alkylmercury species in the equatorial Pacific. Nature 347:457.

Mearns, A.J., and D.R. Young. 1983. Characteristics and effects of municipal wastewater discharges to the Southern California Bight, a case study. Pp. 763-819 in Ocean Disposal of Municipal Wastewater: Impacts on the Coastal Environment, Vol. 2, E.P. Myers and E.T. Harding, eds. Cambridge, Massachusetts: MIT Sea Grant.

Moffett, J.W., and R.G. Zika. 1987. Solvent extraction of copper acetylacetonate in studies of copper (II) speciation in seawater. Mar. Chem. 21:301-313.

Muellenhoff, W.P., A.M. Soldate, Jr., D.J. Baumgartner, M.D. Schuldt, L.R. Davis, and W.E. Frick. 1985. Initial Mixing Characteristics of Municipal Ocean Discharges: Vol. I, Procedures and Applications, EPA/600/3-85/073a; Vol. II, Program Listings, Computer Programs EPA/600/3-85/073b. Narragansett, Rhode Island: U.S. EPA, Environmental Research Laboratory. Available from NTIS as PB86-137478 and PB86-137460.

Nichols, M.M. 1990. Sedimentologic fate and cycling of kepone in an estuarine system: Example from the James River Estuary. The Science of the Total Environment 97/98:407-440.

Nixon, S.W. 1988. Physical energy inputs and the comparative ecology of lake and marine ecosystems. Limnol. Oceanogr. 33:1005-1025.

NRC (National Research Council). 1984. Ocean Disposal Systems for Sewage Sludge and Effluent. Washington, D.C.: National Academy Press.

NRC (National Research Council). 1989. Contaminated Marine Sediments—Assessment and Remediation. Washington, D.C.: National Academy Press.

O'Connor, T.P. 1990. Coastal Environmental Quality in the United States, 1990. Chemical Contamination in Sediments and Tissues, A Special National Oceanic and Atmospheric Administration 20th Anniversary Report. Rockville, Maryland: National Oceanic and Atmospheric Administration.

O'Connor, D.J., J.A. Mueller, and K.J. Farley. 1983b. Distribution of kepone in the James River Estuary. Journal of Environmental Engineering 109(2):396-413.

O'Connor, T.P., A. Okubo, M.A. Champ, and P.K. Park. 1983a. Projected consequences of dumping sewage sludge at a deep ocean site near New York Bight. Can. J. Fish. Aquat. Sci. 40(Sup. 2):228-241.

Officer, C.B., and D.L. Lynch. 1989. Bioturbation, sedimentation and sediment-water exchanges. Estuarine, Coastal and Shelf Science 28:1-12.

Officer, C.B., and J.H. Ryther. 1977. Secondary sewage treatment versus ocean outfalls: An assessment. Science 197:1056-1060.

Okubo, A. 1971. Ocean diffusion diagrams. Deep-Sea Res. 18:789-802.

Olmez, I., E.R. Sholkovitz, D. Hermann, and R.P. Eganhouse. 1991. Rare earth elements in sediments off southern California: A new anthropogenic indicator. Environmental Science and Technology 25(2):310-316.

Palermo, M.R., C.R. Lee, and N.R. Francingues. 1989. Management strategies for disposal of contaminated sediments. Pp. 200-220 in Contaminated Marine Sediments—Assessment and Remediation. Washington, D.C.: National Academy Press.

Paulson, A.J., R.A. Feely, and H.C. Curl. 1989. Separate dissolved and particulate trace metal budgets for an estuarine system: An aid for management decisions. Env. Poll. 57:317-339.

Paulson, A.J., R.A. Feely, H.C. Curl, E.A. Crecelius, and T. Geiselman. 1988. The impact of scavenging on trace metal budgets in Puget Sound. Geochim. Cosmochim. Acta 52:1765-1779.

Roberts, P.J.W., W.H. Snyder, and D.J. Baumgartner. 1989a. Ocean outfalls. I: Submerged wastefield formation. Journal of Hydraulic Engineering, American Society of Civil Engineer 115:1-25.

Roberts, P.J.W., W.H. Snyder, and D.J. Baumgartner. 1989b. Ocean outfalls. II: Spatial evolution of submerged wastefield. Journal of Hydraulic Engineering, American Society of Civil Engineers 115:26-48.

Roberts, P.J.W., W.H. Snyder, and D.J. Baumgartner. 1989c. Ocean outfall. III: Effects of diffuser design of submerged wastefield. Journal of Hydraulic Engineers, American Society of Civil Engineers 115:49-70.

Sanders, J.E. 1989. PCB pollution in the upper Hudson River. Pp. 365-400 in Contaminated Marine Sediments—Assessment and Remediation. Washington, D.C.: National Academy Press.

Sayles, F.L., and W.H. Dickinson. 1991. The ROLAI^{2}D lander: A benthic lander for the study of exchange across the sediment-water interface. Deep-Sea Research 38(5):505-529.

SCCWRP (Southern California Coastal Water Research Project). 1973. The Ecology of the Southern California Bight: Implications for Water Quality Management. Long Beach, California: Southern California Coastal Water Research Project.

SCCWRP (Southern California Coastal Water Research Project). 1991. Annual Report. Long Beach, California: Southern California Coastal Water Research Project.

Schwarzenbach, R.P., and J. Westall. 1981. Transport of nonpolar organic compounds from surface water to groundwater. Laboratory Sorption Studies. Environmental Science and Technology 15:1360-1367.

Sheng, Y.P. 1990. Evolution of a three dimensional curvilinear-grid hydrodynamic model for estuaries, lakes and coastal waters: CH3D. Pp. 40-50 in Estuarine and Coastal Modeling, M. Spaulding ed. New York: American Society of Civil Engineers.

Signell, R.P., and B. Butman. 1992. Modeling tidal exchange and dispersion in Boston Harbor. Journal of Geophysical Research 97(C10):15591-15607.

Silver, M.W., A.L. Shanks, and J.D. Trent. 1978. Marine snow: Microplankton habitat and source of small-scale patchiness in pelagic populations. Science 201:371-373.

Smetacek, V.S. 1985. Role of sinking in diatom life-history cycles: Ecological, evolutionary, and geological significance. Marine Biology 84:239-251.

Stoecker, D.K., W.G. Sunda, and L.H. Davis. 1986. Effects of copper and zinc on two planktonic ciliates. Marine Biology 92:21-29.

Stolzenbach, K.D., K.A. Newman, and C. Wong. 1992. Aggregation of fine particles at the sediment-water interface. Journal of Geophysical Research 97(C11):17889-17898.

Stull, J.K., R.B. Baird, and T.C. Heesen. 1986. Marine sediment core profiles of trace constituents offshore of a deep wastewater outfall. J. Water Pollut. Control Fed. 58(10):985-991.

Sunda, W.G. 1988-1989. Trace metal interactions with marine phytoplankton. Biol. Oceanogr. 6(5-6):411-442.

Sunda, W.G., and R.L. Ferguson. 1983. Sensitivity of natural bacterial communities to additions of copper and curic ion activity: A bioassay of copper complexation in seawater. Pp. 871-891 in Trace Metals in Seawater, C.S. Wong, ed. New York: Plenum Press.

Sunda, W.G., and R.R.L. Guillard. 1976. The relationship between cupric ion activity and the toxicity of copper to phytoplankton. J. Mar. Res. 34:511-529.

Sunda, W.G., and A.K. Hanson. 1987. Measurement of free cupric ion concentration in seawater by a ligand competition technique involving copper sorption onto C-18 SEP-PAD cartridges. Limnol. Oceanogr. 32:537-551.

Sunda, W.G., D.W. Engel, and R.M. Thuotte. 1978. Effect of chemical speciation on toxicity of cadmium to grass shrimps. *Palaemonetes pugio*: Importance of free cadmium ion. Environ. Sci. Technol. 13:213-219.

Sunda, W.G., R.T. Barber, and S.A. Huntsman. 1981. Phytoplankton growth in nutrient rich seawater; importance of copper-manganese cellular interactions. J. Mar. Res. 39:567-586.

Tee, K.T. 1987. Simple models to simulate three-dimensional tidal and residual currents, in three-dimensional coastal ocean models. Pp. 125-147 in Three-Dimensional Coastal Ocean Models, N.S. Heaps, ed. Washington, D.C.: American Geophysical Union.

Thomann, R.V., J.A. Mueller, R.P. Winfield, and C.R. Huang. 1991. Model of fate and accumulation of PCB homologues in Hudson Estuary. Journal of Environmental Engineering 117(2):161-178.

Thorbjarnarson, K.W., C.A. Nittrouer, D.J. DeMaster, and R.B. McKinney. 1985. Sediment accumulation in a back-barrier lagoon, Great Sound, New Jersey. Journal of Sedimentary Petrology 55(6):856-863,.

U.S. GOFS (U.S. Global Ocean Flux Study). 1989. Sediment Trap Technology and Sampling, Planning Report No. 10. Woods Hole, Massachusetts: U.S. GOFS Planning and Coordination Office, Woods Hole Oceanographic Institution. 94p.

van den Berg, C.M.G. 1984. Determination of the complexing capacity and conditional stability constants of complexes of copper (II) with natural organic ligands in seawater by cathodic stripping voltammetry of copper-catechol complex ions. Mar. Chem. 15:1-18.

Van Dover, C.L., J.F. Grassle, B. Fry, R.H. Garritt, and V.R. Starczak. 1992. Stable isotope evidence for entry of sewage-derived organic material into a deep-sea food web. Nature 360:153-155.

Wang, R.F.T. 1988. Laboratory Analysis of Settling Analysis of Wastewater Particles in Sea Water Using Holography, Report No. 27. Pasadena, California: Environmental Quality Laboratory, California Institute of Technology.

Weilenmann, U., C.R. O'Melia, and W. Stumm. 1989. Particle transport in lakes: Models and measurements. Limnol. Oceanog. 34(1):1-18.

Weiner, J.G., W.F. Fitzgerald, C.J. Watras, and R.G. Rada. 1990. Partitioning and bioavailability of mercury in an experimentally acidified Wisconsin lake. Environmental Toxicology and Chemistry 9:909-918.

Wu, Jin. 1983. Sea-surface drift currents induced by wind and waves. J. Phys. Oceanogr. 13:1441-1451.

D

Engineering and Management Options for Controlling Coastal Environmental Water Quality

INTRODUCTION

As discussed in the main body of this report, approaches to wastewater management and engineering in coastal areas should be designed to suit the characteristics of the surrounding environment. This appendix discusses five engineering and management options for wastewater and storm water in coastal urban areas: source control, wastewater treatment systems, disinfection, combined sewer overflows, and nonpoint source controls. Wastewater outfalls are not included here, but are discussed in Appendix C in connection with transport and fates in the coastal environment.

Whereas traditional wastewater management with its single medium focus revolves around treatment and disposal, a fully integrated approach addresses a broader range of considerations including source control; potential impacts on all environmental media (water, air, and land); water, energy, and other natural resource conservation; recycle and reuse; and nonpoint source control.

SOURCE CONTROL

The three basic source control alternatives, which may be practiced independently or concurrently in any municipality, are pollution prevention, pretreatment, and recycle and reuse.

Pollution Prevention

Pollution prevention is the common sense notion of trying to prevent or reduce pollution at the source before it is created. It may include a wide range of activities, programs, and techniques. Elimination or minimization of water pollutants at the source is becoming more important as wastewater treatment plant effluent criteria become more strict. Once a pollutant is discharged into a sewer system, it is diluted by several orders of magnitude and usually much more difficult to remove. Analysis and treatment of these diluted pollutants can be difficult and expensive. Depending on the pollutant's dominant characteristics, it may volatilize into the atmosphere, biodegrade, settle out with the sludge, or pass through into the final effluent.

Pretreatment

Pretreatment refers to the treatment of wastewater at industries or commercial establishments before it is discharged to a sewer system. Pretreatment of wastewater reduces the release of conventional and toxic pollutants into the system. Pretreatment processes include physical and chemical treatment and biological treatment. These processes typically result in some type of cross-media transfer of pollutants from wastewater to land or air. For example, chemical or biological processes produce residuals that contain concentrated levels of pollutants removed in treatment. Thermal and biological processes, however, can destroy all or most of some compounds, but others will concentrate in residuals or escape to the atmosphere.

To date, pretreatment has been the main approach used by the federal government to control the discharge of industrial or commercial waste to publicly owned treatment works (POTWs). Environmental Protection Agency (EPA) effluent guidelines are based on the best available control technology. The enforcement of federal pretreatment standards by POTWs has helped reduce the amounts of contaminants, especially metals and some toxic organics, from being discharged to the nation's waterways.

Recycling and Reuse

Recycling and reuse involve transformation of potential waste materials into products. Internal recycling and reuse occurs when a material that has served its original purpose and could become a waste is recovered and reused at the site of waste generation; the material is controlled by the waste generator. Internal recycling by industry can involve the installation of closed-loop or in-process recycling systems. Internal recycling takes place in the home when, for example, vegetable wastes are composted rather than disposed as garbage. External recycling and reuse occurs away from the site of waste generation. External recycling is a multiple-step process involving separation and collection of the material, transport to a recycling center and/or a reprocessing facility, and resale to a new user.

Pollution Prevention in Municipal Wastewater Management—Background and Definitions

In the recently enacted 1990 Federal Pollution Prevention Act, source reduction (or source control) has been interpreted by the EPA as "any practice that reduces the amount of any hazardous substance, pollutant, or contaminant entering any wastestream prior to recycling, treatment and disposal" (42 U.S.C. 13101 et seq.).[1] Pollution prevention in the context of coastal municipal wastewater management refers to the use of materials, processes, or practices that eliminate or reduce the creation of pollutants, either toxic or conventional, or wastes (e.g., plastics, paper) at the source. Sources can be domestic, commercial, institutional, or industrial. Pollution prevention includes any on-site source reduction or substitution undertaken prior to discharge to a municipal sewer system to reduce the total volume or quantity of pollutants generated or hazardous materials used in order to minimize the impact of the waste itself. It also includes practices that reduce the use of water, energy, or other natural resources at the source. Actions taken away from the source of the waste-generating activity, including off-site treatment of wastes or off-site recycling, are not considered pollution prevention activities by the EPA. These activities may still serve to improve wastewater influent quality.

Pollution prevention options include product changes, technology modifications, raw materials and process changes, and operational changes. In the context of municipal wastewater management, some of the chief pollution prevention activities include source reduction, water conservation, energy conservation, and some approaches to nonpoint source control.

In applying pollution prevention concepts to the field of municipal wastewater management, all sources—domestic, commercial, institutional, and industrial—are potentially significant. Similarly, all chemicals and materials are assessed, including conventional pollutants such as total suspended solids (TSS) and biochemical oxygen demand (BOD), paper and plastics, as well as toxics. For example, at one municipality, the elimination of plastics may be the central pollution prevention project; at another, the minimization of silver and mercury amalgams from dentists' offices may be more effective. The banning of phosphate detergents is another example of source reduction or product substitution that can change wastewater characteristics significantly.

---

[1]References to United States Code are cited with the title followed by "U.S.C." and the section.

Energy Conservation and Energy Recovery

Energy conservation and recovery can be important components of an integrated coastal management program for POTWs. Municipal POTWs generally consume less than one percent of the electrical energy demand of the communities they serve. Nonetheless, optimizing existing treatment processes, limiting sludge production, and using less energy intensive treatment methods have the potential to make a significant reduction in energy demands. Energy recovery using methane gas—a natural by-product of anaerobic sludge digestion—can meet over 50 percent of the electricity needs of a POTW (CSDOC 1989).

Nonpoint Source Control

Nonpoint source control options include a range of activities to limit urban and agricultural runoff and atmospheric deposition into waterways that in turn degrade the coastal environment. Some of these approaches can be considered pollution prevention activities, others are structural and fall under the treatment category. The subject of nonpoint source control options is addressed later in this appendix.

Pollution Prevention Programs

Implementation

At present there is no federal mandate to implement pollution prevention programs in municipalities and/or municipal wastewater management districts. Key ingredients in the implementation of pollution prevention programs in municipalities include 1) setting definitions and goals; 2) conducting an inventory of all resources and pollutants, including, especially, those subject to cross-media transfer; 3) systematically examining each problem pollutant to determine how it can be prevented or minimized; 4) establishing a prioritization, reporting, and tracking system for pollutants; 5) undertaking preventive routine operation and maintenance inspection programs to help eliminate unwanted plant shut-downs and avoidable discharges; and 6) implementing specific projects or actions. Examples of regulatory options that encourage the implementation of pollution prevention practices are shown in Table D.1.

Examples of Pollution Prevention Programs

Orange County, California. A wide range of pollution prevention activities have been instituted and are being incorporated into the existing

TABLE D.1 Local Pollution Prevention Regulatory Options

Indirect Inducements	Direct Requirements
Regulatory flexibility	Mandatory hazardous waste management plans
Aggressive enforcement of local pollution discharge limits	Mandatory pollution prevention requirements incorporated into discharge permits, including the requirement of standard reduction technologies
Development of more comprehensive and stringent limits	Incentives tied to reduced regulatory requirements
Nonprescriptive effluent guidelines	Mandatory recycling requirements for industrial dischargers
Development of control mechanisms for commercial and small industrial dischargers	Investigation of pollution prevention opportunities required by statute prior to planning of new treatment facilities
Option to use mass-based wastewater discharge limits	
Development of waste exchanges and other technical assistance for regional industry to encourage waste reduction	
Low interest loans	

source reduction program at the Orange County Sanitation Districts. Principally, pollution prevention and waste minimization have been used as a tool to assist dischargers in attaining compliance. A program is being developed to train field inspectors in pollution prevention inspections, to coordinate multi-agency pollution prevention work, to hold workshops and other public education events, to mandate wastewater reductions of pretreatment program permittees, to apply mass emission limits instead of concentration limits for permittees, to require the implementation of pollution prevention techniques, and to provide technical assistance to permittees in violation of their permits.

Springfield, Massachusetts. While the pretreatment program at the 67 million gallons per day (MGD) Springfield Regional Wastewater Treatment Plant has effectively controlled the industrial discharge of cadmium, nonindustrial sources and/or nonpoint sources interfere with a Type I classification for the composting and marketing of its sludge. The Springfield pollution prevention pilot project is designed to monitor and quantify the nonpoint source pollutant load. It will be coupled with a pollution prevention educational outreach project targeted at reducing illegal discharges to storm sewers.

Winston-Salem, North Carolina. The city of Winston-Salem is establishing a pilot project at a large POTW by developing a model pollution program designed to meet the needs of both large and small POTWs and integrating pollution prevention evaluation techniques into existing pretreatment program elements.

Cincinnati, Ohio. Cincinnati's Metropolitan Sewer District is developing a pilot program similar to that of Winston-Salem with assistance from the Ohio Environmental Protection Agency. The program will incorporate pollution prevention techniques into the ongoing pretreatment program to reduce loadings to the POTWs.

Economic Advantages of Pollution Prevention

Added regulations, higher industrial treatment and landfilling expenses, and increased liability costs have caused industrial and governmental leaders to reevaluate end-of-pipe pollution control measures in favor of front-end actions. Some of the economic advantages of pollution prevention include reduced on-site capital and operational waste-treatment costs; reduced transportation and disposal costs for wastes sent to an off-site location; reduced compliance costs for permits, monitoring, and enforcement; lower risk of spills, accidents, and emergencies; lower long-term environmental liability and insurance costs; reduced production costs through better management and efficient use of raw materials, transportation, and energy; improvements in process, product quality, and product yield resulting from a reexamination of current practice and the institution of better controls; income derived from the sale or reuse of waste; reduced sewer-use fees; better employee morale; and better public relations.

Economic Advantages of Recycling and Reuse

Because hazardous waste disposal fees can be a major component of an industry's annual operation and maintenance costs, zero sludge production may be attractive on economic as well as environmental grounds. A study sponsored by the California Department of Health Services compared the annual waste management operating costs for two similar circuit board plants for two different treatment plants assuming 10-year lifetimes. Cost data for the first plant were based on an installed conventional treatment system with sludge handling equipment. Cost data for the second plant were based on an installed recovery system with zero sludge production. The results are shown in Table D.2.

In this study, the total annual cost of the zero sludge production alternative is 9 percent higher than the conventional sludge system. However,

TABLE D.2 Comparison of Conventional Sludge Treatment Versus Zero Sludge Production System (Source: Cal-Tech Management Associates 1987. Reprinted, by permission, from Cal-Tech Management Associates, 1987.)

	Conventional Sludge System	Zero Sludge Production System
Total Capital Cost	$450,000	$1,250,000
Capital Recovery[1]	$67,000	$186,000
Operation and Maintenance[2]	$43,000	$75,000
Labor	$75,000	$50,000
Chemicals and Power	$75,000	$54,000
Water - In/Out	$22,000	$2,000
Sludge Disposal & Fees	$48,000	0
Miscellaneous	$10,000	$5,000
Total Annual Cost	$340,000	$370,000

[1]Assuming an 8 percent opportunity cost.
[2]Assuming 12 percent of the original capital cost per annum.

only quantifiable costs have been used in defining the cost parameters. No attempt was made to examine less tangible future costs such as the value of recovered metals, increasing land disposal costs, legal liabilities, or insurance costs. Had these nonquantifiable costs been factored in, it is possible that the zero sludge production system would become the preferred option on economic as well as environmental grounds.

Pollution Prevention or Pretreatment?

An integrated coastal management plan for a given area would compare the environmental and cost benefits of pollution prevention with pretreatment to determine which is most advantageous. The following exercise shows the advantages and trade-offs in each set of activities.

Environmental Benefits—Pretreatment

In the 1991 National Pretreatment Program Report to Congress, the EPA concluded that categorical standards and local industrial discharge limits implemented by POTWs had brought about significant reductions in toxic pollutant loadings from regulated industries (EPA 1991a). The EPA estimated that metals loadings have been reduced by 95 percent to annual loadings of 14 million pounds, and organic loadings have been reduced by between 40 and 75 percent to annual loadings of 65 million pounds. The EPA also concluded that the planned development of additional categorical standards would further reduce loadings of toxic pollutants to POTWs.

In 1988 and 1989, the Association of Metropolitan Sewerage Agencies conducted a survey of its members to determine the effectiveness of their pretreatment programs. It found that the mass discharge of 10 heavy metals had decreased 69 percent and that the mass discharge of cyanide and selected organics had decreased 66 percent (AMSA 1990). These findings confirmed the EPA's Domestic Sewage Study conclusion that "The pretreatment program has been an effective means in reducing the mass discharge of many hazardous constituents to POTWs" (EPA 1986).

This success in the reduction of heavy metals through pretreatment is illustrated by the example of the County Sanitation Districts of Orange County, California. In 1976, the Sanitation Districts adopted a new industrial source reduction ordinance which included numerical limitations on all industrial discharges. The influent heavy metals reductions at the two regional wastewater treatment plants in the 15 years of record are shown in Figure D.1.

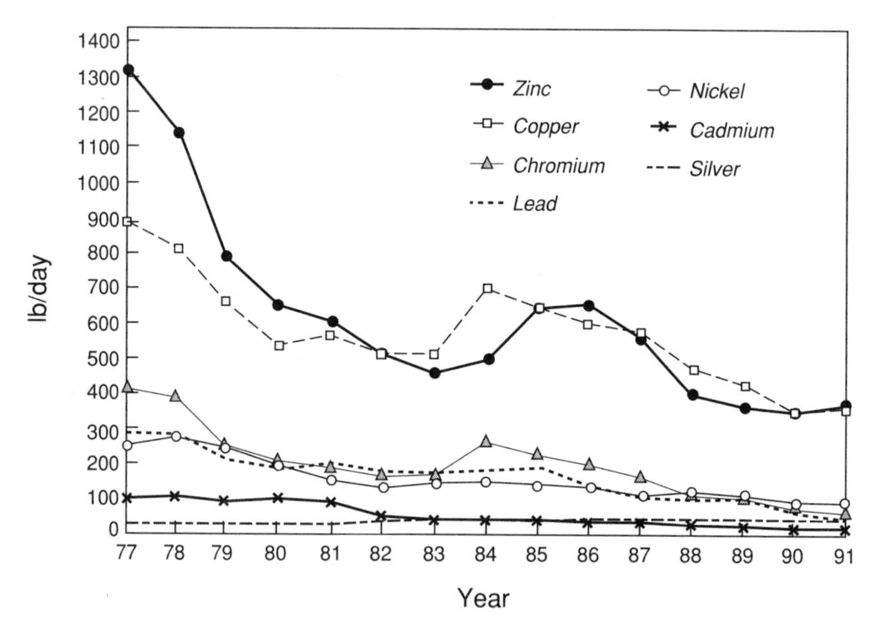

FIGURE D.1 Annual mass inflows of various metals to the County Sanitation Districts of Orange County wastewater treatment plants. (Reprinted, by permission, from County Sanitation Districts of Orange County, California.)

Environmental Benefits—Pollution Prevention

Information on the effectiveness of practices for preventing or reducing discharges to urban wastewater collection systems and receiving waters is available. However, much of it is case-specific and thus complicated to compare with the more extensive information available on pretreatment.

In 1985, the Robbins Company, a metal finishing and plating operation in Attleboro, Massachusetts, was in violation of its water discharge permit and was named a major polluter of the Ten Mile River that empties into Rhode Island's Narragansett Bay. When federal and state officials announced plans to tighten discharge limits further, the company was faced with four options as shown in Table D.3 (Berube and Nash 1991).

The Robbins Company management realized that a pollution prevention approach, through the use of a closed-loop system, although risky, was its best choice. Figures D.2a, D.2b, and D.2c show the Robbins Company's success for the years 1985 through 1990 in water conservation, chemical use, and sludge production after implementing a pollution prevention program. These figures show a 97 percent, 98 percent and 100 percent reduction, respectively, in the use of caustic soda, acid, and chlorine and nearly 100 percent reductions in water use and sludge production. The benefits of such a program are that the reductions can occur in every aspect of the process leading to multiple environmental improvements.

Cost-Benefit Ratios

Although some studies have examined the economics of waste management alternatives for selected industries, such as metal finishers and printed

TABLE D.3 Four Pollution Management Options at the Robbins Company (Source: Berube and Nash 1991. Reprinted, by permission, from the Robbins Company.)

Options	Effect on Compliance	Capital and Operation and Maintenance Cost
Do nothing	Completely out of compliance	Fines up to $10,000 per day
Upgrade present system	In compliance now but probably not in the future	$250,000 capital $120,000/yr O&M
Build a full wastewater treatment plant	Full compliance	$500,000 capital, $120,000/yr O&M
Modify the process and build a closed-loop system	Full compliance	$250,000 capital $21,000/yr O&M

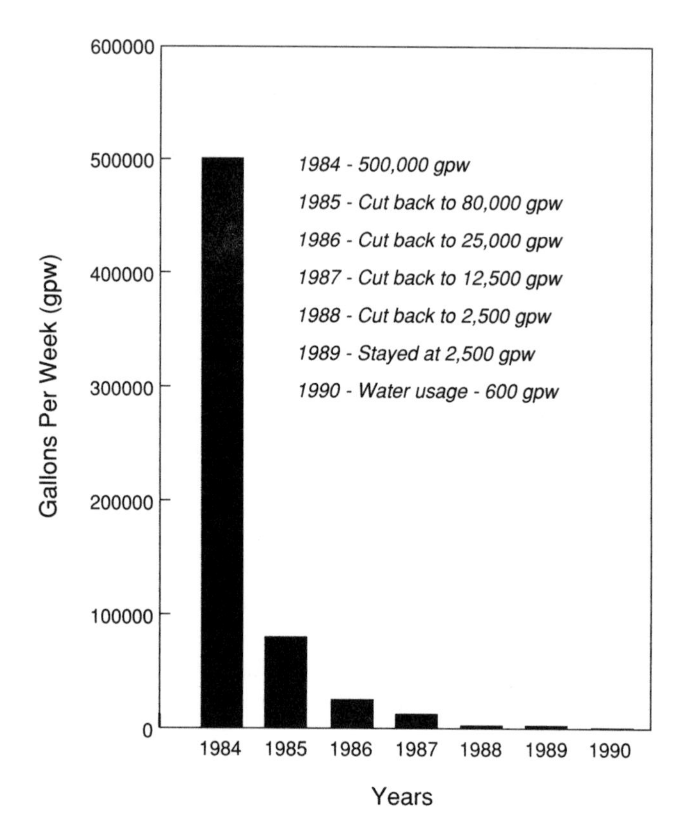

FIGURE D.2a Pollution prevention at the Robbins Company, Attleboro, Massachusetts—Water Usage. (Source: Chatel 1992. Reprinted, by permission, from the Robbins Company.)

circuit board facilities, to date there have been few comparisons of the cost benefits of pretreatment and pollution prevention.

One comparison performed by the EPA looked at the costs of end-of-pipe treatment for an electroplating facility with and without pollution prevention versus a waste recovery system for an electroplating facility (EPA 1979). The three options evaluated were: 1) a system with standard single-stage running rinses and no pollution prevention, 2) a system with pollution prevention in the form of counter-current rinses instead of single-stage rinses, and 3) a system with pollution prevention in the form of recovery units installed after each plating operation in addition to counter-current rinses.

As shown in Table D.4, the cost of the system with counter-current rinses and recovery units provided an annual cost savings of over 50 percent in comparison to the other systems evaluated.

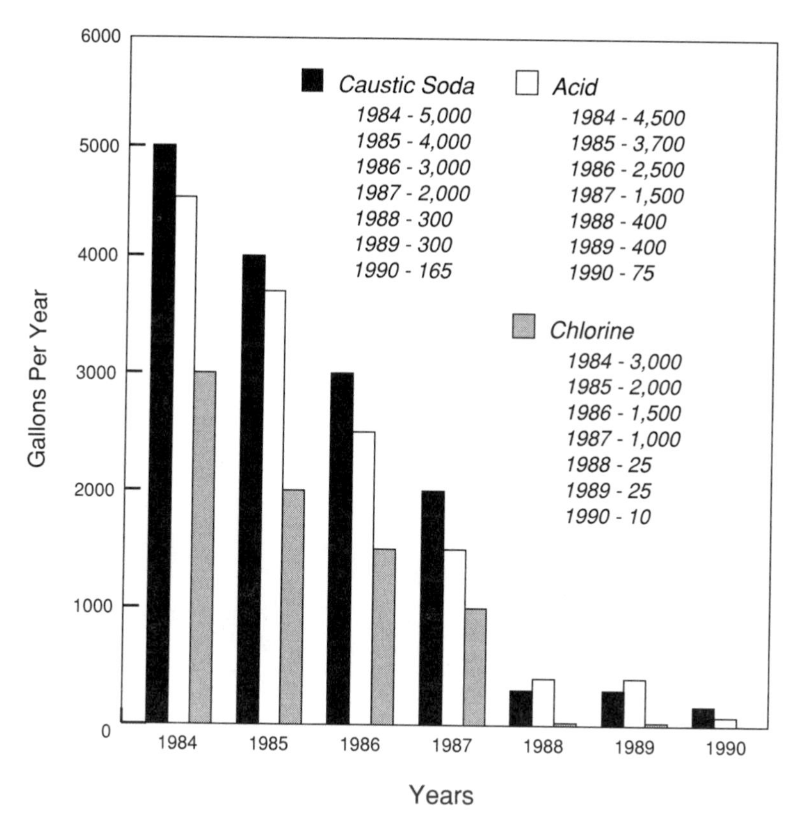

FIGURE D.2b Pollution prevention at the Robbins Company, Attleboro, Massachusetts—Chemical Usage. (Source: Chatel 1992. Reprinted, by permission, from the Robbins Company.)

Grants for Small Business. While many basic pollution prevention practices (e.g., good housekeeping and operational practices, systematic maintenance, and training of personnel) require marginal or no capital investment, other more fundamental practices such as pretreatment systems, production equipment modifications, or raw material substitutions require up-front investment of funds for development, research, engineering, and equipment. Although pollution prevention may represent financial benefits, these benefits may take several years to amortize the original capital investment. Large businesses and corporations are usually able to support the burden of long term capital returns. Small businesses may not be. Consequently, small businesses are often reluctant, and frequently find it impossible, to engage in advantageous pollution prevention or pretreatment programs without federal, state or local financial assistance, such as grants.

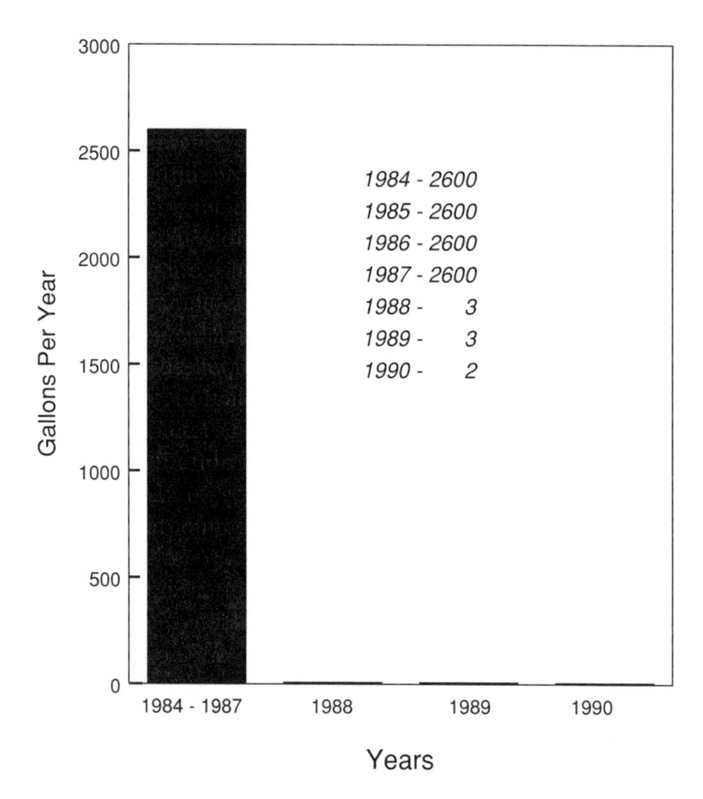

FIGURE D.2c Pollution prevention at the Robbins Company, Attleboro, Massachusetts—Sludge production. (Source: Chatel 1992. Reprinted, by permission, from the Robbins Company.)

TABLE D.4 EPA Comparison of Pretreatment Versus Pollution Prevention Cost (Source: EPA 1979)

	No Pollution Prevention	Pollution Prevention	Pollution Prevention
	Standard Single Stage Rinses	Counter-Current Rinses	Counter-Current Rinses + Plating Recovery Systems
Captial Costs[1]	$192,000	$186,000	$162,000
Annual Costs[2]	$90,000	$87,000	$40,000

[1]The capital costs include equipment for treatment of wastewater and sludge.
[2]The annual costs include annualized capital (10 year life), depreciation, operation and maintenance, energy and power, and sludge disposal costs.

Grants may be used as seed money for small businesses to get pollution prevention and pretreatment programs started.

Quality Certification of Technology. Pollution prevention and pretreatment often involve innovation and new processes and technologies that are unfamiliar to the average business. Many businesses do not have the resources, personnel, or expertise to pursue and obtain good technical information. Unscrupulous vendors and lack of knowledge of technological realities may result in the improper application of a control technology or practice. One problem faced by industry is when a vendor who provides the equipment goes out of business or reneges on guarantees. Then businesses are faced with the problem of replacing the entire pollution control system or paying heavy fines with no redress. In order to provide small business with the necessary resources and confidence to acquire new technologies, a strict quality certification program of technology and vendors could certify the vendors and technology.

Integrated Multi-Media Permitting. The current permitting system responds to single media statues and regulations such as the Clean Air Act, Clean Water Act, and Resource Conservation and Recovery Act. Because of the single media interests addressed by these statutes, the requirements of one ignores impacts on other media. Single media permitting imposes standards and technology that may result in an increase of emissions to another media, and conflict with another set of permitting requirements. An integrated multi-media permitting program would address the balance between the three media—air, water, and land—providing an integrated system for environmental protection. As an incentive for business in general, and small business in particular, it would eliminate some of the duplication of reporting, audits, inspection, monitoring and permitting of the current system. Most importantly, multi-media permitting would allow for an integrated environmental protection program that addresses all media equally and results in the greatest and most cost-effective benefit to the environment.

Conclusions

Pollution prevention programs are generating a different attitude in industries and communities, where pretreatment regulations are based on the command-and-control philosophy of environmental protection. Pollution prevention offers a new type of incentive to better business practice and community relations. Pollution prevention, like labor productivity and energy consumption, can become a measure of productivity and efficiency of industrial operations. For businesses, pollution prevention may be a way to improve profitability and competitiveness.

While enforcement and compliance often provide the impetus for pollution prevention programs, the existing framework of regulations is not sufficient to implement successful pollution prevention programs. There are several key elements necessary for a workable system:

Regulatory Flexibility—e.g., use of alternative requirements such as mass-emission limits as the compliance basis for dischargers and latitude in imposing compliance and enforcement schedules on industry to permit innovative approaches and make allowances for failed technology and development of incentives;

POTWs as Co-Regulators—POTWs must have the flexibility to develop a prevention program tailored to local conditions and the tools to implement the program;

Institutional Structure that Supports a Multi-Media Perspective; and

Parallel Programs—Federal, state, and local governments should initiate grants for small businesses, quality certification of technology, integrated multi-media permitting.

Several factors make the economic determination of pollution prevention benefits a difficult problem. First, because there are innumerable ways to reduce waste, oftentimes each pollution prevention opportunity must be considered on a case-by-case basis. Second, there are a number of nonquantifiable economic benefits to pollution prevention programs, including decreased liability and improved public relations. Nevertheless, some U.S. companies subject to regulation under the Clean Water Act have verified that pollution prevention pays for itself relatively quickly, especially compared with the time needed to comply with regulations (e.g., obtain regulatory permits, and site and construct waste management facilities) (NHSRC 1991, EPA 1992a).

The information presented in this report indicates that the total operating costs of facilities with pollution prevention equipment can be economically competitive with conventional pretreatment systems, even when nonquantifiable costs are not factored into the final result. Recognizing that more research and data collection are essential to a sound analysis of environmental and economic costs and benefits, the following preliminary conclusion can be reached:

- A break-even cost situation between the pollution prevention approach and the pretreatment approach to source control has nearly been reached across a broad spectrum of industries. If higher discharge limitations and/or higher off-site waste disposal costs are applied in the future, businesses with pollution prevention programs in place will have an economic advantage.

MUNICIPAL WASTEWATER TREATMENT

Introduction

Ultimately, after appropriate treatment, wastewater collected from cities and towns must be returned to the land or water. The complex question of which contaminants in wastewater should be removed to protect the environment, to what extent, and where they should be placed must be answered in light of an analysis of local conditions, environmental risks, scientific knowledge, engineering judgment, and economic feasibility.

The total, 20-year capital cost to upgrade U.S. municipal sewerage systems is $110 billion (for the design year of 2010 in 1990 dollars) according to the 1990 Needs Survey Report to Congress (EPA 1991a). The cost for construction of conventional secondary ($37.3 billion) and advanced systems ($11.79 billion) totals $49.0 billion. A decade ago, the United States government paid about 75 percent of the costs to construct new treatment plants. In 1990, the Federal Construction Grants Program was terminated and replaced with a revolving loan program administered by the states for the EPA. Today, more than 80 percent of the cost burden falls on local ratepayers.

Advances in Municipal Wastewater Treatment Systems

A selection of some important advances in municipal wastewater treatment can be organized into five areas: 1) optimization of primary stages of treatment, 2) innovations in biological treatment processes, 3) natural wastewater treatment systems, 4) water reclamation and reuse as an alternative to discharge to receiving waters, and 5) innovations that offer flexibility and/or special capabilities.

Optimization of Primary Stage(s) of Treatment

Economic and space constraints at existing and new sites have provided the impetus to optimize the primary stages of treatment through high-rate settlers, primary filtration, fine screens, and chemically-enhanced primary treatment. Improved primary treatment results in a reduction of larger organic and inorganic particles. Because small (less than 1.0 micrometer) organic particles in wastewater can be biologically degraded about 4 times faster than larger particles (Force 1991), optimization of primary treatment can increase the biodegradation rate in subsequent treatment, which can result in improved performance or cost savings. The first three of these technologies are discussed here; chemically-enhanced primary treatment is covered in the section on Municipal Wastewater Treatment Systems.

High-Rate Settlers. High-rate settlers consist of parallel plates or tubes in sedimentation tanks inclined at an angle of between 30 to 60 degrees from the horizontal. Beginning with an idea first proposed in the early 1900s on the value of maximizing the surface area of a public water supply settling tank (Hazen 1904), the concept was adapted to wastewater applications in the 1970s. High-rate settlers have significantly smaller land-area and basin-size requirements than conventional settlers. They have been operated at increased overflow rates compared with conventional primary treatment facilities, often with chemicals for increased removal efficiency. However, increased maintenance problems caused by the accumulation of grease and other debris must be balanced against the performance advantages. High-rate settlers are used in a number of locations in France, Monaco, the province of Quebec, Canada, and the United States (Forsell and Hedstrom 1975, Leblanc 1987).

Primary Effluent Filtration. Primary effluent filtration makes use of a shallow bed of single-size, fine-grain sand with an underdrain and air pulsing system to filter primary effluent. By reducing the amount of larger-sized organics by filtration, the biodegradation rate of the remaining organic material can be increased in subsequent biological secondary treatment processes. Alternatively, primary effluent filtration can serve as a final treatment step if the prevailing water-quality objectives can be met with the effluent quality achieved by this treatment method. Primary effluent filtration performance efficiencies vary widely for TSS and 5-day biochemical oxygen demand (BOD_5) (Matsumoto 1991). In pilot-scale studies conducted at the County Sanitation Districts of Orange County, California in 1982, average concentrations of 25 mg/l TSS and 104 mg/l BOD_5 were achieved for municipal wastewater (Hydroclear Corporation, unpublished data, January 1983). Full-scale capital and operations and maintenance estimates for this technology on which to determine its relative cost-effectiveness have not been developed.

Fine Screens. Where coarse screens have been used in the preliminary stage of treatment, the development of better screening materials and systems over the last 20 years has led to the use of fine screens as a substitute for or as a means of upgrading conventional primary sedimentation. Fine screen designs include inclined, continuous self-cleaning types, and rotary drum disk types, which are cleaned by spraywater. Screens are made of stainless steel with mesh sizes ranging from 0.001 to 0.25 inches. Suspended solids removals are between 15 percent to 30 percent when the units are used as substitutes for conventional primary treatment. When used as an effluent screen to upgrade primary performance, 15 percent additional removal is typically achieved. Screens have the advantages of being inexpensive, more compact, and having low maintenance requirements (Marshall 1987, WEF/ASCE 1992).

Advances in Biological Treatment Processes

Advances in biological treatment processes have emphasized the importance of space, energy, and cost savings, as well as the need to incorporate further flexibility into these systems. Some of the significant advances include the development of biological aerated filters, sequencing batch reactors, high biomass systems, and biological nutrient removal systems. Further developments in the area of biotechnology are anticipated in the near future. These technologies are discussed here. Nutrient removal systems are addressed in the section on Municipal Wastewater Treatment Systems.

Biological Aerated Filters. Biological aerated filters were developed in France during the early 1980s and are in use today on a full-scale basis at about 100 facilities in Europe, Japan, and Canada. These systems employ shale, aluminum silicate, or expanded polystyrene to foster the growth of high concentrations of bacteria. Depending on the design, wastewater and air are introduced from the top or bottom of the media in a counter-current or co-current manner. The reactors can be used for BOD and/or ammonia removal. Biological aerated filters eliminate the need for final clarifiers and may be cost-competitive relative to other biological systems at low influent concentrations and loadings. This process is compact in areal dimensions, but has significant energy requirements and high operation and maintenance costs.

Sequencing Batch Reactors. Sequencing batch reactors are an elementary form of biological treatment in which the aeration, settling, and decant phases of each treatment cycle occur in a single reactor. The steps of operation include:

- fill - the reactor is filled to a predetermined level (with or without aeration),
- react - the introduction of air allows the aerobic degradation of carbon, ammonia, and other degradable compounds,
- settle - solids are allowed to settle to the bottom of the reactor without mixing or aeration,
- decant - clarified effluent is withdrawn from the reactor, and
- idle - waste sludge is removed (as necessary).

Prior to 1984, there were four sequencing batch reactor facilities operating in the United States. Since then, over 150 new sequencing batch reactor facilities have been built in this country, 80 percent of which have flows less than 1 MGD. The largest U.S. plant has a flow of 9.2 MGD flow and is located in Cleveland, Tennessee. The advantages of a sequencing batch system compared with a continuous flow system include the ability to absorb both hydraulic and organic shock loads, to hold wastewater until regulatory limits are met, and to remove phosphorus during the anoxic fill period. N-removal is possible through long react and idle periods. Sequencing batch reactor systems contain no secondary clarifiers or associated return sludge facilities (Heidman 1990). However, post-reactor treatment facilities either must be sized excessively or use flow equalization because of the intermittent, high-flow discharges.

High Biomass Systems. High biomass systems include various forms of inert media within aeration basins, which support the growth of fixed film organisms as a supplement to the suspended biological growth. This technology can lower the solids loading rate on subsequent clarifiers, thereby improving the treatment capacity within existing basins. The media can increase the solids retention time and lower the food-to-microorganism (F/M) ratio, thus permitting or improving nitrification, reducing aeration tank volumes for new facilities, and reducing secondary clarifier surface area requirements. It can be incorporated in a new treatment facility and can also increase the capacity of existing activated sludge systems, thereby forestalling or avoiding new tank construction. Significant long-term, full-scale data on these systems are currently limited to facilities in Europe and Japan and are scant.

Natural Wastewater Treatment Systems

Natural wastewater treatment systems include various methods of wastewater application to land (e.g., crop irrigation, landscape irrigation, and groundwater recharge). They also include the use of constructed wetlands and floating aquatic plant systems. The majority of natural land-based wastewater treatment systems are preceded by a minimum of primary treatment. Slow-rate, rapid infiltration, and overland flow processes are the predominant municipal natural treatment systems in use today. Table D.5 summarizes effluent quality of these three types.

Constructed wetlands can be free water surface or subsurface flow systems, in which the treatment takes place as the water flows through the stems and roots of the wetlands vegetation. The former type may be designed with the goal of establishing new wildlife habitats or enhancing adjacent natural areas. The use of floating aquatic plants is a variant on the wetlands theme. Typical species used include water hyacinth and duckweed.

TABLE D.5 Comparison of Average Effluent Quality of Three Natural Treatment Systems (Source: WPCF 1990, EPA 1992b)

in mg/l:	Slow-Rate	Rapid Infiltration	Overland Flow
TSS	< 2	2	10 - 30
BOD_5	< 1	2 - 10	15 - 30
Total Phosphorus	< 1	< 1 - 5	4 - 6
Total Nitrogen	1 - 8	10 - 20	5 - 15
Ammonia Nitrogen	< 0.5	< 0.5	> 4

Land-area requirements for natural wastewater treatment systems can be quite large relative to more conventional systems, and thus these systems are generally not feasible for use in urban areas. Under an integrated coastal management plan, natural wastewater treatment systems could be used as a complement to conventional in regions where the required land area is available. Table D.6 presents a range of the land-area requirements for all of the systems described here.

Solar Aquatic Systems. Aquaculture is the growth of fish and other aquatic organisms for food production. Wastewater has been applied to such systems around the world where the production of biomass remains the chief objective. In the United States, a new aquaculture/wastewater concept modification has been developed. Solar aquatic systems combine alternative energy use with the selective use of aquatic plants and animals for the purpose of obtaining a clean wastewater effluent. To date, five demonstration facilities have been built in New England, and several other facilities have been designed or built in other parts of the United States and in Swe-

TABLE D.6 Land-Area Requirements for Land-Based Wastewater Systems (After Primary Treatment) (Source: Metcalf and Eddy 1991. Reprinted, by permission, from McGraw-Hill, Inc., 1991.)

	Area Required (Acres/MGD)
Slow-Rate	50 - 550
Rapid Infiltration	5 - 60
Overland Flow	6 - 50
Natural or Constructed Wetlands	20 - 60
Floating Aquatic Plants	20 - 60
Conventional Activated Sludge	0.4

den. The technology is currently applied only to small flow facilities of less than 1 MGD because of cost and land requirement considerations.

The solar aquatics design makes use of a series of aerated tanks to accept the pretreated and flow-equalized effluent. After the first bank of tanks, the flow is passed to a constructed wetland, then on to another series of aerated tanks. As wastewater flow progresses through the system, phytoplankton, bacteria, and stress-tolerant plants give way to zooplankton and finally fish and other higher life forms. The animals, including the fish, snails, clams, and crayfish feed on sludge after microbial treatment. Water polishing and final nutrient removal is completed by flowers, shrubs, and trees. Retention times vary from 4 days for sewage up to 11 days for septage lagoon supernatant, each at constant flow. These times do not include required pretreatment systems.

Most of the treatment achieved in these systems is attributed to bacteria attached to floating aquatic plants, with little evidence that fish contribute directly to treatment. The health risks associated with the use of aquatic organisms requires further investigation. Controlled pilot-scale studies have reported removal efficiencies for TSS and BOD_5 of over 98 percent and nitrogen from 85 to 95 percent, with 14 of 15 EPA volatile organic compounds found in the influent but none in the effluent and effluent fecal coliforms of less than 1,000 per 100 milliliters (Guterstam and Todd 1990, Teal and Peterson 1990).

Water Reclamation and Reuse as an Alternative to Wastewater Discharge

Wastewater reclamation and reuse has become increasingly important in water resources planning as the demand for water increases due to increased population, drought conditions, and decreasing supplies of high quality surface water and ground water. The use of reclaimed water allows municipalities to meet specific water needs while increasing their long-term water supply reliability. Coastal states, as shown in Figures D.3a and D.3b, are among the largest wastewater producers and also have high water demands. Therefore, significant water reuse projects have been implemented in water-short areas such as California, Florida, and Texas. Currently California recycles over 400,000 acre feet of water per year. This volume is expected to increase by 50 percent by the year 2010 (see Table D.7) (Arora 1992).

Wastewater reuse provides the option of putting constraints on discharge to surface waters and the marine environment. Recognition of environmentally sensitive conditions in coastal and estuarine waters, such as those of western Florida and northern California, has resulted in strict discharge limits or prohibition of wastewater discharge. For example, waste-

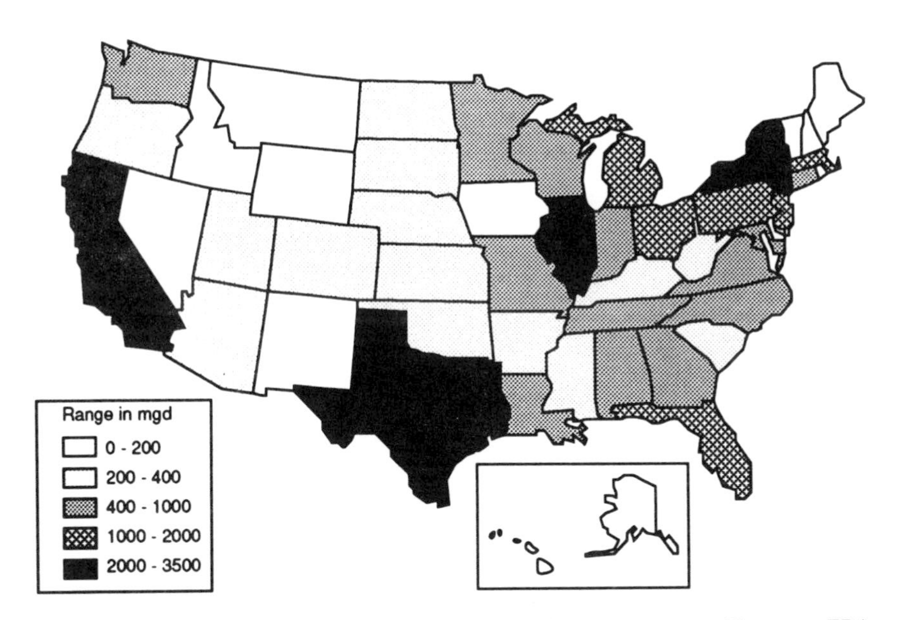

FIGURE D.3a Total treated wastewater design flows by state. (Source: EPA 1992b)

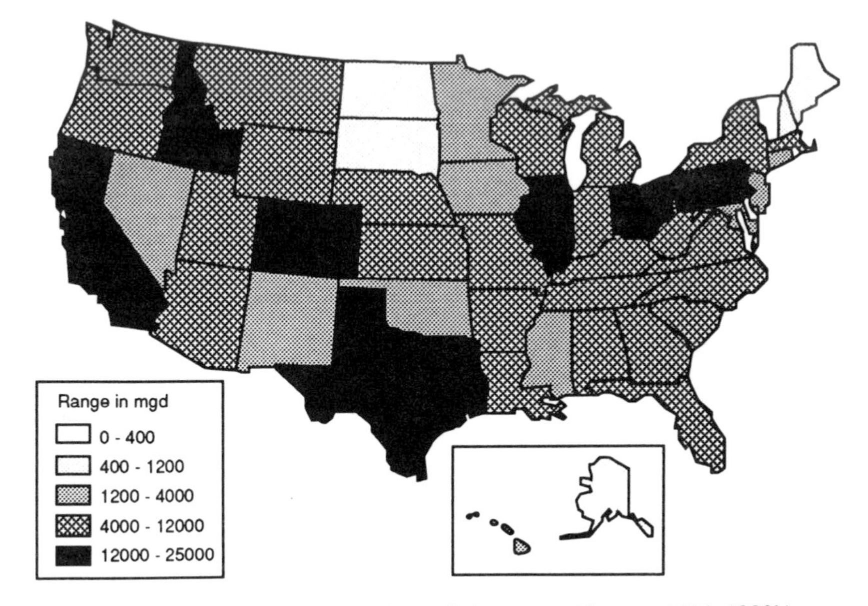

FIGURE D.3b Total fresh water demands by state. (Source: EPA 1992b)

TABLE D.7 Present and Projected Annual Use of Reclaimed Wastewater in California in 1,000s of Acre-Feet (Source: Arora 1992. Reprinted, by permission, from Water Environment Federation, 1992.)

Hydrologic Area	Year 1980	1990	2000	2010	Increase 1980-2010
North Coast	9	10	10	10	1
San Francisco Bay	10	11	13	15	5
Central Coast	9	25	27	27	18
Los Angeles	59	101	196	267	208
Santa Ana	29	47	73	78	49
San Diego	9	43	55	55	46
Sacramento Basin	21	22	23	25	4
San Joaquin Basin[1]	23	25	29	33	10
Tulare Lake Basin[2]	67	78	86	99	32
North Lahontan	6	6	7	8	2
South Lahontan	4	13	15	15	11
Colorado River Basin	4	20[2]	33[2]	45[2]	41[2]
Total	250	401	567	677	427

[1]Does not include planned reclamation of agricultural drainage water.

[2]Includes reclaimed agricultural return flows (normally lost to the Salton Sea) for power plant cooling.

water discharge into the Russian River in northern California is prohibited during the summer months.

Increasingly stricter discharge limits have made many communities, such as Vero Beach, Florida, implement water reuse. In anticipation of these stricter discharge limits, there is an increasing need for coastal communities to include water reclamation and conservation in local and regional planning. An integrated coastal management approach could facilitate environmental and economic priority-setting to address discharge and reuse issues.

Economic Feasibility of Water Reuse. Wastewater reclamation costs, including treatment and distribution range widely from $200 per acre foot to almost $1,500 per acre foot in total cost. Thus, to understand the costs of wastewater recycling correctly, it is important to be aware of the various assumptions and factors used in developing cost estimates. Costs associated with secondary treatment of municipal wastewater are normally considered as water pollution control costs and are not included in wastewater reclamation and reuse costs (Asano 1991).

To aid in determining cost-effectiveness of water reclamation and reuse programs, the EPA has formulated "Cost Effectiveness Analysis Guidelines"

(40 CFR, Part 35). Care must be taken in accounting for factors such as environmental impacts and quality of life, which are not traditionally accounted for in cost-benefit analyses. In November 1991, the Florida Department of Environmental Regulation produced "Guidelines for Preparation of Reuse Feasibility Studies for Applicants Having Responsibility for Wastewater Management" in order to standardize economic evaluations (FDER 1991). These guidelines consider two alternatives 1) no action, and 2) the implementation of a water reuse system. The criteria set forth in the guidelines gives recycled water equal value to the potable water conserved through its use (EPA 1992b).

Cost effectiveness and economical feasibility of water recycling programs are difficult to evaluate due to the complex nature of the water supply issue. Unrealistically low potable water rates are a major obstacle in establishing the economic feasibility of water recycling projects. Increasing potable water prices will aid in making water recycling economically feasible.

Water Conservation

In the United States, water historically has been a cheap commodity, demand for it has generally been met, and there has been little incentive to curb its use. With the mounting environmental costs of developing new supplies and with shortages occurring in drought-prone and/or densely populated areas, water conservation is gaining increasing credibility. In Massachusetts, obstacles to the development of new supplies has led to rigorous leak detection programs, public education campaigns to encourage voluntary conservation, domestic device retrofitting efforts, and cooperative programs with industry. These combined efforts have led to the reduction of upstream water use at the Massachusetts Water Resources Authority by 17 percent over 5 years (M. Conner, Harbor Studies, Massachusetts Water Resources Authority, personal communication, 1992). In California, the drought of 1986-1992, the second longest of the century, has taxed an already burdened water-delivery system and led to drought survival efforts that include mandatory rationing, conservation programs for domestic and industrial water use, water waste restrictions, and excess water-use penalties. These efforts have resulted in decreases in wastewater flows of between 10 percent and 40 percent at various municipal wastewater treatment plants (Bruskin and Lindstrom 1992). Reduced wastewater flows can lower operating costs at existing facilities and lower costs, postpone, and reduce the capacity of future wastewater treatment facilities where inflow and filtration are minimized. In coastal cities where infiltration and inflow are not controlled, water conservation alone will not reduce flows enough to cause a reduction in facility needs.

Advances that Offer Flexibility and/or Special Capabilities

Membrane Filtration. Membranes made from cellulose acetate, polyamides, or combinations of polymers separate suspended, colloidal or dissolved particles from wastewater when the liquid flows through the membrane. The smaller the membrane pore, the smaller the rejected species and the more costly the membrane technology. Membrane processes can be used for removing fine particles, turbidity, trihalomethane (THM) precursors, specific organics or for disinfection. The five membrane processes with the greatest potential for wastewater applications are reverse osmosis, nanofiltration, ultrafiltration, microfiltration, and electrodialysis.

Magnetite Process. The magnetite process begins with a rapid-mixing step in which metal salts and/or a polymer are added to the wastewater to flocculate the solids. The floc is then seeded with magnetite (Fe_3O_4), a highly magnetic material. The mixture is passed over a magnet, on which the particles collect. The magnetite particles can be reused repeatedly by regeneration with alkali, which strips off the adsorbed contaminants and reactivates the surface. This concept has been studied on a limited scale in the United States under EPA support, and, without chemical addition, is currently being tested in Australia (D. Dallis, Sala, Inc., Wellesley, Massachusetts, personal communication 1990; WBNSWG 1991).

Natural Chemical Coagulants. Research is being conducted on the use of a variety of new natural substances such as seeds, carrageenan (a seaweed product from algae), vegetable gums, chitosan (shell extract), ashes, starches, bark resins, and other biodegradable, renewable, and/or non-petrochemical-based substances capable of removing constituents from wastewater, either selectively or comprehensively. Seeds from two *moringa* species native to the sub-Himalayan region of India have out-performed metal salts in municipal wastewater and water treatment tests (Folkard 1986). Chitosan has comparable performance characteristics to metal salts at dosages 5 to 10 times less than those typically used in chemically enhanced treatment (Murcott and Harleman 1992b). Insoluble starch xanthate, developed by the U.S. Department of Agriculture, adsorbs heavy metals from wastewater and has been found to be especially effective in removing aluminum, zinc, chromium, cadmium, mercury, nickel, copper, tin, silver, and gold (Hauck and Masoomian 1990). Recycled sludge from drinking water treatment processes has also been used as a wastewater coagulant.

Institutional Barriers to Innovation

The EPA Innovative and Alternative Technology Program, phased out in 1990, has been the primary federal effort to promote innovation in wastewater treatment technology. The program was most successful in small communities but had little impact on large municipalities in coastal urban areas. Any future federal program to promote innovation should address existing barriers to steady progress in improving wastewater treatment technologies. Some of these barriers are

1. Reluctance of design engineers to try new technologies when there is no incentive to do so;
2. Reluctance of decisionmakers in local municipalities to spend public funds for technologies involving a substantial degree of technical risk;
3. Reluctance of state agencies to approve the first installation of a new technology in their state;
4. Delayed implementation: There is an unreasonable delay between the demonstration of new technology and its recognition in the regulatory process. This may be due to inertia in the process or an inability to adopt new technology or science in an incremental manner;
5. Potential performance problems of new technologies experienced after implementation;
6. Reputations of engineers and of state and local water officials may be damaged if a project fails;
7. Conservative state design standards: Most states continue to be governed by the conservative "Ten State Standards," which may preclude optimal design;
8. Lack of performance specifications which would permit more competition in design choices; and
9. Many engineering firms are simply not familiar with new technologies.

Municipal Wastewater Treatment Systems

A large number of technically feasible wastewater treatment technologies are currently available. Ten representative systems, arranged roughly from the simplest to the most complex, have been selected to demonstrate the wide range of treatment capabilities and costs. The ten systems are all proven technologies in full-scale operation in the United States. The data presented below for these ten systems can be used to make comparative judgments regarding performance, to estimate the approximate costs of meeting various effluent discharge standards, and to compare the costs of point and nonpoint source treatment options.

The ten wastewater treatment systems are as follows:

1. Primary (PRI)
2. Chemically enhanced primary (CEPT)

a. low-dose chemically-enhanced primary (CEPT)
b. high-dose chemically-enhanced primary (HD-CEPT)

3. Conventional primary + biological treatment (BIO)
4. Chemically-enhanced primary + biological treatment (CEPT-BIO)
5. Primary or chemically enhanced primary + nutrient removal (NUTR)
6. System 5 + gravity filtration (NUTR-FILT)
7. System 5 + high lime + filtration (NUTR-HI-FILT)
8. System 5 + granular activated carbon + filtration (NUTR-FILT-GAC)
9. System 5 + high lime + filtration + granular activated carbon (NUTR-HI-FILT-GAC)
10. System 9 + reverse osmosis (NUTR-HI-FILT-GAC-RO).

Common elements for all treatment systems presented in this report include influent pumping, preliminary treatment (bar screens and grit removal), effluent disinfection (chlorination and dechlorination), effluent pumping, and a sludge processing system consisting of dissolved air flotation of biological sludges, and anaerobic digestion of combined primary and thickened biological sludges, followed by belt press dewatering.

Description of Ten Wastewater Treatment Systems

1. Primary Treatment

Primary treatment is a physical process that involves gravity separation of settleable and floatable solids from the influent wastewater stream. Removal of settleable solids takes out some associated pollutants, including organic matter, nutrients, heavy metals, toxic organics, and pathogens. Other physical separation processes, such as fine screens and filters can be included in this treatment step.

2. Chemically-Enhanced Primary Treatment

Chemically-enhanced primary treatment is a modification of the primary clarification process through the use of chemical coagulants, typically metal salts and/or organic polyelectrolytes. By varying the chemical dose, the performance of chemically-enhanced clarification systems can be adjusted to increase the removal of suspended solids, BOD, and/or total phosphorus. Chemically-enhanced primary treatment facilities can be divided into two categories:

a. Low-dose chemically-enhanced primary treatment is used mainly for increasing the removal of suspended solids, BOD, metals, and toxic organics. A low-dose chemically-enhanced primary treatment plant is defined as the addition of a metal salt or other primary coagulant in concentrations between 5 mg/l and 100 mg/l, with or without the application of a polymer, prior to primary clarification.

b. High-dose chemically-enhanced primary treatment is used mainly to increase the removal of suspended solids, BOD, metals, and toxic organics in addition to the removal of phosphorus. The added metal salts react with soluble ortho-phosphate in the influent wastewater, producing a precipitate that is removed in the waste sludge. High-dose chemically-enhanced treatment is defined as the addition of a metal salt or other primary coagulant in concentrations greater than 100 mg/l, with or without the application of a polymer, prior to primary clarification.

Chemically-enhanced primary treatment provides opportunities for size reduction of follow-on treatment; if iron salts are used, the control of hydrogen sulfide, a major cause of odor problems; and in some cases, the potential for production of extra methane as a fuel source.

3. Conventional Primary + Biological Treatment

Conventional biological treatment systems, often classified as either suspended (e.g., activated sludge) or attached growth systems (e.g., trickling filters), use a diverse culture of microorganisms to break down organic matter in the wastewater, oxidizing a portion and converting the remainder into biological solids. Organic contaminants are removed by biodegradation and volatilization. Nondegradable suspended contaminants are removed by physical entrapment and subsequent removal with the generated biomass. Some soluble constituents (i.e., heavy metals) are removed by adsorption on the biomass. Some nutrient removal occurs through incorporation into the generated biomass. Biological treatment systems convert some influent organic nitrogen and urea to ammonia, thereby increasing the ammonia concentration making it more biologically available upon effluent discharge. Some biological systems are operated to convert ammonia to nitrate. Gas-liquid mass transfer is required to supply oxygen to the biological process, and this often results in stripping and gaseous discharge of volatile compounds. With disinfection, this effluent is sometimes used for irrigation of agricultural lands. In fact, effluent from all of the following treatment trains can be reclaimed for certain uses.

4. Chemically-Enhanced Primary + Biological Treatment

Chemically-enhanced primary + biological treatment involves the use of metal salts and polymers with either a conventional or innovative biological treatment system. Chemically-enhanced primary treatment has three major effects on a biological system—enhanced removal of phosphorus (HD-CEPT), potential improved BOD removal by the biological system due to enhanced efficiency of the primary treatment stage, and increased sludge quantity. Raw sludge production is increased in the chemically-enhanced primary stage and decreased in the biological stage such that the overall amount of raw sludge is greater by 10 percent to 20 percent compared with a conventional primary + biological treatment system. However, there is less than a 10 percent difference in the overall amount of digested sludge produced by the two systems due to the destruction of the volatile content in the chemically-enhanced primary stage (Chaudhary et al. 1991).

5. Nutrient Removal

Wastewater treatment systems can be configured to remove the nutrients nitrogen and/or phosphorus. Nitrogen removal is accomplished by an extension of the conventional biological system to incorporate the biochemical processes of nitrification and denitrification. Nitrification is the oxidation of ammonia and organic nitrogen to nitrate nitrogen. The process is mediated by the activity of a specialized class of autotrophic bacteria that can be grown in conventional activated sludge biological systems by extending the biological solids residence time resulting in more complete biodegradation of organic matter. Nitrogen removal is subsequently obtained by denitrification whereby the nitrate nitrogen is reduced to nitrogen gas and then released into the atmosphere.

Phosphorus removal can be accomplished by chemical or biological means. High-dose metal salts addition, as described in the section on chemically enhanced treatment, results in phosphorus removal. Alternatively, biological phosphorus removal can be accomplished through the selection of high phosphorus content microorganisms, resulting in a greater mass of phosphorus in the excess biological solids removed. Biological phosphorus removal systems are more capital cost-intensive and less operations and maintenance cost-intensive than chemical phosphorus removal systems and their efficiency can vary depending on a number of factors. Consequently, biological phosphorus removal systems typically incorporate some degree of chemical addition (usually for polishing) to ensure reliability and low phosphorus concentrations in the effluent.

6. Nutrient Removal with Gravity Filtration

This alternative includes a filtration system in addition to the nutrient removal system. This combination will remove additional quantities of TSS, along with other contaminants associated with the TSS (such as BOD_5, nitrogen, phosphorus, and heavy metals). The capability to add chemicals to the effluent filters is also provided, allowing further removal of phosphorus and other pollutants. Sludge production is increased slightly with this alternative relative to the nutrient removal system alone due to the increased removal of pollutants. Effluent filtration will increase the removal of pathogenic organisms, metals, and toxics from the treated effluent and enhance the performance of downstream disinfection processes. This system is used in some areas to produce water for use in urban irrigation.

7. Nutrient Removal with High Lime

This alternative follows a nutrient removal system with a high lime treatment system. High lime treatment involves the addition of lime (CaO) to elevate the pH of the treated effluent to over 11.0. This process provides considerable removal of phosphorus and heavy metals, removal of high molecular weight residual organics, and disinfection of the treated effluent. Solids produced as a result of lime addition are separated in a downstream clarifier and the pH of the clarifier effluent is adjusted back to neutral using either carbon dioxide or acid. Alternatively, two pH adjustment steps (referred to as two-stage recarbonation) can be used to remove excess calcium and magnesium from the treated effluent. In the first step, the pH of the clarified effluent is reduced to 9.3 to allow precipitation of calcium carbonate, which is removed in a downstream clarifier. The pH of the effluent from this second clarifier is subsequently reduced to 7.0. Because a high level of phosphorus removal will be obtained independent of influent phosphorus concentration, upstream removal of phosphorus by either biological or chemical means is not necessary.

Large quantities of chemical sludge are produced in the high lime treatment process. The chemical sludge can either be dewatered and land-filled or it can be processed to thermally regenerate lime, suitable for reuse in the same process. In either case, processing of the lime sludge is relatively difficult and expensive. For the purposes of this analysis, thermal regeneration of the lime is assumed.

8. Nutrient Removal and Granular Activated Carbon

This alternative adds the granular activated carbon (GAC) system to the NUTR-FILT system. GAC is a physical process for the removal of residual organic materials, including toxic organics, from a treated wastewater effluent. Some heavy metals may be removed as well. Treated effluent is applied to downflow packed beds containing granular activated carbon. Residual organic materials are removed as the treated effluent passes through the packed beds. GAC has a fixed capacity for removal of organics and, when this capacity is fully utilized, the beds must be removed from service and the exhausted GAC regenerated. Thermal regeneration is typically used and is assumed for this case.

9. Nutrient Removal with High Lime and Granular Activated Carbon

This alternative, the first of two potable reuse alternatives described, adds high lime and GAC to the nutrient removal system. This process with disinfection has typically been applied to wastewater that is reclaimed for indirect potable reuse.

10. Nutrient Removal with High Lime, Granular Activated Carbon, and Reverse Osmosis

This alternative, the second of two potable reuse alternatives described, involves an additional treatment step, a reverse osmosis (RO) system, after System 9. The RO process is applied typically when dissolved solids removal is required to prevent salt build-up within a recycling system. RO involves application of highly treated effluent under high pressure to a membrane. It allows the water to flow through but is not permeable to (i.e., rejects) dissolved solids. Dissolved solids are concentrated as a brine in a reject stream. RO's best application is as a complementary process with GAC to accomplish complete removals of a broad spectrum of pollutants. Operating costs for the RO process are high due to the energy costs of maintaining high pressure, limited membrane life due to fouling, and the high cost of brine disposal. Owing to its high cost, RO has been applied primarily for water reuse applications in areas where water is scarce and expensive. In many instances, only a portion of the treated effluent is processed through RO.

Matrix of Performance and Cost Summary Tables

Data for the ten systems are summarized in the Matrix of Performance and Cost Summary Tables D.8a, D.8b, D.8c, and D.8d.[2] Performance comparisons are made on the basis of conventional parameters.

Performance and Costs

The performance of each of the 10 wastewater treatment systems was assessed based on two surveys of over 100 U.S. POTWs undertaken in 1990

[2] These tables subdivide the ten systems into two categories. Systems 1-4 are those for which significant performance data exist, based on the two above-mentioned surveys. Average effluent concentration and average percentage removal values are based on these data. Systems 5-10 are ones for which fewer data exist. Average effluent concentration and average percentage removal values for these systems are based on the technical literature and professional judgment.

The assumptions used to standardize results for Systems 5-10 are as follows:

Plant size = 20 MGD
TSS influent concentration = 250 mg/l
BOD_5 influent concentration = 250 mg/l
Total phosphorus influent concentration = 8 mg/l
Total nitrogen influent concentration = 35 mg/l

TABLE D.8a Average Influent/Effluent Concentrations and Percentage Removals for Systems 1-4

	Primary (1)	Low-Dose Chemical Primary (2a)	High-Dose Chemical Primary (2b)	Biological (3)	Chemical Primary + Biological (4)
TSS (mg/l)	214/93	182/52	177/13	234/14	186/10
BOD_5 (mg/l)	202/139	168/80	146/33	203/16	174/9
TP (mg/l)	6/4	6/2	5/0.3	6/3	5/1
TN (mg/l)	30/23	30/19	30/—	28/19	16/—
NH_4 (mg/l)	14/—	15/13	—	—	—
Oil & grease (mg/l)	41/20	42/12	36/6	50/<	—
TSS (%)	55	71	92	93	93
BOD_5 (%)	30	55	78	92	95
TP (%)	38	63	93	38	87
TN (%)	15	37	40	31	31
Oil & grease (%)	51	71	82	98	—

NOTE: TSS = total suspended solids, BOD_5 = 5-day biochemical oxygen demand, TP = total phosphorus, TN = total nitrogen, NH_4 = ammonia nitrogen, — = data insufficient or unavailable.

and 1991[3] and also on the technical literature and the professional experience of members of the Committee on Wastewater Management in Coastal Urban Areas and its Panel on Source Control and Treatment Technologies.

Costs for the ten wastewater treatment systems were estimated by a professional engineering firm with confirmation provided by the two surveys. Many factors can affect the cost of a wastewater treatment system and, as a consequence, the cost of a specific wastewater treatment facility may vary significantly from the general costs presented in this section.

Costs are expressed as capital cost, operation and maintenance cost, and total cost. Capital cost is expressed in two sets of units: dollars per gallon per day of installed capacity, and dollars per mission gallons.[4] Operation and maintenance costs, and total costs are expressed in dollars per million

[3]Information on the two nationwide surveys conducted to obtain data on these ten candidate systems and the screening criteria used to select them has been published as a separate document: Performance and Innovation in Wastewater Treatment—Technical Note #36, January, 1992, by Murcott, S., and Harleman, D., Parsons Laboratory, Massachusetts Institute of Technology, Cambridge, Massachusetts.

[4]Total capital cost is calculated as follows:

$$\text{Total capital cost} = (\text{annual cost in \$/MG}) \times (20\ \text{MGD}) \times (365\ \text{days/year}) \times (\text{uniform series present worth factor})$$

Uniform series present worth factor = $[(1 + i)^n - 1]/[i(1 + i)^n]$, where i = 8% and n = 20 years.

TABLE D.8b Costs for Systems 1-4

	Primary (1)	Low-Dose Chemical Primary (2a)	High-Dose Chemical Primary (2b)	Biological (3)	Low-Dose Chemical Primary + Biological (4)
Capital Cost ($/gpd)	0.9-1.1	1.1-1.4	1.2-1.8	2.4-2.6	2.6-2.9
Capital Cost ($/MG)	245-310	320-400	400	610-720	750-870
O & M Cost ($/MG)	205-240	230-280	250-350	320-410	350-450
Total Cost ($/MG)	450-550	550-680	650-750	930-1,130	1,050-1,1 50

gallons. All costs are annualized costs. Assumptions used in computing these costs include an 8 percent interest rate for a 20 MGD facility with a design period of 20 years. Land costs are not included. For systems 5 to 10, lime recalcination and other increased sludge production were internalized into the cost so that the additional sludge shows up as an increased cost rather than as increased sludge.

TABLE D.8c Average Influent/Effluent Concentrations and Percentage Removals for Systems 5-10

	Nutrient Removal (5)	Nutrient Removal + Filtration (6)	Nutrient Removal + High Lime + Filtration (7)	Nutrient Removal + Filtration + GAC (8)	Nutrient Removal + High Lime + Filtration + GAC (9)	Nutrient Removal + High Lime + Filtration + GAC + Reverse Osmosis (10)
TSS (mg/l)	250/15	250/5	250/3	250/2	250/2	250/0
BOD_5 (mg/l)	250/15	250/5	250/3	250/3	250/2	250/1
TP (mg/l)	8/1.5	8/1	8/0.1	8/0.5	8/0.1	8/<0.1
TN (mg/l)	35/3	35/2	35/2	35/1.5	35/1.5	35/<1
NH_4 (mg/l)	—/0.5	—/0.5	—/0.4	—/0.5	—/0.4	—/<0.1
TSS (%)	94	98	99	99	99	100
BOD_5 (%)	94	98	99	99	99	100
TP (%)	81	88	99	94	99	100
TN (%)	91	94	94	96	96	97

NOTE: TSS = total suspended solids, BOD_5 = 5-day biochemical oxygen demand, TP = total phosphorus, TN = total nitrogen, NH_4 = ammonia nitrogen, — = data insufficient or unavailable.

TABLE D.8d Costs for Systems 5-10

	Nutrient Removal (5)	Nutrient Removal + Filtration (6)	Nutrient Removal + High Lime + Filtration (7)	Nutrient Removal + Filtration + GAC (8)	Nutrient Removal + High Lime + Filtration + GAC (9)	Nutrient Removal + High Lime + Filtration + GAC + Reverse Osmosis (10)
Capital Cost ($/gpd)	2.9-3.3	3.5-3.9	5.2-5.6	4.5-4. 9	6.1- 6.7	8.5-9.5
Capital Cost ($/MG)	750-870	890-1,140	1,300-1,700	1,150-1,450	1,500-1,800	7,000-2,500
O & M Cost ($/MG)	500-580	560-660	1,100-1,300	850-950	1,350-1,650	2,500-3,000
Total Cost ($/MG)	1,250-1,450	1,450-1,800	2,400-3,000	2,000-2,400	2,900-3,500	4,500-5,500

Qualitative Comparisons

Ten Systems. The data presented in the Matrix of Performance and Cost Summary Tables have been used to develop cost-performance relationships for the removal of various pollutants. Figures D.4 and D.5 plot the data, providing TSS, BOD, P, and N performance-cost relationships for all the systems. The figures show ranges both of performance and cost. The numbering of the ten systems on these figures follows the same order as given above.

Several observations are evident from an examination of these four figures:

• Unit total treatment costs increase exponentially as a higher level of treatment is provided. In general, the point at which treatment costs begin to increase quite rapidly and produce only small improvements in improved effluent concentrations occurs at a unit total treatment cost on the order of \$800 to \$1,200 per million gallons treated.

• Smooth curves are indicated for TSS and BOD_5 performance-cost plots (Figures D.4a and D.4b). However, smooth curves are not obtained when nutrient control (total nitrogen and/or total phosphorus) is taken into consideration (Figures D.5a and D.5b). Here, the cost for conventional primary + biological treatment (System 3) is significantly higher than the cost for either primary (System 1) or low-dose chemically-enhanced pri-

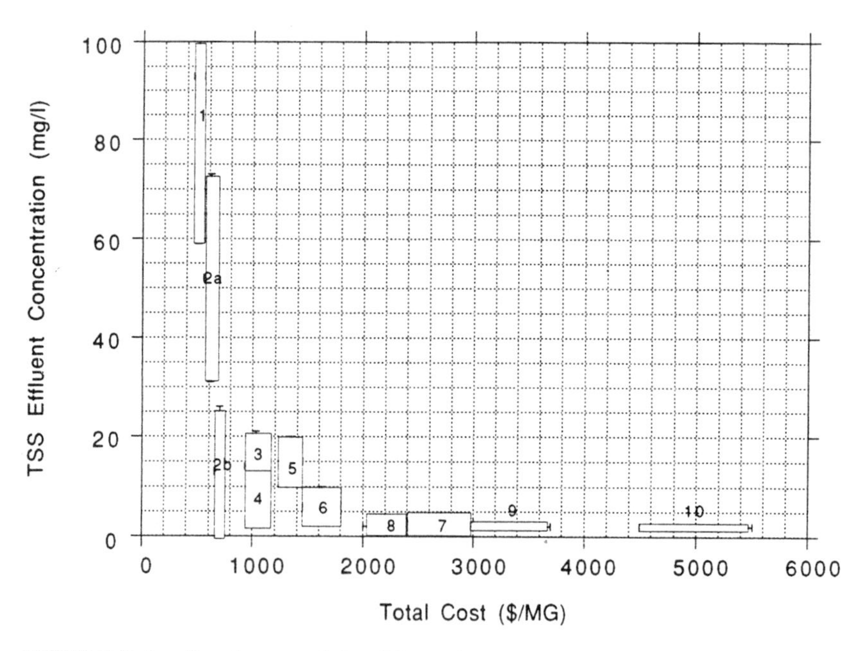

FIGURE D.4a Total suspended solids performance and cost relationship.

mary (Systems 2a and 2b), even though there is no advantage in terms of total phosphorus effluent concentration. In other words, conventional primary + biological treatment is not a particularly cost-effective strategy for phosphorus removal if phosphorus removal is the only treatment objective. The same can be said for conventional primary + biological treatment with regard to total nitrogen removal.

- The more advanced technologies such as HI-LIME, GAC, and RO significantly increase costs while producing little additional removal of TSS, BOD_5, or nutrients. These technologies generally lie on the *other side* of the cost and performance curve for these parameters. Their major purpose is for the removal of specific toxic organic compounds, heavy metals, and other unconventional pollutants, which must be controlled for specific reuse or discharge applications.

Comparison of Low-Dose Chemically-Enhanced Primary with Conventional Primary + Biological Treatment. Data were collected on 34 primary and 19 low-dose chemically-enhanced primary treatment plants through the two above-mentioned national surveys. These data are used to compare low-dose chemically-enhanced primary treatment performance with conventional primary and conventional primary plus biological treatment. In most

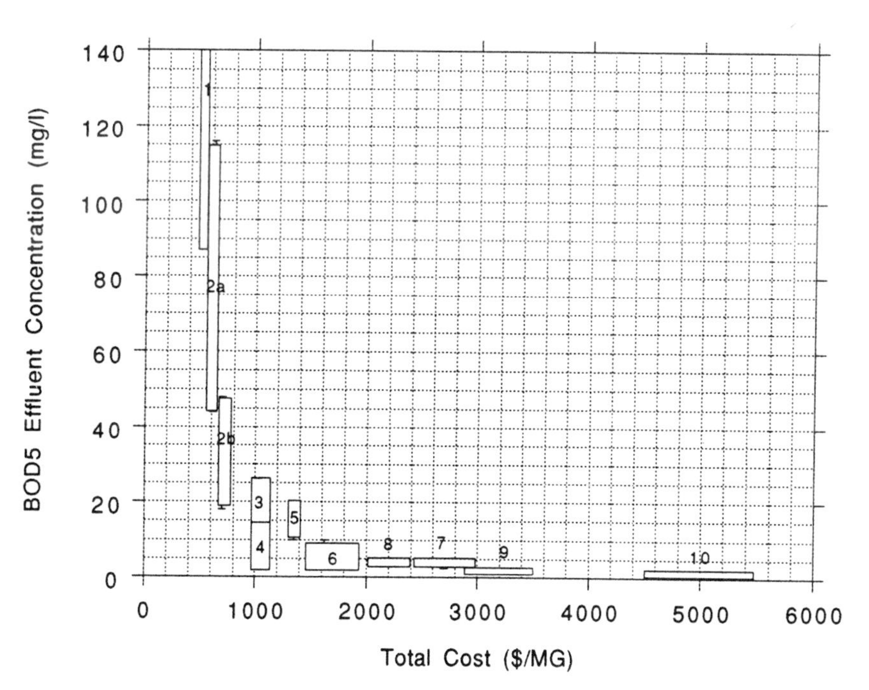

FIGURE D.4b Five-day biochemical oxygen demand (BOD_5) performance and cost relationship.

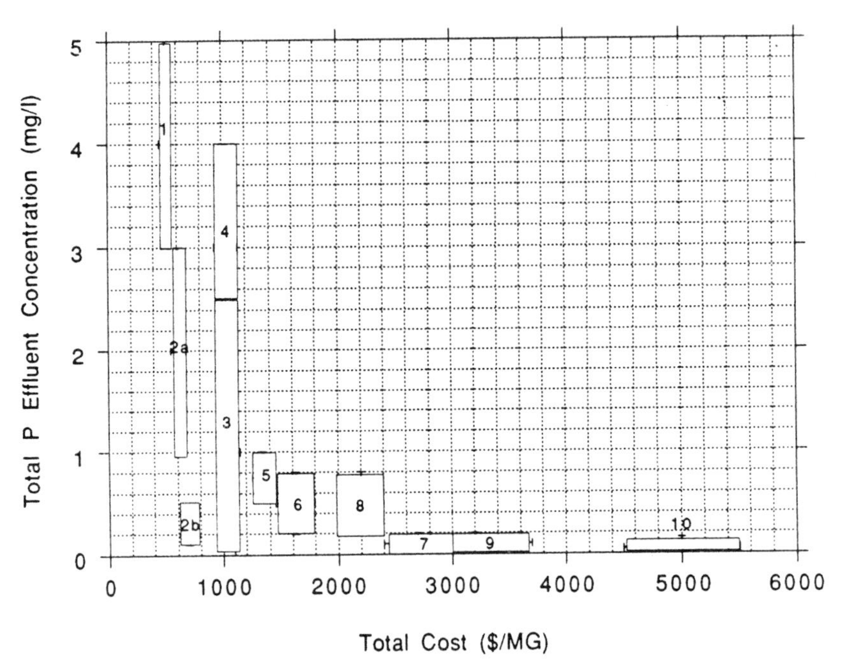

FIGURE D.5a Total phosphorus performance and cost relationship.

TABLE D.9 Comparison of Low-Dose Chemically-Enhanced Primary Treatment with Conventional Primary Treatment Plus Activated Sludge

Constituent	Low-Dose Chemically-Enhanced Primary	Conventional Primary + Biological
TSS Effluent (mg/l)	52	14
BOD_5 Effluent (mg/l)	80	16
Total Phosphorus Effluent (mg/l)	2	3
Total Nitrogen Effluent (mg/l)	19	19
Oil & Grease Effluent (mg/l)	6 - 12	0 - 1
Raw Sludge (lb solids/lb TSS removed)	1.3 - 1.5	1.8
Cadmium Effluent (μg/l)	6 - 20	6 - 20
Copper Effluent (μg/l)	130 - 220	90 - 150
Chromium Effluent (μg/l)	150 - 300	140 - 280
PAH[1] Effluent (μg/l)	1 - 6	1 - 6
Pathogens (% removed)	< 99%	< 99%
Total Energy Use (kWh/yr $\times 10^{-3}$)	290	700
Capital Cost ($/gpd)	1.1 - 1.4	2.4 - 2.6
O & M Cost ($/MG)	230 - 280	320 - 410
Total Cost ($)	550 - 680	930 - 1130

[1]PAH = polycyclic aromatic hydrocarbons.

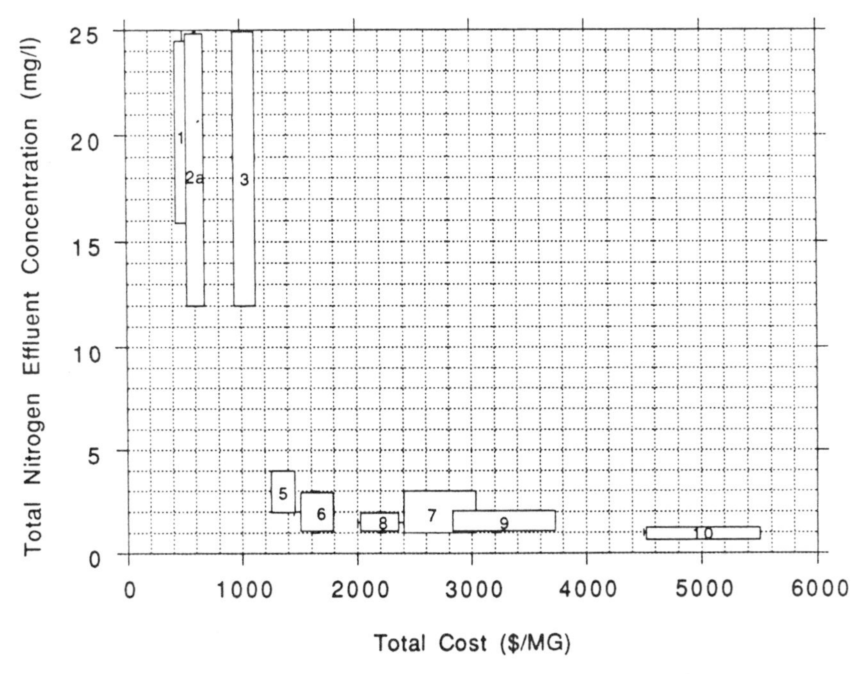

FIGURE D.5b Total nitrogen performance and cost relationship.[5]

cases, significant removals of organic materials and metals are achieved only by the most complex treatment trains.

Table D.9 summarizes performance characteristics for two systems, low-dose chemically-enhanced primary (System 2a) and conventional primary treatment + biological treatment (System 3), for the purpose of showing a side-by-side comparison of their respective efficiencies. This comparison shows that, on average, chemically-enhanced primary treatment is comparable to conventional primary + biological treatment in phosphorus, nitrogen, cadmium, and PAH removal, and there is little difference in their removals of copper and chromium. Conventional primary + biological treatment removes TSS, BOD_5, and oil and grease to lower levels. Low-dose chemically-enhanced treatment produces roughly 20 percent less sludge. Neither treatment method without add-on disinfection processes adequately removes pathogenic organisms. Conventional primary + biological treatment uses approximately 2.5 times as much total energy. Total cost of chemically-enhanced primary treatment is approximately half that of conventional primary + biological treatment. This is the average capability of these two systems based on full-scale yearly operations data in the United States in 1989 and 1990.

[5] Data were not available for Systems 2b and 4.

Comparison of Conventional Primary Treatment and Low-Dose Chemically-Enhanced Primary Treatment. Figures D.6a and D.6b show graphs of average TSS and BOD_5 effluent concentration versus overflow rate at full-scale POTWs in the United States. Overflow rate, a measure of treatment process efficiency, is the wastewater flow divided by the surface area of the treatment tank. Figure D.6a compares TSS effluent concentration versus overflow rate for primary with low-dose chemically-enhanced primary. The average primary TSS effluent concentration is 93 mg/1 with a standard deviation of 34. The chemically-enhanced primary TSS effluent concentration averages 52 mg/1 with a standard deviation of 21 mg/1.

Figures D.7a and D.7b are similar plots comparing BOD_5 effluent concentration versus overflow rate for primary and low-dose chemically-enhanced primary treatment. The average primary BOD_5 effluent concentration is 139 mg/1 with a standard deviation of 50. The average low-dose chemically-enhanced primary BOD_5 effluent concentration is 80 mg/l with a standard deviation of 36 mg/l.

These figures show the advantages of low-dose chemically-enhanced primary over conventional primary treatment. They also show no obvious decrease in removal efficiency with increasing overflow rates for plants with overflow rates up to 2,400 gpd/sf (except for primary effluent BOD_5). These data suggest that design overflow rates may be unnecessarily conservative for primary and low-dose chemically-enhanced primary treatment, but more controlled studies are necessary to verify this suggestion.

Seven-day average data on the performance of low-dose chemically-enhanced primary treatment come from two plants: Point Loma Treatment Plant, San Diego, California and Hyperion Wastewater Treatment Plant, Los Angeles, California. These two southern California low-dose chemically-enhanced primary facilities remove approximately 80 percent TSS and 55 percent BOD_5 at overflow rates between 1,600 and 2,100 gpd/sf. Comparing the performance at these well-operated plants with the average performance treatment as shown in the Matrix of Performance and Cost Summary Tables, a 48 percent increase in TSS performance and an 77 percent increase in BOD_5 performance over conventional primary treatment is observed.

Toxic Organics and Metals

The ten representative wastewater treatment systems provide different capabilities in terms of their ability to remove toxic organics and heavy metals. Table D.10 lists a number of toxic organics and heavy metals and

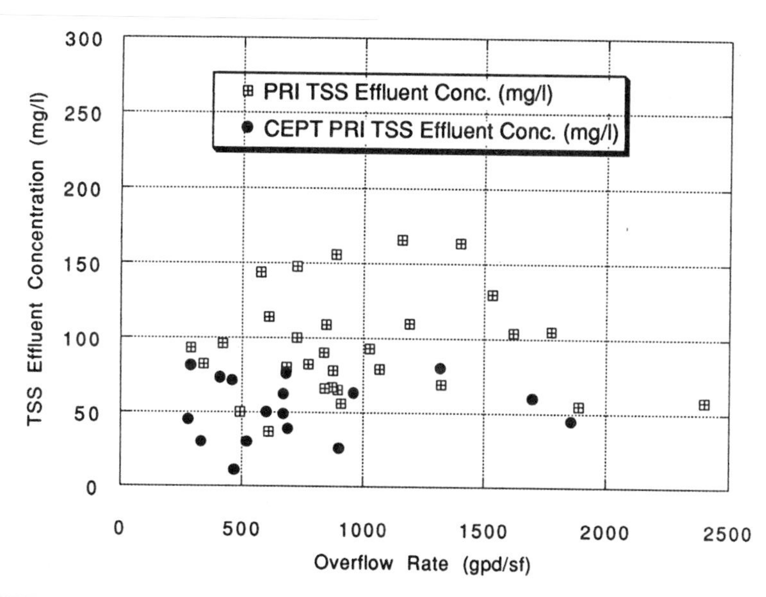

FIGURE D.6a TSS removal efficiency for average primary and chemically-enhanced primary treatment.

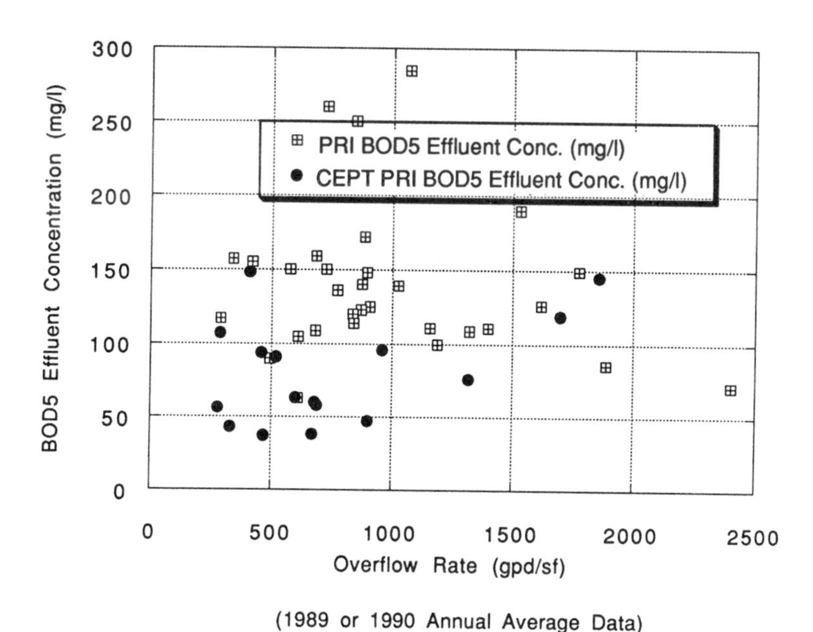

FIGURE D.6b BOD_5 removal efficiency for average primary and chemically-enhanced primary treatment.

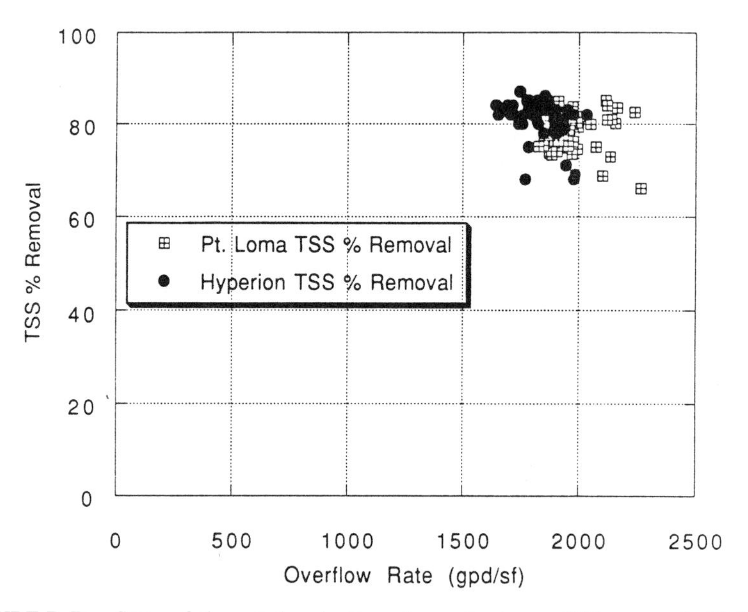

FIGURE D.7a State-of-the-art chemically-enhanced primary TSS removal efficiency at San Diego (Point Loma) and Los Angeles (Hyperion) POTWs.

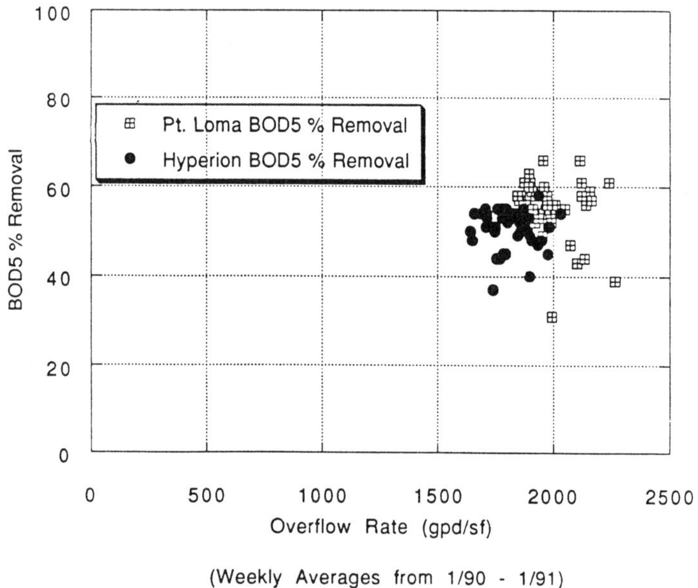

FIGURE D.7b State-of-the-art chemically-enhanced primary BOD_5 removal efficiency at San Diego (Point Loma) and Los Angeles (Hyperion) POTWs.

shows a range of expected effluent concentrations attainable by each of the ten treatment systems.[6]

Fats, Oil, and Grease

The wastewater parameter known as *fats, oil, and grease* (FOG) refers to a number of compounds. The term grease generally includes fats, oils, and waxes, while fats and oils are compounds (esters) of alcohol or glycerol (glycerin) with fatty acids. In domestic wastewater, fats and oils typically come from butter, margarine, lard, and cooking oil. However, oils are not only of animal or vegetable origin but also come from petroleum products. FOG can cause problems in wastewater plant operations, both in the sewers and in the plant itself. It can contribute to the loss of hydraulic capacity, clogging of screens, poor grit separation, poor settling in sedimentation tanks, and interference with biological processes. It also causes aesthetic and toxicity problems if discharged into receiving waters (Metcalf and Eddy 1991).

The parameter measured in U.S. wastewater treatment plants is termed *oil and grease*. Data on oil and grease removal efficiency for Systems 1-3, based on the two above-mentioned POTW surveys is presented in Table D.11. The table shows an increasing improvement in removal of FOG with the increasing level of treatment. This trend is consistent with most literature reports.

---

[6] Procedure for determining the treatment performance levels:

1. Begin with extensive data base for activated sludge (AS) effluent to determine the range for typical AS effluents (column 3).
2. Review metals data base (Esmund et al. 1980) and toxic organics data base (McCarty et al. 1980) to determine typical influent concentration levels and AS treatment efficiencies. Using a consistent AS treatment efficiency applied to column 3, determine range for typical influent concentration.
3. For column 1 and 2, apply treatment efficiency factors from Hannah et al. (1986). Infer from filtration data from Esmund et al. (1980) when not available in Hannah et al. (1986).
4. For columns 4 and 5, apply a chemical treatment efficiency factor: Esmund et al. (1980) for metals, McCarty et al. (1980) for toxic organics, to column 3 and 4.
5. For column 6, apply a filtration efficiency factor: Esmund et al. (1980) for metals, McCarty et al. (1980) for toxic organics, to column 5.
6. For column 7, apply a high lime and filtration factor: Esmund et al. (1980) for metals, McCarty et al. (1980) for toxic organics, to column 4.
7. For column 8, apply a GAC efficiency factor: Esmund et al. (1980) and McCarty et al. (1980), to column 7.
8. For column 9, apply a GAC factor to column 8.
9. Column 10 was obtained from McCarty et al. (1980). Inferences were made when needed.

TABLE D.10 Typical Effluent Concentrations of Organics and Metals for Selected Treatment Trains

	Treatment Train Effluent Concentration (micrograms/liter)				
Constituent	Influent	1	2	3	4
Chloroform	7-60	7-60	5.6-48	1.0-9.0	1.0-9.0
Bromodichloromethane	0.3-1.7	0.3-1.7	0.3-1.7	0.1-.05	0.1-0.5
Dibromochloromethane	1.0-6.0	1.0-6.0	1.0-6.0	0.1-0.7	0.1-0.7
Bromoform	0.3-1.2	0.2-1.0	0.2-1.0	0.1-0.4	0.1-0.4
Carbon Tetrachloride	1.0-8.0	1.0-8.0	1.0-8.0	0.2-2.0	0.2-2.0
1,2-Dichloroethane	5.0-15.0	5.0-15.0	3.9-11.7	0.8-2.4	0.8-2.4
1,1,1-Trichloroethane	7.5-12.5	7.5-12.5	7.5-12.5	3.0-5.0	3.0-5.0
Tetrachloroethylene	1.0-4.0	1.0-4.0	1.0-4.0	0.5-2.0	0.5-2.0
Trichlorothylene	1.0-2.0	1.0-2.0	1.0-2.0	0.5-1.0	0.5-1.0
Xylene	0.06-0.2	0.06-0.2	0.06-0.2	0.03-0.1	0.03-0.1
Chlorobenzene	1.0-25.0	0.8-20.0	0.7-18.0	0.1-2.5	0.1-2.5
1,2-Dichlorobenzene	1.0-8.0	0.8-6.4	0.7-5.6	0.1-0.8	0.1-0.8
1,3-Dichlorobenzene	1.0-8.0	0.8-6.4	0.7-5.6	0.1-0.8	0.1-0.8
1,4-Dichlorobenzene	15.0-25.0	12.0-20.0	10.0-17.5	1.5-2.5	1.5-2.5
1,2,4-Trichlorobenzene	1.0-5.0	0.8-4.0	0.7-3.5	0.1-05	0.1-0.5
Ethylbenzene	0.4-15.0	0.3-13.0	0.3-9.0	0.04-1.5	0.04-1.5
Naphthalene	1.0-20.0	0.2-17.4	0.2-15.4	0.03-0.6	0.03-0.6
1-Methylnaphthalene	0.33-30.0	0.29-26.1	0.25-23.1	0.01-0.9	0.01-0.9
2-Methylnaphthalene	033-30.0	0.29-26.1	0.25-23.1	0.01-0.9	0.01-0.9
Dimethylphthalate	33-106	21-67	5.0-16.0	5.0-16.0	3.2-10.4
Diisobutylphthalate	20-33	12-21	3.0-5.0	3.0-5.0	1.9-3.2
Bis-[2-ethylhexyl] phthalate	66-200	41-126	10.0-30.0	10.0-30.0	6.5-19.5
PCBs	5.0-33	3.1-20.7	0.55-3.6	0.5-3.3	0.3-2.6
Arsenic	9-22	9-22	9-22	8-20	5.6-14.0
Barium	120-160	120-160	120-160	60-80	60-80
Boron	300-500	300-500	300-500	300-500	300-500
Cadmium	6.6-22.2	5.8-19.5	5.8-19.5	3.0-10.0	2.2-7.3
Chromium	160-320	149-297	137-275	40-80	12-24
Copper	167-267	134-214	94-150	50-30	31-50
Iron	600-1600	600-1600	300-800	300-800	150-400
Lead	100-150	70-105	50-80	40-60	32-48
Manganese	41-81	37-73	33-65	30-60	21-42
Mercury	0.25-2.5	0.2-2.0	0.2-2.0	0.1-1.0	0.08-0.8
Nickel	93-147	88-140	79-126	70-110	60-95
Selenium	4.2-15.0	3.8-13.5	3.8-13.5	1.0-3.5	0.9-3.1
Silver	0.4-6.7	0.4-6.7	0.4-6.7	0.2-3.0	0.2-3.0
Zinc	250-400	225-360	225-360	100-160	70-112

NOTE: Influent values attempt to be representative of concentrations entering POTWs. However, values can be quite variable depending on the nature of the service area.
MDL=minimum detection level

5	6	7	8	9	10
1.0-9.0	1.0-9.0	1.0-9.0	1.0-9.0	1.0-9.0	0.1-1.0
0.1-0.5	0.1-0.5	0.1-0.5	0.04-0.2	0.04-0.2	0.02-0.1
0.1-0.7	0.1-0.7	0.1-0.7	0.03-0.2	0.03-0.2	0.01-0.08
0.1-0.4	0.1-0.4	0.1-0.4	0.02-0.08	0.02-0.08	0.01-0.03
0.2-2.0	0.2-2.0	0.2-2.0	0.1-1.6	0.1-1.6	0.01-0.16
0.8-2.4	0.8-2.4	0.8-2.4	0.2-0.6	0.2-0.6	0.02-0.06
3.0-5.0	3.0-5.0	3.0-5.0	0.1-1.2	0.1-1.2	0.01-0.1
0.5-20	0.5-2.0	0.5-2.0	0.05-0.2	0.05-0.2	0.05-0.2
0.5-1.0	0.5-1.0	0.5-1.0	0.35-0.7	0.35-0.7	0.35-0.7
0.03-0.1	0.03-0.1	0.03-0.1	0.01-0.03	0.01-0.03	0.01-0.03
0.1-2.5	0.1-2.5	0.1-2.5	0.01-0.02	0.01-0.02	0.01-0.02
0.1-0.8	0.1-0.8	0.07-0.6	0.03-0.3	0.03-03	0.02-0.2
0.1-0.8	0.1-0.8	0.05-0.4	0.05-0.4	0.02-0.2	0.01-0.1
1.5-2.5	1.5-2.5	0.9-1.5	0.4-0.7	0.4-0.7	0.3-0.6
0.1-0.5	0.1-0.5	0.03-0.15	0.01-0.05	0.01-0.05	0.01-0.05
0.04-1.5	0.04-1.5	0.04-1.5	0.03-1.1	0.03-1.1	0.03-1.1
0.03-0.6	0.03-0.6	0.02-0.5	0.01-0.02	0.01-0.02	0.01-0.02
0.01-0.9	0.01-0.9	0.01-0.9	0.01-0.9	0.01-0.9	0.004-0.36
0.01-0.9	0.01-0.9	0.01-0.9	0.01-0.9	0.01-0.9	0.004-0.36
3.2-10.4	3.2-10.4	3.2-10.4	1.1-3.7	1.1-3.7	0.46-1.5
1.9-3.2	1.9-3.2	1.9-3.2	0.24-0.41	0.24-0.41	0.17-0.29
6.5-19.5	6.5-19.5	6.5-19.5	5.9-17.7	5.9-17.7	2.2-6.5
0.3-2.6	0.3-2.6	0.3-2.6	0.1-0.3	0.1-0.3	0.1-03
5.6-14.0	5.0-12.6	1.4-3.6	5.0-12.6	1.4-3.6	<MDL
60-80	60-80	60-80	60-80	60-80	2.0-5.0
300-500	300-500	300-500	300-500	300-500	100-300
2.2-73	2.2-7.3	1.4-4.7	2.1-6.9	1.3-4.5	0.7-2.0
12-24	9-18	8-16	5.4-10.8	4.8-9.6	0.2-2.0
31-50	31-50	15-24	15-25	7.0-12.0	1.0-10.0
150-400	120-320	30-80	84-224	21-56	20-30
32-48	27-41	18-27	16-25	11-16	1.0-3.0
21-42	17-34	5.6-11.2	13.6-27.2	5.0-10.0	1.0-4.0
0.08-0.8	0.08-0.8	0.07-0.7	0.06-0.6	0.05-0.5	<MDL
60-95	60-95	49-77	50-79	41-64	4.0-10.0
0.9-3.1	0.7-2.6	0.6-2.1	0.35-1.3	0.3-1.1	<MDL
0.2-3.0	0.2-3.0	0.12-1.8	0.2-3.0	0.12-1.8	0.1-1.2
70-112	70-112	40-64	45-73	34-54	5.0-30.0

TABLE D.11 Oil and Grease Influent and Effluent Concentrations and Percent Removal for Systems 1-3 (Source: Murcott and Harleman 1992a)

	Primary (1)	Low-Dose Chemical Primary (2a)	High-Dose Chemical Primary (2b)	Biological (3)
Oil & grease inf/eff (mg/l)	41/20	42/12	36/6	50/1
Oil & grease (% removal)	51	71	82	98

Sludge

Sludge Quantity

The quantity of sludge produced depends mainly on the amount of TSS and BOD_5 removed. As performance efficiency in terms of their removal increases, sludge quantity generally increases. This relationship is reflected in Table D.12 which gives sludge production values for Systems 1 to 10.

Sludge Treatment Costs

The cost of sludge treatment is calculated as the cost following sludge stabilization and dewatering.[7] In this analysis, four alternative sludge management options were considered: 1) land disposal in a dedicated landfill or a refuse landfill, 2) composting with the give-away of the finished product, 3) incineration with the disposal of ash to landfill, and 4) direct land application. In all cases, it is assumed that sludges of 20 percent and 30 percent solids are produced by dewatering the sludge from the treatment plant, thus giving a band of costs. The incineration and composting operations are assumed to be at the plant site, while the ultimate land-disposal facility is assumed to be 100 miles from the treatment plant. This distance is not excessive for large cities. Transport costs are considered for all alternatives. In addition, anaerobic digestion costs were not factored into the

[7] The costs are mainly derived from the Handbook for Estimating Sludge Management Costs (EPA 1985a), which contains cost curves for the various sludge treatment and disposal processes, all in 1983 dollars. Conversion from 1983 dollars to 1991 dollars is by using the Engineering News Record Construction Costs Index. This index was 4,006 in 1983 and is presently at 4,818. Therefore all of the 1983 costs are converted to 1991 dollars by multiplying by 1.2. Where the cost curves did not cover the full range of sludge production for larger plants, the curves were extrapolated. The costs thus derived were checked against the results of the National Sewage Sludge Survey (EPA 1989a) and found to be in agreement.

TABLE D.12 Sludge Production in Systems 1 to 10[1] (Source: Murcott and Harleman 1992a)

Sludge Production for Systems 1-4					
Primary (1)	Low-Dose Chemical Primary[1] (2a)	High-Dose Chemical Primary[1] (2b)	Biological (3)	Chemical Primary + Biological (4)	
1 lb solids/lb TSS removed	1.3 - 1.5 lb solids/lb TSS removed	1.3 - 2 lb solids/lb TSS removed	1 lb solids/lb BOD_5 removed (bio sldg only) or 1.8 lb solids/lb TSS rem (conv pri+bio sldg)	1.2 lb solids/lb BOD removed (bio sldg only) or 2.2 lb solids/lb TSS rem (chem pri+bio sldg)	
Sludge for Systems 5-10 (expressed as lb/million gallons)					
Nutrient Removal (5)	Nutrient Removal + Filtration (6)	Nutrient Removal + High Lime + Filtration (7)	Nutrient Removal + Filtration + GAC (8)	Nutrient Removal + High Lime + Filtration + GAC (9)	Nutrient Removal + High Lime + Filtration + GAC + Reverse Os. (10)
2500	2750	2750	2750	2750	2750

[1]All chemical primary sludge data in this table has been computed according to the formulas described in Murcott and Harleman (1992a).

incineration alternative. Further assumptions are listed under each alternative.

All of the sludge management options listed include final disposal. In the case of landfilling, the sludge is deposited in a dedicated landfill and covered with soil or deposited in a municipal solid waste landfill. In the latter case, the sludge assists in the anaerobic decomposition of the solid waste and may be considered a beneficial addition to the landfill. Composting is an aerobic decomposition process, usually carried out in either open piles or windrows or within closed vessels. In the case of sludge composting, a bulking agent is required to make it possible for oxygen to penetrate the compost piles. Typical bulking agents are sawdust, leaves, bark, or shredded paper. The product of sludge composting is typically an excellent soil additive or conditioner, and a useful fertilizer. Incineration may be considered the ultimate oxidation process, resulting in a mostly inorganic ash, which also must be disposed of, usually in a dedicated landfill. Finally, direct land application of sludge involves the addition of sludge to either agricultural or nonagricultural land such that the sludge is assimilated into the soil.

Land Disposal

If sludge at either 20 percent or 30 percent solids is transported to a dedicated landfill, the cost includes the land cost. This is assumed to be not very different from the disposal of sludge in a refuse landfill, where the cost of sludge disposal would be prorated. Liquid sludge disposal is not considered because of the high cost of truck haul for large treatment plants and the inability of most to pass the paint filter test required for such sludges to be permitted at a landfill.

The transport and land disposal calculated for this alternative would be the minimum sludge-handling cost for any treatment plant. The cost is calculated as the sum of the land disposal and transportation costs, as shown in Table D.13.

Composting

It is assumed that composting is by the static aerated pile method with wood chips or similar materials used as the bulking agent. No credit is given for the sale of the compost, which is often the case. Exceptions usually yield less than $10 per ton. The published curves did not extend to the required plant size and were extrapolated. Costs are calculated for both 20 percent and 30 percent feed sludge. Transportation figures are for dewatered sludge, assuming a 60 percent solids concentration of the finished compost. The cost results are shown in Table D.14.

TABLE D.13 Cost of Land Disposal

Dry Tons of Sludge Disposed Per Day	Annual Cost of Land Disposal ($ million)	Annual Cost of Transport ($ million)	Total Annual Cost ($ million)
	Dewatered sludge of 20 percent solids		
20	0.7	1.8	2.5
60	1.5	4.0	6.4
120	3.0	9.1	12.1
180	4.2	13.3	17.5
	Dewatered sludge of 30 percent solids		
30	0.7	1.8	2.5
90	1.5	4.0	6.4
180	3.0	9.1	12.1
270	4.2	13.3	17.5

TABLE D.14 Cost of Composting

Dry Tons of Sludge Composted & Disposed of Per Day	Annual Cost of Composting ($ million)	Annual Cost of Transport ($ million)	Total Annual Cost ($ million)
	Dewatered sludge feed of 20 percent solids		
3	1.2	0.1	1.3
9	2.7	0.3	3.0
18	4.3	0.6	4.9
30	6.5	0.9	7.4
60	20.1	1.8	21.9
	Dewatered sludge feed of 30 percent solids		
3	1.1	0.1	1.2
9	2.3	0.3	2.6
18	3.7	0.6	4.3
30	5.4	0.9	6.3
60	14.6	1.8	16.4

NOTE: If the compost cannot be given away or sold, it must be disposed of on land. Compost may be used in landfill disposal as the daily cover requirement.

Incineration

A fluidized bed incinerator is assumed. Feed solids are assumed as both 20 percent and 30 percent suspended solids with 70 percent volatile. The process operates for 24 hours per day, 360 days per year. A reduction of 70 percent by weight is assumed and the ash is destined for a land disposal facility 100 miles from the treatment plant. The land disposal costs and transport costs are assumed to be the same as for the land disposal, reduced by 70 percent.

The incineration cost curves in the *Handbook for Estimating Sludge Management Costs* (EPA 1985a) are limited to sludge quantities less than 30 dry tons per day, and the higher values are therefore extrapolated.

If chemically-enhanced primary treatment is used in a wastewater treatment facility, the quantity of sludge is increased relative to primary treatment and the unit heating value of the sludge incinerated is decreased. It is assumed here that the heating value for sludge produced in chemically-enhanced treatment plants is 25 percent less than in conventional primary + biological plants and that the amount of auxiliary fuel needed to run the incinerator is increased by 25 percent. The cost of fuel oil is assumed at $1.50 per gallon.

Since incineration of sludge precludes the necessity for digestion, credit is given for the savings in digestion costs. The results of the calculations are shown in Table D.15.

TABLE D.15 Cost of Incineration

Dry Tons of Sludge Incinerated Per Day	Annual Cost of Incineration (incl. Transport & Disposal) ($ million)	Total Annual Cost of Incineration + Credit Given for Digestion ($ million)
	Dewatered sludge of 20 percent solids	
9	3.7	2.1
18	5.8	3.6
30	7.8	5.3
90	22.7	19.0
270	67.2	59.9
	Dewatered sludge of 30 percent solids	
9	2.8	1.2
18	4.7	2.5
30	6.1	3.7
90	21.9	18.2
270	62.2	54.9

TABLE D.16 Cost of Incineration of Chemically-Enhanced Primary Sludge

Dry Tons of Sludge Incinerated Per Day	Total Annual Cost[1] of Incineration for a Dewatered Sludge Feed of 20% Solids ($ million)	Total Annual Cost[1] of Incineration for a Dewatered Sludge Feed of 30% Solids ($ million)
9	2.3	1.4
18	3.9	2.8
30	5.8	4.1
90	20.3	19.6
270	6.4	59

[1]Total annual cost includes transport and disposal, with credit given for digestion and added cost of fuel for low BTU sludge.

If chemically-enhanced wastewater treatment is employed, the costs are as shown in Table D.16. Note that the quantity of sludge to be incinerated is greater for chemically-enhanced primary treatment plants than for conventional primary treatment.

Direct Land Application

In direct land application, treated sludge is hauled to a site in a liquid or dewatered state and injected into, or spread and incorporated into the soil. The nutrient-rich organic matter in the sludge provides a food source for microbiological organisms and earthworms, which provide nutrients for crop uptake and benefit soil structure. The beneficial reuse of treated sewage sludge increases crop production and reduces the potential pollution from the use of chemical fertilizers. The organic material in the sludge also increases the soil's ability to store water.

The costs of managing sludge by direct land application typically range from $100 to $150 per dry ton. In some urban areas on the east coast, direct land application costs can be as high as $750 per dry ton. The costs include either the purchase or contracting of dedicated farm land and the cost of transportation to the sites. Costs of transportation for land application are a function of the distance between the farm land and the treatment facility. Cost estimates for transportation range from a low of approximately $.50 per dry ton per mile to $2.50 per dry ton per mile. Application costs vary greatly, but generally are less than $50 per dry ton. These transportation and land application rates do not include the potential for crop and backhaul rebates.

Energy Use in Municipal Wastewater Treatment

Although the efficiency of energy recycle and recovery in municipal sewage treatment and disposal has improved since the passage of the Clean Water Act of 1972, U.S. POTWs remain net consumers of energy. U.S. POTWs consumed an estimated 257×10^{12} Btu/yr of energy in 1990, which represents 0.32 percent of total national energy use or approximately 4 percent of total electricity use (Jones 1991). This increase represents a doubling of energy used by POTWs since 1972.

Wastewater facilities typically account for 15 percent or more of a municipal energy budget (Jones 1991). The amount of energy consumed at POTWs and the inefficient use of that energy means that there are significant opportunities for energy conservation and demand-side management. Minimizing pumping and improving the delivery of process air are examples of where significant energy savings can be obtained.

Energy in POTWs can be described by the terms primary energy and secondary energy. Primary energy is the energy employed in the operation of a facility, such as electricity used in various processes and in space heating. Secondary energy can be defined as the energy needed in the manufacture of materials to construct a POTW facility; the construction of the facility itself; and energy associated with chemical use, labor, and transportation. The main energy sources for POTWs are electric power, natural gas or propane, and diesel fuel or gasoline.

Approximately 25 percent to 40 percent of the annual costs of running U.S. POTWs are primary energy costs, with the operation of the facilities accounting for the major share of the energy consumption. The energy use associated with operating treatment plants is a function of level of treatment, plant size, location, and pumping needs. Pumping can be a substantial energy-consuming process, especially in the headworks and outfall. The costs associated with energy use are financed by user charges. Table D.17 gives the energy requirements for Systems 1-4.

Energy consumption can be reduced by any of several means, including reduction of energy demand by process optimization and energy efficient installations, innovation in design and operation of treatment systems, and improvements in energy recovery from digester gas or incineration. Recovery and use of digester gas as a fuel source for space heating, steam generation, electricity, incineration, and other internal needs is the major form of energy reclamation in sewage treatment. Sludge volatile solids serve as an energy source in incineration and in anaerobic digestion.

DISINFECTION

Sources of pathogens include discharges from sewage treatment plants, combined sewer overflows (CSOs), and stormwater runoff. Water fowl and

TABLE D.17 Energy Requirements for Treatment Systems 1-4 for a 1 MGD Plant, Energy Usage (kWh/yr × 10^{-3}) (After Tchobanoglous and Schroeder 1985. Tchobanoglous/Schroeder, *Water Quality* © 1985 by Addison Wesley Publishing Company, Inc. Reprinted with permission of the publisher.)

	Primary Energy		Secondary Energy			
	Electricity	Fuel	Plant Construction	Chemicals	Parts & Supplies	Total
Primary	50	90	44	—	8	192
Low-Dose Chemical Primary[1]	58	95	54	70	15	292
Biological						
—Activated Sludge	237	209	156	60	34	696
—Trickling Filter	130	180	165	60	30	565
Chemical Primary + Biological[2]	177	200	132	100	32	641

[1]Low-Dose Chemical Primary includes increased energy requirements due to mixing, flocculation, additional sludge collection, and transport.

[2]Chemical Primary + Biological assumes a 50 percent reduction in the size of the biological system due to the enhanced removal from the Chemical Primary stage.

animal wastes, failing septic tank leaching systems, discharge from boats, and storm drain systems are other potential sources.

Disinfection Methods

Methods of disinfection include 1) the addition of chemicals; 2) the use of physical agents such as light or heat; 3) mechanical methods; and 4) exposure to electromagnetic, acoustic, or particle radiation. Disinfection technologies are considered as add-ons to any of the ten treatment trains already discussed and to other treatment systems. They can also be used for treating combined sewer overflow discharges. A variety of the present and future disinfection methods, the major advantages and disadvantages of each method, and selected costs are given below.

Chlorination

Chlorine (Cl_2), calcium hypochlorite ($Ca(OCl)_2$), or sodium hypochlorite (NaOCl) are the most widely used disinfectants in the United States and the rest of the world. Chlorination is inexpensive, widely available, and has a long history of proven effectiveness. However, organic compounds typically present in treatment wastewater can combine with chlorine to form toxic chloro-organic compounds and excess free-chlorine is toxic to many aquatic species. Chloroform, the best-known trihalomethane, is a documented animal carcinogen. Chloramines, especially monochloramine, form in wastewater in the presence of ammonia. These compounds are stable and do not dissipate. Because the high chlorine dose required for break point chlorination are not used, chloramine disinfection is less expensive than free chlorine disinfection. However, chloramines are also less reactive than free-chlorine. Some chloro-organics are potential carcinogens, mutagens, or toxins (WPCF 1986). Recent Occupational Safety and Health Administration standards for safety require intensive training and other expensive measures for systems employing chlorine gas. Sodium hypochlorite does not have the same safety issues associated with it as does gaseous chlorine.

Chlorine dioxide (ClO_2) has been used in the past for wastewater treatment disinfection where phenolic wastes are present. It is used as an alternative disinfectant of raw water supplies. Its advantages are that it does not produce THMs and is a very effective bactericide and viricide over a broad pH range. Because it does not react with ammonia, it can provide effective disinfection at relatively low applied concentrations. Its disadvantages include high capital cost for the ClO_2-generating equipment, and a lesser understanding of its toxicity. Chlorine dioxide produces chlorate and chlorite which may have subsequent impacts on the environment.

Capital costs of chlorination systems can range from minor for cases where adequate contact time can be attained in the outfall to substantial, where a contact tank and mechanical mixing must be provided. Operation and maintenance costs for chlorine systems are about $15 per million gallons treated (at a dose rate of 5 mg/l).

Dechlorination

Dechlorination is the process of removing chlorine residuals after chlorination. Dechlorination typically is done by either chemical means or with granular activated carbon (GAC). Chemical methods involve the injection of sulfur dioxide(SO_2), sodium sulfite ($NaSO_3$), sodium bisulfite ($NaHSO_3$), or sodium metabisulfite ($Na_2S_2O_5$) following the chlorine contact tanks. Like sulfur dioxide, the salts produce the same active ion, sulfite (SO) upon dissolution in water. Due to cost, GAC for dechlorination generally is limited to those instances where a GAC process is already in place at a wastewater treatment plant to remove toxic organics. The GAC can be a gravity or pressure bed system.

Dechlorination by chemical means is a simple, relatively inexpensive, and effective method. However, while reduced sulfur ions react rapidly with free ions and combined chlorine residuals, they do not always completely remove organic chloramines or chloro-organics. Additional treatment may be needed to further reduce total organic halogens (TOX, where X is chlorine, bromine, or iodine), which are potential carcinogens, mutagens, or toxins. It is estimated that municipal and industrial treatment plants using some 100,000 to 200,000 tons of chlorine per year produce several thousand tons of chloro-organics per year (WPCF 1986).

GAC is effective in removing many residuals. However, it is also expensive and therefore used only in those instances where high levels of organic removal are required. Operations and maintenance costs total about $20 per million gallons treated.

Ozone Disinfection

Ozonation of drinking water has a long history, especially in Europe. However, its application to wastewater has not been widely accepted. Ozonation produces fewer toxic by-products than chlorination and, because it reduces the need to store large quantities of chlorine in urban areas, it is safer to operate than a chlorine system. Because ozone is the strongest oxidizing agent and disinfectant used in wastewater treatment, only small doses and short contact times are needed. The efficiency of ozone disinfection is also independent of pH in a range between pH 6-10 and of temperature in a range between 36-86 degrees Fahrenheit.

Ozone use has been limited in the United States, however, because of its relatively high cost. Ozonation has higher capital and operating costs than chlorination. It is energy intensive, requiring 16-24 kilowatt hours of electricity per kilogram of ozone. Ozone disinfection of wastewater for a city of 500,000 could require 10,000 kWh/day of electricity. Also, ozone does not maintain a residual concentration in the treated water which allows for the possible regrowth of microorganisms after disinfection.

Ultraviolet Irradiation

Ultraviolet (UV) irradiation disinfection is a developing technology and is receiving increasing attention and application in the United States. UV disinfection has been demonstrated to be effective on a variety of organisms, particularly on viruses. It is as effective or more effective than chlorination or ozonation (WPCF 1986) and leaves no toxic residue. UV irradiation disinfection equipment is inexpensive and occupies little space, and is relatively easy to maintain and operate.

As with ozonation, however, UV disinfection leaves no residual inactivation agents. Thus, the regrowth of microorganisms after disinfection is possible. Also, some microorganisms may be able to repair damage done by UV disinfection if exposure is not fatal. Investigators have found that suspended and dissolved matter and water itself absorb UV radiation; thus the efficacy of this process is compromised when particulate matter is elevated (Qualls et al. 1985). Because UV disinfection is a relatively new technology and is generally more widely applicable to smaller POTWs, it requires further development for larger installations.

Electron Beam

This technology uses high speed electrons to kill microorganisms and toxic organic compounds. The technology, used for many years to preserve food and disinfect medical supplies, is now being tested as to its feasibility in treating wastewater (Jones 1991) but has been successful in treating sludges in limited studies (EPA 1989c).

Efficacy of Disinfection Methods in Pathogen Inactivation

Disinfection is the term used for the selective destruction of pathogenic organisms in order to protect the health of people and other animal life. The efficacy of different disinfection methods is a function of the concentration of the chemical agent or intensity and nature of the physical agent, the contact time, the temperature, and the number or types of pathogenic organisms present. It also depends on water quality factors (e.g., the amount of solids, dissolved organic material, inorganic compounds, and pH), treatment plant design, and level of treatment. Generally, the relative resistance of microorganisms can be listed in the following order, from the most resistant to least: parasitic cysts and acid fast bacteria, bacterial spores, viruses, vegetative bacteria.

Viruses are less easily inactivated than bacterial indicator organisms. These differences in sensitivity are one reason indicator bacteria do not adequately specify the effectiveness of disinfection measures in the inactivation of parasites and viruses (see Appendix B). Table D.18 shows the relative resistance of waterborne microorganisms for chloramines, chlorine, chlorine dioxide, and ozone (Sobsey 1989). The ranges of *CTs*, the product of the disinfection concentration and the contact time, for 99 percent inactivation reflect the variety of water disinfection conditions during experimentation as well as the variety of viruses, cysts, and bacterial cultures used.

The inactivation of microorganisms in chlorinated activated sludge effluents is highly variable. Disinfection reduces pathogen levels, especially when the effluent is free of suspended solids, but effectiveness is reduced in the presence of higher concentrations of suspended and colloidal solids. Depending on the level of nitrification, chlorination results in 0.8 to 1.3 $\log_{10}$ reduction of fecal streptococci and 0.1 and 0.5 $\log_{10}$ reduction of F-specific coliphages, used as surrogates for human enteric viruses. Table D.19 summarizes these data.

Table D.20 summarizes several surveys that have evaluated the occurrence of enteric viruses in chlorinated and unchlorinated primary + biologi-

TABLE D.18 Ranges of CT[1] for 99 Percent Inactivation (Source: Sobsey 1989)

	E. coli	Enteric Viruses	Giardia Cysts
Chloramines	113	345 - 2,100	430 - 1,400
Chlorine	0.6 - 2.7	0.3 - 12.0	12 - 1,012
Chlorine Dioxide	0.5	0.2 - 6.7	2.7 - 15.5
Ozone	0.006 - 0.02	0.006 - 0.72	0.53 - 4.23

[1]CT is the product of the disinfectant concentration (mg/l) and the contact time (minutes).

TABLE D.19 Log_{10} Reduction of Microorganisms in Chlorinated Activated Sludge Effluents (Source: Nieuwstad et al. 1988)

	Fecal Streptococci	F-specific Phage
Activated Sludge	0.8	0.1
Moderately Nitrified	1.4	0.2
Nitrified	1.3	0.5

TABLE D.20 Enteric Virus Levels in Treated Wastewater (Source: Rose and Gerba 1991, Asano et al. 1992)

	% of Positive Samples	Average PFU/100 liters
BIO + Chlorination	31	13
	70	130
BIO + Unchlorinated	67	79
BIO + Filtration + Chlorination	8	1.25
	0.8	not detected
	7	0.13

cal treatment effluents (System 3) and in chlorinated activated sludge effluents with filtration (Rose and Gerba 1991, Asano et al. 1992). Viruses were recovered from 31 percent to 70 percent of the samples at concentrations averaging between 13 and 130 plaque forming units per 100 liter in chlorinated and unchlorinated primary + biological treatment effluents. Filtration with chlorination reduced the prevalence and concentrations by 10- to 100-fold. This reduction was the result both of the improved physical removal of viruses due to the filtration and the enhanced inactivation due to the higher quality effluent.

COMBINED SEWER OVERFLOW CONTROLS

History and Problems

In many older cities across the United States—particularly the northeast seaboard cities and Seattle, Portland, and San Francisco on the west coast—combined sewer systems were constructed to provide drainage and wastewater disposal services. As shown in Figure D.8, combined sewers are designed to convey both wastewater and surface drainage from residential and business areas to a discharge location. During dry weather, an interceptor sewer accepts wastewater from the combined sewer and conveys it to a treatment plant. During rain events, the limited capacity of the interceptor sewer allows only a portion of the wastewater/stormwater mixture to be carried to a treatment plant. The remainder discharges from the combined sewer into nearby creeks, open channels, and rivers. The discharged mixture of wastewater and stormwater is called a combined sewer overflow (CSO). The construction of combined sewer systems in the United States ceased for the most part after 1945.

Technical uncertainties and difficulties in developing cost-effective measures to mitigate CSOs has plagued regulators since the beginning of the century. The first federal legislation to address this problem, the River and Harbor Act of 1899, exempted CSOs from regulation. The Clean Water Act of 1972 removed this exemption, but even after the July 1, 1988 deadline for municipal compliance with the Clean Water Act, there had been only partial elimination of some dry weather overflows and limited elimination of wet weather overflows.

The uncertainty on how to address CSOs emanates from concerns regarding the adequacy of technical and financial resources available to address them compared with their priority as a water quality problem. Federal uncertainty and lack of a consistent national program to fund and administer CSO control programs has resulted in uneven application of the law in the United States. Even so, several cities (Chicago, San Francisco, and Milwaukee) have undertaken large-scale programs to mitigate overflows from

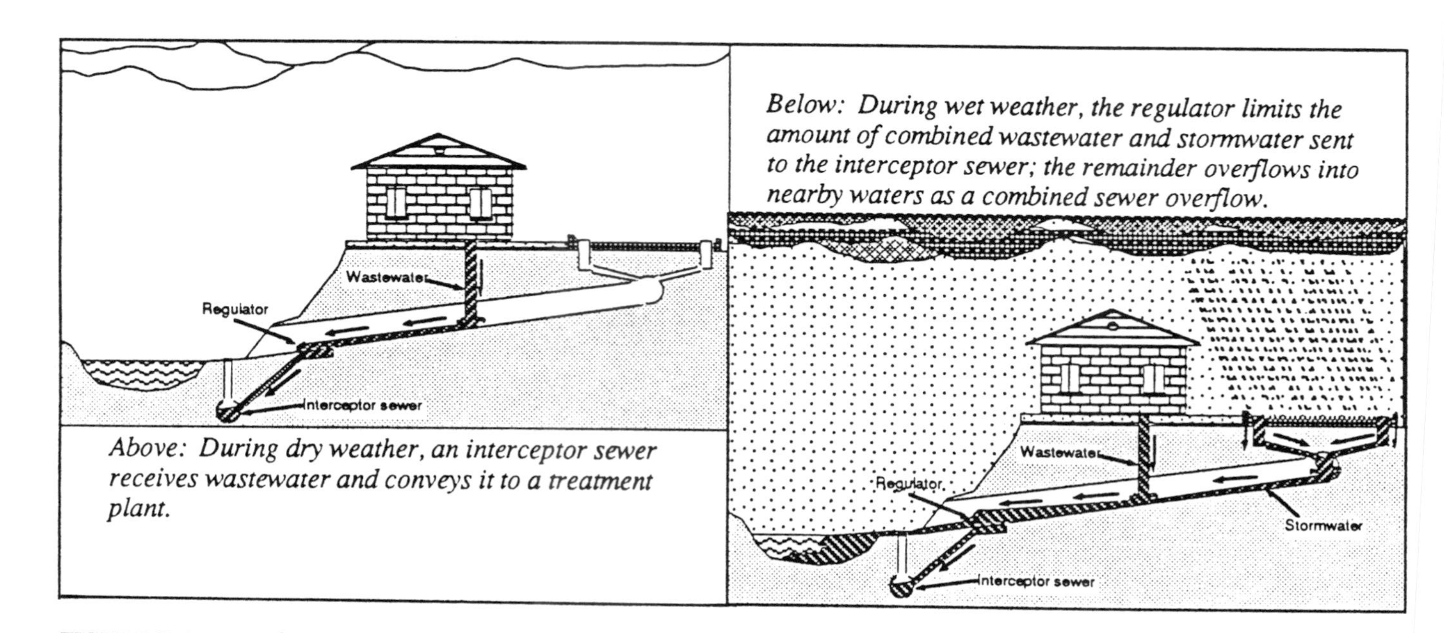

FIGURE D.8 Combined sewer operation. (Source: Camp Dresser & McKee, Inc. 1991. Reprinted, by permission, from Camp Dresser & McKee, Inc., 1991.)

their combined sewer systems. But many of the more recent advances in CSO control technology come from western Europe. In Europe, research and development for promoting new CSO controls and creating ways and methods to design system-wide holistic controls flourishes. For example, in a 1990 CSO research and development program (Project Rainfall), the West German government spent more in one year than the U.S. EPA did in the past eight years. Similar programs of this scale are under way in the United Kingdom, France, The Netherlands, and Canada.

The U.S. government, however, has now initiated a formal CSO regulatory program in response to the Water Quality Act Amendments of 1987. On September 8, 1989, the U.S. EPA published its National Combined Sewer Overflow Strategy (Federal Register 1989). This strategy required six technology-based limitations as a minimum best control technology and best available technology, which is established on a best professional judgement basis:

1. Proper operation and regular maintenance programs for the sewer system and combined sewer overflow points,
2. Maximum use of the collection system for storage,
3. Review and modification of pretreatment programs to assure CSO impacts are minimized,
4. Maximization of flow to the POTW for treatment,
5. Prohibition of dry weather overflows, and
6. Control of solid and floatable materials in CSO discharges.

The following section describes 5 categories of CSO controls that represent the state-of-the-art of this technology.

1. Source Controls,
2. Flow System Optimization,
3. Sewer Separation,
4. Satellite Treatment, and
5. Off-Line Storage.

The first 3 categories apply, essentially, to the EPA's minimum requirements 1 to 5. The fourth and fifth categories: satellite treatment and off-line storage, relate to the EPA's item 6.

CSO Technologies

Source Controls

The source controls cited here deal with reducing the amount of pollutants that accumulate during dry weather on the land surface, streets, and within sewer systems. Minimizing these accumulations means that during rainstorms there will be a smaller pollutant mass discharged from the urban land areas to the receiving waters. While it is straightforward to postulate that this is the *cleanest* and most obvious class of controls, in real practice, the opposite is generally true. Long-term dependency on a labor force to perform these types of controls simply has not worked.

Street Sweeping. Although the major objective of street sweeping is to enhance roadway appearance, periodic removal of surface accumulations of litter, debris, dust, and dirt also reduces the transport of such materials into the sewer system. Common methods of street sweeping include manual sweeping and the use of mechanical sweepers utilizing brooms or vacuums. Most communities now practicing street sweeping rely mainly on one of the mechanical methods, which loosen debris from the street surface, pick it up, and store it for later disposal. Street sweeping cannot be performed during wet weather or during periods of ice and snow accumulation.

The technology of street sweeping and its pollutant removal effectiveness has been assessed extensively in the last decade. Almost all manuals for urban stormwater water-quality management cover the topic (FHA 1985). The extensive statistical reviews of this practice (11 sites and 5 pollutants) during the EPA's National Urban Runoff Project indicated that no significant reductions in pollutant concentrations are realized, although they could occur in certain site-specific cases (EPA 1983).

The EPA concludes that street sweeping in most of the United States is appropriate for aesthetic purposes but has limited water-quality benefits. Street flushing, which is only practiced in a few cities in the United States but is common in Europe, provides better removal of fine particles than street sweeping but is only feasible in combined sewer areas.

Catch-Basin Cleaning. Catch-basins are commonly cleaned with mechanical bucket devices or vacuums. Catch-basin cleaning removes heavy solids, eliminating possible sources for silt deposition problems in downstream sewers. When properly designed and maintained, catch-basins are effective in removing both coarse solids and floatables. However, the finer solids are not effectively removed. Catch-basins accumulate solids and liquids which, when flushed by runoff, can contribute significant pollutant loads. Frequent vacuum or suction-type (eductor) cleaning of catch-basins can remove accumulated pollutant material and maintain catch-basins' removal efficiency. However, catch-basins are not frequently cleaned because of high labor requirements (CH_2M Hill 1989, FHA 1985).

Sewer Flushing. First flush includes the wet weather scouring of wastewater solids deposited during dry weather in combined sewer systems. First flush

TABLE D.21 Effectiveness of Single 300-Gallon Manual Flush (Tanker) to Scour, Entrain, and Transport Materials within 12" to 18" Laterals

Pollutant removed in a length of:	250 feet	700 feet	1,000 feet
Organic & Nutrient Deposits (BOD_5, TP, TN)	75% - 90%	65% - 75%	35% - 45%
TSS Deposits	75%	55% - 65%	18% - 25%

also includes the transport of loose solid particles from the urban ground surface to the sewerage system. These particles settle out in the combined system and are flushed during periods of larger flows. The magnitude of these combined loadings during runoff periods has been estimated to be as much as 30 percent of the total daily dry weather sewage loadings (CH_2M Hill 1989).

Sewer flushing involves scouring and transport of deposited pollutants to the wastewater treatment plant during dry weather when there is sufficient interceptor capacity to convey these flows. In 1979, a three-year research and development program sponsored by the EPA was conducted in the Dorchester area of Boston to determine the pollution reduction potential of flushing combined sewer laterals (Pisano 1978, 1979). It was concluded that small volume flushing would transport organics, nutrients, and heavy metals sufficient distances to make the option feasible and attractive. Relevant conclusions are listed in Table D.21.

Sewer flushing of large diameter combined sewers was investigated in the Elizabeth, New Jersey, CSO Facility Plan (Clinton Bogert Associates 1991, Kaufman and Lai 1978). The plan concluded that daily flushing of troublesome deposition sections within seven subareas using 12 automatic flushing systems reduced the first flush overflow pollutant loadings by about 28 percent.

Flow System Optimization

This method involves adjusting the flow controls within existing pipe systems to maximize the carrying capacity of interceptors or to take advantage of unused large pipe storage during wet weather. This control is very efficient in the capture of small storms (less than 0.1 inch), costing under $200 per acre.

Combined sewer systems have flow regulation structures that divert wet weather flow to the POTW through the interceptor at rates normally from 1.5 to 3 times average dry weather flow. Any greater flow escapes and discharges to the environment. When the wet weather flow exceeds this rate, the POTW sees only a small fraction of the potential first flush. Many North American communities have increased their POTW capacity in order to maintain higher ratios of wet to dry flow to better capture the first flush. New types of flow controllers are needed to increase the first flush capture without overloading the POTW during the remainder of the storm in cases where interceptors can accept the increased flows.

Enhanced Flow Regulation and Static In-Line Control. Static control regulators have the operational advantage of no moving parts. Interception is limited to preset levels. Typically, static control regulators cannot maximize intercepted flow before a spill. The standard type of controllers include: fixed orifices, drop inlets, leaping and side-spill weirs, siphons, and manually operated gates.

In the last ten years a new type of static flow controller called a *vortex throttle* has improved in-line storage of combined sewage at regulator chambers for later drainback to the POTW. Vortex devices work using no moving parts or external energy supply. Besides providing a more positive degree of flow control than orifices or spill weirs, they require less maintenance than mechanical float-operated controllers.

The largest U.S. system-wide configuration of vortex flow controllers for combined sewer system control is in Saginaw, Michigan. In the mid-1980s, the city converted 21 flow regulators to static vortex types at a cost of $200 per acre served. In the spring of 1988, Saginaw reported that the amount of wet sludge processed during rainfall events had increased by about 12 tons per day during runoff events preceded by several days of antecedent dry weather (J. Anderson, Supt. of Wastewater Treatment, Department of Public Utilities, City Saginaw, Michigan, personal communication, 1988).

The Marigot project in Laval, Quebec includes 13 new regulators controlling combined sewer flow to a new tunnel more than three miles long with pumpage to a new POTW located on the southerly side of the island of Laval, adjacent to Montreal, Canada. This project is the largest in North America in terms of number of units and the scale of the technology. Construction was completed in 1987.

Dynamic In-Line Storage and Real Time Control. Variable control is provided by regulators that control a gate or similar structure opened in response to an external signal. This controller can adjust the amount of interception and is connected to a central control that optimizes selection of overflow time and location in response to actual system flow and rainfall conditions. Most variable controls require constant maintenance to ensure proper operation.

Seattle, Washington, maintains the largest computerized in-system control program in the United States. It monitors pumping stations, regulator stations, POTWs, and rain gages and uses programmed information to provide control.

The Northeast Ohio Region Sewer District has operated and maintained in-line storage systems in Metropolitan Cleveland for nearly 20 years. Over 140 in-line systems are in operation using motorized sluice gates and inflatable plastic fabric dams (fabridams) controlled by computers. Cleveland's system fabridam failure rate initially approached 30 percent because of design and installation problems. As a result, the Northeast Ohio Region Sewer District continues to upgrade design and installation quality control. It is estimated that the automated program eliminates 98 percent overflows for a 0.12 inch rain. Ohio's experience demonstrates that a slow and progressive correction approach is preferable to one of immediate full-scale implementation (Hudson 1990).

Sewer Separation

Sewer separation is a method of minimizing the amount of street runoff that mixes with sanitary sewage. However, complete separation is difficult and prohibitively expensive to achieve and street runoff can be quite polluted. Thus, in contrast to 20 or 30 years ago, the practice of sewer separation within combined-sewered areas is not practiced on a large scale today. Separation is still practiced to solve pollution problems within small portions of combined sewer areas connected to separated systems or to solve flooding problems within combined systems where there is inadequate flow capacity.

Conventional Full and Partial Separation. Sewer separation has several different meanings. *Complete* separation means strict separation of all sanitary, commercial, and industrial sewage flows into a system that is separated from a storm sewer system serving the same area. *Partial* separation generally means construction of a new storm sewer system to handle street runoff load, i.e., pipes connecting to all the catch-basins. The effectiveness of stormwater inflow reduction for complete separation systems is about 95 percent. Partial separation effectiveness ranges from 50 percent to 85 percent.

Separation is viewed as a viable CSO control in some limited circumstances, where substantial separation already exists. However, the federal stormwater permit program promulgated in 1991 has lead most CSO-impacted communities to reject prior separation as a CSO control approach and to express serious concern about separating large, new areas. Areas impacted by basement flooding will continue to separate in order to reduce health hazards and property damage.

The cost of a separation project is dependent on the degree of prior separation; the configuration of the existing system; the size, population density, and geography of the area; and the design objective. A complete separation cost was developed for the CSO catchment area in Boston (CH_2M Hill 1988). Total construction costs varied from $60,000 per acre for partially separated residential neighborhoods to $190,000 per acre for entirely combined downtown areas.

Flow Slipping. Flow slipping involves the use of inlet control in urbanized areas to manage the stormwater entering existing combined sewer systems. It has its roots in Scandinavia and the United Kingdom where the concept has been used on undersized combined sewers to relieve basement flooding and to mitigate the volume and frequency of overflows. This practice is widely used for basement flooding and drainage control in Ohio, Maine, Illinois, and Quebec and is used as a CSO control throughout Ontario Province in Canada.

Flow restrictors placed within stormwater catch-basins are widely used to induce overland flow away from sensitive areas to more attractive capture/storage locations (Smisson 1981; Pisano 1982a, b; Wisner 1984; Walesh 1985; Havens & Emerson Engineers 1987; TWA and HRD 1987). In the last several years, Boston has conducted a long term field evaluation of this concept and has noted no adverse flooding or deposition problems. It intends to utilize this concept to optimize on-going separation projects (S. Shea, Head of Sewer Construction, Boston Water & Sewer Commission, personal communication, 1991).

In Hartford, Connecticut, preliminary CSO plans for the Franklin Avenue District (a 1,000 acre CSO catchment) included extensive use of flow slipping within combined sewer areas, and new or existing storm drain outlets. Average costs for conventional street load sewer separation within the district are about $56,000 per acre. Flow slipping separation costs ranged from $4,000 to about $21,000 per acre.

High-Rate Satellite Treatment

Often it is impossible to capture and treat all combined sewage even from low production rainstorms. Most POTWs can treat no more than about 0.003 inches per hour (on the average) of rainfall. Federal and some state CSO control agencies are considering requiring capture and secondary treatment of one-year storm events, equalling approximately one inch per hour, which could represent more than 300 times the capacity of a POTW. The costs of near-surface retention, deep storage facilities, and treatment facilities to process these returned flows from storage can be enormous.

For example, the current CSO policy for metropolitan Toronto and area municipalities restricts the system to one overflow per recreational season for critical receiving water areas and 90 percent volumetric control (annual) for all other areas. Up-system near-surface tanks and waterfront tunnels will cost about $350 million, while the balance of the estimated $1.3 billion program is associated with additional new conveyance pipes and treatment capacity at existing POTWs to handle new growth and additional wet weather flows. These flows are gradually sent to the POTW over an average of three days. Toronto is considering satellite treatment to decrease the amount of retained storage and to decrease new POTW capital investment. Satellite treatment is usually distributed throughout a collection system at the outfalls of large combined sewer trunks.

There are only a few practical high-rate treatment processes in modern CSO control design practice. These include screening, vortex separation, and vortex separation with storage. Treatment units such as dissolved air floatation, dual-media high-rate filtration, high-gradient magnetic separation, and powdered activated carbon-alum coagulation (all preceded by screening) are operations that have been used with some success in experimental demonstration projects. For a variety of technical reasons, these unit operations are not practical in full-scale operation where intermittent, heavy shock CSO flows, debris, and pollutant loadings occur.

Combinations of vortex separation with conventional near-surface storage are being used in several projects. Several recently constructed facilities in Decatur, Illinois, using new German vortex separators coupled with relatively small volumes of storage, provide higher degrees of pollutant removal than conventional high-rate treatment schemes. At the same time, they reduce maintenance requirements and reduce the amount of required storage relative to conventional *near surface* storage. For an investment of roughly $2,000 to $4,000 per acre, the storage and vortex treatment concept consumes very little land space. During peak design flow conditions, this system can achieve solids removals at levels between those obtained in preliminary and primary treatment.

Screening Facilities. Screening technology has an inconsistent classification system, using such terms as bar, coarse, fine, or micro, depending on the screen construction and spacing. Bar and coarse screens are used to remove gross floatable and settleable materials. Coarse screens are used as a pre-treatment *protective* measure for vortex separation or off-line storage facilities, especially where pumping is required. Fine screens and micro screens are usually used at POTWs and centralized CSO treatment facilities. Types include static screens, hydraulic sieves, drum screens, and vertical rotating screens.

In practical terms, because it operates reliably, mechanical screening of the type used in POTW headworks is the only screening method that is viable for remote satellite operations. Micro strainers require constant maintenance and frequently are blinded with solids. They are recommended only at continuously staffed operations.

Five major screening and disinfection facilities (200-300-MGD range) were recently designed in Atlanta, Georgia. Each satellite plant includes a series of parallel screening channels with preset overflow diversion weirs for bypass. Flows within the design capacity of the facility pass through coarse and then fine screens, followed by chlorination and discharge. The facilities are intended to reduce fecal coliform and remove floatable and settleable solids larger than 3/8 inch in diameter. Presently, the $90 million construction program is under way. Average cost per million gallons treated per day for the 5 facilities is estimated at $10,000, which is approximately one-half of one percent of the cost of conventional wastewater treatment.

Vortex Solids Separators. Vortex solids separators coupled with storage facilities can provide a significant degree of coarse solids removal that is cost-effective and requires low maintenance. A vortex solids separator is a small, compact, solids separation device with no moving parts. If the unit is used as a combined sewer regulator, dry weather sewage passes through unimpeded. If the device is intended to operate only as an off-line treatment unit, then storm flows are deflected by gravity or by pumpage into the unit. During wet weather the unit's outflow is restricted, causing the unit to fill and inducing a swirling vortex-like operation. Settleable grit and floatable matter are rapidly removed. Concentrated foul matter is sent to the POTW, while the cleaner, treated flow discharges to the receiving waters or into temporary storage for later treatment at the POTW.

A type of vortex separator developed in the early 1970s, the *swirl concentrator* has been extensively tested in the United States (Drehwing 1979, Sullivan et al. 1982, Wordelman 1984, Heinking and Wilcoxon 1985, Hunsinger 1987), and the performance results are extremely mixed. In many early applications, the device was intended to remove substantial amounts of suspended solids, while its original design was only intended to remove coarse grit and floatables at regulation chambers. Other deficiencies relate to excessive vessel turbulence at design flow. Also, combined sewage contains finer materials that do not readily separate.

There are 19 swirl concentrators in the United States, which have a total design flow capacity of 888 MGD, including two under construction in Euclid, Ohio (EPA 1989b). With the exception of Decatur, Illinois, all installations are stand-alone, off-line devices. The largest swirl concentrator complex in the United States is the Robert F. Kennedy facility in Washington D.C. The site contains three 57-foot diameter units with a total design flow of 400 MGD. In the mid-1980s, a vortex separator was developed which appears to achieve better removal efficiencies than the swirl concentrator (Brombach 1987, Pisano et al. 1990). Presently there are 13 such units in seven U.S. projects, which have a total design capacity of 1.2 billion gallons per day.

In general terms, properly designed U.S. vortex separators remove 15 percent to 35 percent of settleable solids, with higher removals associated with first flush. Construction costs of vortex vessels range from $5,000 to about $8,500 per acre treated. A recent full-scale vortex separator investigation in Tengen, Germany, has concluded that this technology achieves settleable solids removals of at most 60 percent (Brombach et al. 1992).

Off-Line Storage

Storage facilities, basins, or tunnels have been extensively used to capture excess runoff during storm events. Storage allows the maximum use of existing dry weather treatment facilities and is often the best low-cost solution to CSO problems. Combined sewage flow is stored until the treatment facilities can treat the excess flows. Off-line storage requires detention facilities, basins or tunnels, and the facilities for either draining by gravity or pumping flow to and from storage.

Utilization of relatively small volumes of retention storage can be effective in retaining pollutant loadings from small, frequent, storm events. Loadings from the larger storms usually represents only a small fraction of the total annual CSO. This is sometimes overlooked. Small amounts of system storage (in-line or off-line), on the order of 200 to 400 cubic feet per acre, can effectively address much of the CSO problem.

There are a number of different storage strategies in use. The first is called *upstream stormwater hybrid*. The concept is a British and Scandinavian *inlet control* practice, where catch-basins are restricted to force street loads to move over land to new catch-basin intakes. These intakes discharge into shallow off-line storage tanks which have throttled outlets leading back into the sewer system. This practice is widely used to solve basement flooding problems and is often viewed as a CSO control. The value of this approach is limited because it typically is not possible to provide large volumes of surface storage that can be gravity-drained back into sewer systems. This approach benefits congested residential areas because it can be used as a last resort in situations where there is no other good, cost-effective way of creating near-surface storage. This approach is used in the Cleveland and Chicago areas and has recently been proposed in Hartford, Connecticut.

The second type is *near-surface contaminated upstream storage* and is popular in Europe. This technique includes small volume storage that captures first flush flows in excess of several times the average dry weather flow. The captured flow is returned to the sewer system for treatment after the storm passes. Over 12,000 such tanks in Germany, Switzerland, and Holland presently exist with a median size of about 150,000 gallons. Because the trend in the United States has been toward much larger downstream facilities, very few of these types of tanks are used.

The third type is *near-surface contaminated downstream storage*. There are only about 20 in the United States, and they are usually of 1 to 4 million gallons. The largest facility, in Sacramento, California, has a storage volume in excess of 25 million gallons. The greatest concentration of large near-surface retention and detention facilities is in Michigan, where very stringent performance standards of the one-year, six-hour storm with secondary treatment have been adopted. Cincinnati has included in its CSO Master Plan a number of such tanks. New York City is currently designing several near-surface storage tanks with storage volumes on the order of 10 to 40 million gallons.

One negative feature of large near-surface storage tanks is the large amount of land required. However, if properly designed, the covered surface of the tanks can be used for tennis courts, basketball courts, or parking. Some designs include soil covering and a planted park area. Another negative feature is the problem of cleaning the large tanks of accumulated sediments and organic material. Because the solids loadings on downstream treatment can be significant, cleanout timing is important. Capture performance of such tanks is high if total retention is provided.

The fourth and fifth methods of storage are deep tunnels and reservoirs in bedrock, either decentralized or consolidated. Large caverns have been excavated in Chicago, Illinois, Milwaukee, Wisconsin, and Rochester, New York for storage and subsequent treatment. Such systems are also proposed for Cincinnati, Ohio and Boston, Massachusetts. These approaches are becoming more attractive because the costs of tunneling have dramatically decreased in the last decade.

The last method described is the *moat* storage as employed by the city of San Francisco. This system will eliminate 85 percent of the city's CSOs by adding storage in large underground conveyance boxes that effectively ring the city along the shoreline. The construction cost for a service population of 727,000 was $1.35 billion. Flexible *bags* located in the receiving water have been demonstrated in small scale applications represent a low-cost variation of this approach if it proves to be successful on the large scale.

The costs of near-surface storage typically ranges between $5,000 to $15,000 per acre. There are a number of factors that influence overall cost. In general, the most expensive facilities are underground, rectangular tanks resting on poor soils and requiring odor treatment in urban settings. Maintenance cost is also significant since the tanks must be visited by a maintenance crew to flush settled solids from the tanks after every storm. Such tanks in New York City presently cost about $4 to $5 per gallon of storage generated. Storage systems can be very effective in CSO control. Annual overflow reductions on the order of 80 percent to 90 percent are possible, but maintenance is costly because of the need to remove heavy solids regularly from the storage facilities.

Integrating CSO Control Techniques

A $4 million full-scale demonstration project in Metro Toronto, begun in May 1991 is exploring alternative CSO control strategies including a vortex solids separator, detention tanks, chemical addition, and alternative disinfection schemes (i.e., conventional chlorination and dechlorination versus ultraviolet treatment). The central idea of this demonstration project is to ascertain what portion of the CSO flows can be "safely treated and acceptably discharged" in a satellite context such that POTW costs can be reduced. The project is funded by Metro Toronto Works, Environment Canada, and the Ontario Ministry of Environment.

This new idea of storage and high-rate treatment can be appreciated by a review of the Decatur, Illinois, CSO control projects. Projects at 3 sites: McKinley Ave (40 MGD), 7th Ward (113 MGD), and Lincoln Park (416 MGD) are discussed here.

The concept is to direct all flow through coarse screens. A portion of the first flush is captured in retention storage tanks and the rest passes through vortex separators. The underflow from the vortex separators is pumped into the retention storage tank. When the retention storage tank is full, the vortex separator continues to operate, but underflow is set to zero. Some separation occurs, but efficiency is greatly diminished. The retention storage tanks are provided with aerators and mixers to prevent odors and facilitate cleanout. Washdown is performed using water cannons. Following a storm event, the vortex separator pump station dewaters the tank with pumpout sent to the nearby interceptor. Typical ratios of wet weather flow volume to dry weather flow volume interception ranges from 3 to 5.

Storage provided by these projects ranges from 113 to 217 cubic feet per acre. The 416 MGD Lincoln Park facility covers 3.4 acres. A conventional near-surface storage facility would occupy at least three times that area and, under certain storm conditions, still have overflow. The facility consists of two screening buildings, two flow deflection chambers, one vortex flow divider, four vortex separators, one pump building, and the first flush storage tank. The 113 MGD 7th Ward facility, which provides preliminary treatment (15 percent settleable solids removal at a design flow of 174 cubic feet per second), covers 2.2 acres. A number of unit costs are presented in Table D.22.

The cost of vortex separators per MGD design flow are $4,125/MGD

TABLE D.22 CSO Project—Decatur, Illinois, Construction Costs Only (1991)[1] (Source: Pisano and Wolf 1991. Reprinted, by permission, from Water Environment Federation, 1991.)

Parameter	McKinley	7th Ward	Lincoln Park
Tributary Area (acre)	661	860	2,491
Design Area (MGD)	40	113	416
Site Area (acre)	1.5	2.2	3.4
Cost ($ in million)	1.755	3.947	7.84
Cost ($/acre)	2,655	4,134	3,147
First Flush Storage (cf/acre)	101	117	73
Storage Underflow (cf/acre)	27	100	40
Total Storage (cf/acre)	138	217	177[1]
Cost Storage ($/gal)	1.09	0.76	N/A
Diameter Vortex Sep (ft)	25	44	44
Cost Vortex Sep ($1,000)	171	492	N/A

N/A = not available.
[1]Additional in-line storage = 60 cf/acre.

and $3,125/MGD for similar projects in Burlington, Vermont and Saginaw, Michigan. The unit volumetric cost of storage within the vortex separators is $1.66 and $2.43 per gallon respectively.

It is believed that providing vortex separators with conventional facilities would provide the greatest operational flexibility. The future of satellite treatment will likely couple near-surface storage with high-rate vortex solids separation treatment devices. This approach provides the flexibility to expand facilities to meet more stringent regulatory control requirements. There is a limit to the removal of solids by physical means. Chemical addition to detention storage is an option to increase removal. If the requirement is to obtain fewer overflows per year, then adding more retention storage is about the only option available.

Summary of Comparative Performance of CSO Control Technologies

The efficacy and cost of the various CSO control technologies is summarized in Tables D.23, D.24, and D.25. Table D.23 compares the relative advantages, Table D.24 presents the pollutant removal capability, and Table D.25 shows the comparative costs of each CSO technology.

TABLE D.23 Comparative Advantages of Each Technology

Attribute	a	A	B	C	D	E	F	G	H
Source Controls	○	○	◒	○	○	○	○	○	○
Catchbasin Cleaning	○	○	●	○	○	○	○	○	○
Street Sweeping	○	○	◒	○	○	○	○	○	○
Sewer Flushing	○	○	○	◒	○	○	○	○	○
Flow System Optimization	○	◒	◒	◒	○	○	◒	●	◒
Enhanced Flow Regulation and Static In-Line Control	○	●	●	○	○	○	●	●	◒
Dynamic In-Line Storage	○	◒	○	●	○	○	◒	◒	◒
Real Time Control	○	○	○	●	○	◒	○	○	◒
System Flow Reduction	●	●	●	○	○	●	●	○	●
Conventional Full Separation	●	●	●	○	○	●	●	○	●
Conventional Partial Separation	◒	●	●	○	○	●	●	○	●
Flow Slipping	○	●	◒	○	○	◒	◒	◒	◒
High Rate Satellite Treatment	◒	◒	◒	●	◒	◒	◒	◒	◒
Screening	◒	○	○	●	○	◒	◒	○	○
Vortex Solids Separators	◒	●	●	○	◒	◒	◒	◒	◒
Vortex Separators and Storage Combination	◒	◒	◒	◒	◒	◒	○	●	●
Off-Line Storage	●	●	◒	◒	●	●	●	◒	●
Upstream Stormwater Hybrid	◒	◒	●	○	○	◒	○	○	○
Near-Surface Contaminated Upstream Storage	●	◒	○	◒	◒	◒	◒	◒	◒
Near-Surface Contaminated Downstream Storage	●	◒	○	●	●	●	●	●	●
Decentralized Deep Storage	◒	●	●	◒	◒	●	●	○	●
Consolidated Deep Storage	◒	●	●	◒	◒	●	●	●	●

Key to Table D.23

Column: Category	An Empty Circle Means...	A Partially Filled Circle Means...	A Filled Circle Means...
a: Construction disruption	Little to slight disruption	Mid-level disruption	Major community disruption
A: Reliability	Low robustness	Average reliability	Excellent reliability
B: Maintainability	Difficult to maintain	Mid-level maintenance effort	Easy to maintain
C: Level of expertise	Low skill level needed	Moderate skill level needed	High skill level needed
D: Land area required	Small area requirements	Medium area requirements	Large area requirements
E: Capital cost	Low capital costs	Medium capital costs	High capital costs

TABLE D.23 *Continued*

Key to Table D.23—*continued*

F: Previous effectiveness	Unproven effectiveness	Moderate effectiveness	Proven effectiveness
G: Wide use applicability	Low potential for wide use	Medium potential for wide use	High potential for wide use
H: Removal effectiveness	< 25% control of overflow volume and solids removal	25% - 60% control of overflow volume and solids control	> 60% control of overflow volume and solids control

TABLE D.24 Comparative Pollutant Removals

Technology	Pollutant: A	B	C	D	E	F	G	H
<u>Source Controls</u>	◓	○	○	○	○	○	○	◓
Catchbasin Cleaning	○	○	○	○	○	○	○	◓
Street Sweeping	◓	○	○	○	○	○	○	○
Sewer Flushing	◓	○	○	○	○	○	○	○
<u>Flow System Optimization</u>	○	◓	○	○	○	○	○	○
Enhanced Flow Regulation and Static In-Line Control	○	◓	○	○	○	○	○	○
Dynamic In-Line Storage	○	◓	○	○	○	○	○	○
Real Time Control	○	◓	○	○	○	○	○	○
<u>System Flow Reduction</u>	○	○	○	○	○	○	○	○
Conventional Full Separation	○	○	○	○	○	○	○	○
Conventional Partial Separation	○	○	○	○	○	○	○	○
Flow Slipping	○	○	○	○	○	○	○	○
<u>High Rate Satellite Treatment</u>	◓	◓	○	○	○	◓	◓	◓
Screening	○	○	○	○	○	○	○	○
Vortex Solids Separators	○	◓	○	○	○	○	○	○
Vortex Separators and Storage Combination	◓	◓	◓	◓	◓	◓	◓	◓
<u>Off-Line Storage</u>	●	●	●	◓	◓	◓	◓	◓
Upstream Stormwater Hybrid	◓	◓	○	○	○	○	○	○
Near-Surface Contaminated Upstream Storage	◓	◓	◓	◓	◓	◓	◓	◓
Near-Surface Contaminated Downstream Storage	●	●	●	●	●	●	●	●
Decentralized Deep Storage	●	●	●	●	●	●	●	●
Consolidated Deep Storage	●	●	●	●	●	●	●	●

KEY:
A: Metals (Cu, Cd, Cr)
B: TSS (suspended solids)
C: Organics (PAHs)
D: Organics (surfactants)
E: Pathogens (coliform)
F: Nitrogen
G: Phosphorus
H: Aesthetics (oil and grease)

○ - 25% or less removal effectiveness
◓ - 25% - 50% removal effectiveness
● - greater than 60% removal effectiveness

TABLE D.25 Comparative Capital Costs for CSO Control Options

CSO Control Options	Capital Costs ($ per acre)
Source Controls	
Catchbasin Cleaning	Not Applicable
Street Sweeping	Not Applicable
Sewer Flushing	$100 - $500
Flow Svstem Optimization	
Enhanced Flow Regulation and Static In-Line Control	$100 - $500
Dynamic In-Line Storage	$500 - $1,000
Real Time Control	$500 - $1,000
Svstem Flow Reduction	
Conventional Full Separation	$50,000 - $100,000
Conventional Partial Separation	$10,000 - $50,000
Flow Slipping	$5,000 - $10,000
High Rate Satellite Treatment	
Screening	$2,000 - $5,000
Vortex Solids Separators	$2,000 - $5,000
Vortex Separators and Storage Combination	$5,000 - $10,000
Off-Line Storage	
Upstream Stormwater Hybrid	$10,000 - $50,000
Near-Surface Contaminated Upstream Storage	$5,000 - $10,000
Near-Surface Contaminated Downstream Storage	$5,000 - $10,000
Decentralized Deep Storage	$10,000 - $50,000
Consolidated Deep Storage	$10,000 - $50,000

NONPOINT SOURCE MANAGEMENT OPTIONS

Introduction

According the joint EPA/National Oceanic and Atmospheric Administration document Proposed Development and Approved Guidance—State Coastal Nonpoint Pollution Control Programs, nonpoint source pollution has become the largest single factor preventing the attainment of water quality standards nationwide (EPA 1991b). For estuarine waters, current best estimates show that of the approximately 75 percent of waters assessed, 10 percent are threatened and 35 percent are impaired. Nonpoint source pollution is an important component of these threats and impairments. The leading sources of nonpoint pollution in estuarine waters are urban and agricultural runoff. More specifically, these include erosion from construction sites; runoff from urban areas; erosion from agricultural lands, streambeds, and roadways; runoff from livestock production areas; and runoff from farmland contaminated by fertilizers, pesticides, and herbicides. Pollutants from nonpoint sources are carried to the coast primarily by surface water through the action of rainfall runoff, snow melt, and ground water seepage. Principal pollutants of concern include sediment, nutrients, bacteria, metals, organics, and oil and grease.

Nonpoint sources include atmospheric deposition. Atmospheric pollutants deposit directly onto coastal water bodies in the form of precipitation or dry deposition. In the Great Lakes, the principal source of PCBs is through atmospheric deposition. In the Chesapeake Bay, atmospheric deposition is a major source of nitrogen to the bay.

Direct atmospheric deposition on coastal waters is probably a small source in the mass balance of most pollutants in most cases. However, deposition on land areas and subsequent delivery to the shore by storm channels and rivers can be significant, as, for example, in southern California, where urban storm runoff was the main pathway for lead transport before the use of unleaded gasoline was mandated.

Nonurban sources, especially agriculture and forestry, can generate large quantities of nonpoint source pollution. Indeed, in bays fronted by farms or harvested forests, nonurban activities may be the predominant source of pollutants such as pesticides; and naturally occurring substances, such as nitrogen and minerals. The largest source of nitrogen in the Chesapeake Bay comes from upstream agriculture in the fertile river valleys. The largest source of nitrogen and copper in Puget Sound is forestland runoff, the result of heavy logging and subsequent erosion. The high acidity of urban precipitation likely contributes to the elevated levels of metals found in urban runoff.

The Problem of Characterizing and Controlling Nonpoint Sources

Unlike sewage treatment plants and other point sources which discharge at relatively constant rates, nonpoint sources deliver pollutants in pulses linked to storm events. The quantity and type of pollutant contained in nonpoint sources depends on the human activity, the intensity and duration of precipitation, and the time between storms. The combination of the randomness of rainfall with the varying level of human activity makes controlling nonpoint sources relatively difficult.

In part because of the complex nature of nonpoint sources and in part because nonpoint sources were not historically recognized as a significant pollution source, nonpoint source pollution control efforts have taken a back seat to point source control efforts. From the Clean Water Act's enactment in 1972 until its reauthorization in 1987, total national spending on nonpoint source pollution controls totaled only 6 percent of the amount spent on point source pollution (EPA 1990). In the 1987 Clean Water Act reauthorization, Congress required that the EPA establish a national nonpoint source pollution control program. However, funding for nonpoint source pollution control efforts has been limited. In 1990, the EPA still allocated less than 6 percent of its water pollution control budget to nonpoint sources (GAO 1990).

Because of the relative lack of attention devoted to nonpoint sources, information about their character and the effectiveness of control measures is wanting. This increases the difficulty of implementing control measures.

The Composition of Urban Runoff

Tables D.26 and D.27 show ranges of contaminant levels found in urban runoff for several contaminants of concern, plus lead and suspended solids, as reported in the technical literature. Table D.26 presents average pollutant concentrations. Table D.27 shows annual pollutant loadings from runoff per hectare of land area for overall urban land and four specific land uses: residential, commercial, industrial, and highway. These loadings vary greatly from location to location. Tables D.26 and D.27 show that there is considerable range in the reported data.

Where do these pollutants come from? How does land use affect the quantity of pollution generated? Which pollutants represent the greatest risk?

Urbanization creates nonpoint source pollution problems by increasing the volume of runoff and by broadening the spectrum of substances that accumulate on land. Pavement covers large percentages of urban surface areas and reduces the opportunity for stormwater to filter into the ground. In rural areas or places with large pervious surfaces, surface runoff events are generated only during large storms and snow melt. In urban areas, even relatively small storms (a few tenths of an inch) can create significant runoff.

Urbanization also creates nonpoint source pollution problems by increasing the number of contaminant sources and expanding the variety of contaminants. Any by-product of human activity deposited on an impervious surface and not removed by street cleaning, wind, or decay eventually ends up in surface runoff. The sources of contaminants encompass all human urban activities and include the following:

Traffic—Traffic is directly responsible for the deposition of substantial amounts of toxic hydrocarbons, metals, asbestos, and oils from exhaust

TABLE D.26 Pollutant Concentrations in Urban Runoff Ranges of Values Reported in Technical Literature

Contaminant Class	Contaminant	Reported Concentrations	Reference
Nutrients	Nitrogen (mg/l)	5.6 - 7.1	EPA 1983
	Phosphorus (mg/l)	0.4 - 0.5	EPA 1983
Aesthetics	Oil and grease (mg/l)	4.1 - 15.3	Stenstrom et al. 1984
		10	Wakeham 1977
		13	Eganhouse and Kaplan 1981
Suspended Solids	TSS (mg/l)	71 - 1,194	SCCWRP 1990
		141 - 224	EPA 1983
Metals	Cadmium (μg/l)	3.3 - 4.2	SCCWRP 1990
	Chromium (μg/l)	11 - 43	SCCWRP 1990
	Copper (μg/l)	17 - 138	SCCWRP 1990
		19	Marsalek 1986
		38 - 48	EPA 1983
	Lead (μg/l)	23 - 242	SCCWRP 1990
		90	Marsalek 1986
		161 - 204	EPA 1983
Pathogens	Fecal coliforms (MPN/ 100 ml)	1,000 - 21,000	EPA 1983
		> 2,000	Olivieri et al. 1977

pipes, tire wear, solids carried on tires and vehicle bodies, and lubrication fluid loss.

Litter—Litter deposits contain items such as cans, broken glass, bottles, pull tabs, papers, building materials, plastic, vegetation, dead animals and insects, and animal waste that eventually wash into storm sewers.

Atmospheric deposition—Cadmium, strontium, zinc, nickel, lead, nutrients and many organic hazardous chemicals are transported with atmospheric fallout from local or distant combustion sources and city dust. Atmospheric pollutants find their way into runoff by either settling directly to the ground or becoming entrained in falling rain drops. In larger cities, the deposition rate of atmospheric particulates in wet and dry fallout ranges from 63 kg/hectare/month to more than 270 kg/hectare/month (Novotny and Chesters 1981).

TABLE D.27 Annual Pollutant Loadings from Urban Runoff (pollutant loading units = kg/ha-year)

Class	Contaminant	Residential	Commercial	Industrial	Highway	Urban Land	Reference
Metals	Cadmium	0.013 - 0.016	0.016	0.024	N/A	N/A	Marsalek 1978
		0.026 - 0.028	0.028	0.044	N/A	N/A	Marsalek 1978
		0.09 - 0.11	N/A	N/A	N/A	N/A	Whipple et al. 1978
	Copper	0.045 - 0.049	0.049	0.077	N/A	N/A	Marsalek 1978
		0.03	0.07 - 0.13	0.29 - 1.3	N/A	0.02 - 0.21	Sonzogni 1980
	Lead	0.157 - 0.174	0.174	0.269	N/A	N/A	Marsalek 1978
		0.01- 0.90	2.70	2.7	4.96	N/A	Bannerman et al. 1984
		0.06	0.17 - 1.1	2.2 - 7	N/A	0.14 - 0.5	Sonzogni 1980
		1.0 - 2.3	N/A	N/A	N/A	N/A	Whipple et al. 1978
	PAHs	2.7×10^{-6}	5.9×10^{-6}	8.4×10^{-5}	1.8×10^{-4}	2.2×10^{-5}	Hoffman et al. 1984, Hoffman 1985
Nutrient	Nitrogen	5.0 - 7.3	1.9 - 11	1.9 - 14	N/A	0.2 - 18	Sonzogni 1980
		9 -11.2	11.2	7.8	N/A	N/A	Marsalek 1978
		N/A	N/A	N/A	N/A	5	Beaulac and Reckhow 1982
		N/A	N/A	N/A	N/A	2 - 9	Uttormark et al. 1974

	Phosphorus	0.04 - 1.12	1.49	1.49	1.04	N/A	Bannerman et al. 1984
		1.6 - 3.4	3.4	2.2	N/A	N/A	Marsalek 1978
		1.2 - 8.0	N/A	N/A	N/A	N/A	Whipple et al. 1978
		N/A	N/A	N/A	N/A	0.3 - 4.8	Sonzogni 1980
		N/A	N/A	N/A	N/A	1	Beaulac and Reckhow 1982
		N/A	N/A	N/A	N/A	1.1	Uttormark et al. 1974
Aesthetics	Oil & Grease	3.67	65.4	26.9	31.6	12.4	Stenstrom et al. 1984
		1.8	5.8	140	78	21	Hoffman et al. 1983
		N/A	N/A	N/A	N/A	12.8	Eganhouse and Kaplan 1981
Sus. Solids	TSS	11 - 487	957	957	979	N/A	Bannerman et al. 1984
		360 - 390	360	672	N/A	N/A	Marsalek 1978
		620 - 2,300	50 - 830	450 - 1,700	N/A	200 - 4,800	Sonzogni 1980
		N/A	N/A	N/A	N/A	141 - 224	EPA 1983

N/A = Not available

Plant debris—In residential areas, fallen leaves and grass clippings dominate street refuse composition. In the fall, a mature tree can produce from 15 to 25 kilograms (dry weight) of organic leaf residue containing significant amounts of nutrients (Heaney and Huber 1973). The fallen leaves are about 90 percent organic and contain from 0.04 percent to 0.28 percent phosphorus, and thus if washed into receiving waters, may produce excess nutrients and oxygen-demand as the leaves decay.

Lawn chemicals—Lawn chemicals can potentially contaminate runoff with pesticides, herbicides, and excess nutrients.

Deicing chemicals—The road salt used in most of the northern United States is composed mainly of sodium chloride with added calcium chloride and calcium sulfate, but sand-salt mixtures applied to snow-covered roads may also contain significant amounts of phosphorus, lead, and zinc (Oberts 1986).

Erosion—Stream channel erosion in an urban watershed is two to three times that under predevelopment conditions. Except at construction sites, urban pervious surfaces in humid areas are usually well protected from sheet and rill erosion by vegetation. However, in arid areas where droughts have forced a transition from lawns to less water-demanding landscapes using native arid plants and wood or stone mulches, erosion of pervious lands can be quite extensive.

Septic systems—Septic systems, though not a problem in sewered urban areas, are a significant source of suburban pollution in the coastal zone. Pollution from septic systems enters water via two potential pathways: subsurface transport of mobile pollutants (primarily nitrate) and effluent surfacing from failing systems.

Cross-connections and illicit discharges into storm sewers—Nonstormwater discharges in storm sewers originate from vehicle maintenance activities and from sewage and industrial wastewater leaking from sanitary sewers and failing septic systems into storm-sewered areas. However, almost any type of pollution may find its way into urban storm sewers by illicit discharges and accidental spills. Deliberate dumping of used oil or waste paint into storm sewers and catch basins is especially common and troublesome. Leaking underground storage tanks and leachate from landfills and hazardous waste disposal sites might also infiltrate sewers. The detection, identification, and elimination of such discharges is a major focus of the EPA's new stormwater permit program.

Combined sewer/sanitary systems—Without special control measures, overflows from combined systems occur on average between 40 to 80 times per year. CSOs harbor all the pollutants found in municipal wastewater, including pathogenic microorganisms, trash, and unpleasant odors, and may carry objectionable debris such as the medical waste found on east coast beaches in recent years. Today, between 15,000 and 20,000 CSO discharge points remain in operation (AMSA 1988), serving a total population of about 37 million. They are contained in 1,100 to 1,300 distinct combined collection systems (EPA 1978).

The Effect of Different Land Uses on Urban Pollutant Loading

Micro-level studies have reported that pollutant concentrations in runoff vary significantly with land use. That does seems logical—in industrial areas and roadways, more combustion occurs and more anthropogenic chemicals are used, so one should expect higher concentrations of combustion byproducts and industrial chemicals in that runoff. However, this logic has not been borne out in the most comprehensive national study of urban runoff carried out to date.

The study, the EPA's National Urban Runoff Project (NURP), evaluated contaminant concentrations in runoff in 28 cities between 1978 and 1983. When NURP researchers compiled the national data and analyzed it statistically, the result was surprising: there appeared to be no statistical correlation between pollutant concentrations and three typical urban land uses—residential, mixed, and commercial (EPA 1983). Only pollutant concentrations from open/nonurban lands were significantly different from the three land-use types.

Having determined that land-use category appears to be of little utility in explaining overall site-to-site, storm-to-storm variability of urban runoff, NURP researchers concluded that for estimating pollutant concentrations at unmonitored sites, the best general characterization may be obtained by pooling data for all land uses (other than the open, nonurban ones).

As shown in Table D.27, the NURP studies do not disprove that land use influences unit loadings of pollutants (expressed as mass per unit land area). Pollutant loadings could vary with land use even if one assumes a constant pollutant concentration for all land uses. The smallest loadings are typical for suburban areas with natural surface drainage, which allows much runoff to percolate through the soils. The highest pollutant loadings are emitted from highly paved, urban, industrial centers in which very little runoff is absorbed by the ground.

And so, the question of how urban land use influences pollutant levels in runoff is unresolved.

The Most Significant Contaminants in Urban Runoff

Tables D.26 and D.27 show that urban runoff may transport measurable quantities of all of the pollutants of concern. Which categories of contaminants are most significant in urban runoff?

Metals and Organics

In addition to analyzing the effect of land use on urban runoff, the NURP studies evaluated nationwide data to determine which pollutants occur most commonly. They found that toxic metals are by far the most prevalent priority pollutants in urban runoff (EPA 1983). Copper, lead, and zinc were present in 91 percent of the samples. Other frequently detected inorganic pollutants included arsenic, chromium, cadmium, nickel, and cyanide. Hazardous organic pollutants were detected less frequently and at lower concentrations than metals. The most commonly found organic waste was the plasticizer bis (2-ethylhexyl) phthalate (found in 22 percent of runoff samples), followed by the pesticide α-hexachloro-cyclohexane (α-BHC) (found in 20 percent of the samples). An additional 11 organic pollutants were reported with detection frequencies between 10 percent and 20 percent. Nutrients and oxygen-depleting organics were also detected but in less significant quantities than both toxic metals and toxic organic chemicals.

Pathogens

The NURP and other studies found levels of fecal coliform bacteria in urban runoff high enough to signal a *potential* health risk, as shown in Table D.26. However, the *degree* of health risk posed by these coliform levels is unknown. Since fecal coliforms in urban runoff come mainly from animal waste, fecal coliform counts may not be useful in identifying human health risks. Some researchers have reported the presence of human pathogens in runoff. For example, Olivieri et al. (1977) found that Baltimore storm runoff samples with fecal coliform densities greater than 2,000/100 milliliters were 95 percent positive for *Salmonella*. Ranges of *Salmonella* densities in urban runoff from Baltimore were less than 1 to more than 11,000 per 10 liters. These researchers also analyzed six stormwater flows for viruses; all tested positive (Olivieri et al. 1977).

Suspended Solids: Pollutant Transporters

The NURP study found that urban runoff carries high quantities of sediment. This is significant from a water quality standpoint because other studies have shown that a large portion of pollutants carried in runoff are bound to sediment. For example, Marsalek (1986) measured much higher concentrations of metals and PAHs in runoff-borne sediment than in the runoff water itself. Table D.28 shows pollutant concentrations in runoff solids as measured by Ellis (1986) for different land uses.

That a large quantity of pollutants is bound to solids is important for

TABLE D.28 Pollutant Concentrations in Solids from Urban Runoff (After Ellis 1986. Reprinted, by permission, from Springer-Verlag New York, Inc., 1986.)

Class	Pollutant	Residential	Commercial	Industrial	Highways
Metals	Cadmium (μg/gram)	2.7 - 3.2	2.9	3.6	2.1 - 10.2
Metals	Lead (μg/gram)	1,570 - 1,980	2,330	1,390	450 - 2,346
Pathogens	Fecal coli (MPN/gram)	25,621 - 82,500	36,900	30,700	18,768 - 38,000
Nutrients	Nitrogen (μg/gram)	460 - 610	410 - 420	430	223 - 1,600

reducing pollutant loads delivered by runoff. It means that control measures designed to remove solids will also remove other contaminants. However, a complicating factor is that most pollutants are associated with smaller particles. Studies by Sartor et al. (1974) indicated that more than 50 percent of the sediment-borne phosphorus is contained in the 6 percent of particles smaller than 43 microns. Consequently, systems designed to remove only larger solids will be ineffective for removing most solids-borne pollutants.

Management Options

Introduction

In comparison to wastewater treatment technology, urban runoff treatment technology is in its infancy. By 1991, only the states of Florida and Maryland required urban runoff treatment. There are only a few cities with such requirements, notably Seattle and Bellevue, Washington. Thus, it is from these areas that the field has gained most of the knowledge about the design and performance of urban runoff quality controls. There are few data on the performance of these controls because monitoring of facility performance is not required. For information on removal efficiencies, it is necessary to rely on data gathered in the National Urban Runoff Program conducted in the late 1970s and early 1980s (EPA 1983) and miscellaneous research studies conducted since. For the most part, these data were collected on existing prototype facilities, most of which had not been designed with stormwater treatment in mind but rather simply as drainage facilities. Some had to be retrofitted with other controls to increase detention time. This circumstance limits the usefulness of the data for developing design criteria.

Still, a lot has been learned. Sufficient information is available to design structures into the drainage systems that will definitely improve the quality of urban runoff. Exactly how much improvement will result from a given system is difficult to predict, but estimates are possible. The following sections present an overview of current knowledge of urban runoff quality controls. They address structural and nonstructural types of control, design hydrology and pollutant removal efficiencies for structural controls, cost for implementation, and next steps required in applications and research.

Types of Controls and Control Philosophy

There are two basic controls of pollution in urban runoff: source reduction and structural control. Source reduction prevents the pollutants from ever coming in contact with rainwater or runoff. When it is cost-effective, source reduction is the better approach because the pollutant never gets into the runoff and therefore never enters receiving waters. Source reduction practices include street sweeping, and mitigation of illicit connections and illegal dumping. Many also consider land-use regulations and restrictions as a form of source reduction. Most source reductions can be used in existing as well as in new developments.

Structural controls are those that remove pollution from urban runoff by either reducing the amount of runoff or by providing some type of treatment. Controls that reduce the amount of runoff include reducing impervious areas and increasing infiltration. Treatment controls include sedimentation and biological removal. Typical structural controls include grassy swales, buffer strips, infiltration devices, detention basins, and wetlands.

Source Reduction of Pollution in Urban Runoff

Source reduction of pollution in urban runoff prevents or minimizes the potential of pollutants into contact with rainfall or runoff. Source reduction controls are viewed by most nonpoint source experts as the most cost-effective runoff quality control for developed urban areas, especially for developments more than 20 or 25 years old. The most common source reduction measures are

- elimination of illicit connections,
- mitigation of illegal dumping,
- coverage of chemical storage areas,
- prevention and containment of spills,
- minimization of chemical applications,
- street sweeping and catch basin cleaning, and
- erosion control.

Illicit connections are defined by the EPA as connections to storm drains that contain discharges other than surface runoff. They may or may not be illegal, but they simply do not belong there. Illicit connections might be direct or indirect. A direct connection would be a sanitary sewer connected directly to a storm drain or the connection of a floor drain to the storm drain. A study in Michigan (Murray 1989) revealed that over 60 percent of service stations and auto repair stores had connections directly to storm drains. These connections accounted for 35 percent of all illicit connections to stormsewers in the study area.

In this report, illegal dumping means the intentional disposal of chemicals and waste materials into a storm drainage system. Some illegal dumping is done in known violation of existing laws; other illegal dumping is done simply in ignorance. Examples of the former are septage haulers emptying their trucks into storm sewers and chemical users washing their excess or waste supplies into storm drains. Ignorance is generally the case when an individual changes oil over a catch-basin, dumps waste household chemicals into the gutter, or sweeps yard trash into a catch-basin.

Covering chemical storage areas, loading docks, and other areas at industrial or commercial facilities where contact with rainfall or runoff occurs is a cost-effective control. Spill prevention and containment is a housekeeping measure. Storing chemicals away from storm drains and diking chemical storage areas can prevent contaminants from entering storm sewers.

Control of chemical application rates in areas exposed to rainfall and runoff is another source reduction measure. This control applies to fertilizer, pesticide, and herbicide application in private and public properties for gardening and pest control. There has been significant progress over the past 10 years in developing technology to minimize highway deicing application rates without compromising safety (Lord 1989).

Street sweeping is a debated control practice because it is difficult to determine its effectiveness, while more benefits are accruable to catch-basin cleaning.

Surface erosion from developed watersheds is generally less than that under natural conditions. However, during construction, it can be a major problem. The principal impact on receiving water is due to sedimentation since erosion from construction sites usually consists of natural soil. Although the technology for erosion control on construction sites is well-developed (VSCC 1984), erosion control requirements for construction sites are rarely enforced.

Land-use controls include zoning to minimize pollutant loads to receiving waters. Runoff pollution is directly related to the intensity of development. Low-density residential development is characterized by having a smaller impervious area and smaller amount of human activity than higher density residential and/or commercial areas. Thus, where runoff controls are difficult to implement due to terrain or topography such as steep slopes or shorefront development, zoning for low density development becomes a source reduction practice. While density limitations have not been used much in the past due to pressure from developers, land-use controls are likely to be used more extensively in shoreline areas, especially estuarine waters, for runoff quality management, in the future.

Source reduction is implemented by two methods: education and ordinances. Education is an important component of source reduction strategies because the public generally has little idea about where storm drains go. Many communities are embarking on public education programs that include workshops, videotapes, and flyers alerting the public to runoff pollution and what they can do to reduce it. The city of Ann Arbor, Michigan, has undertaken a program to paint a fish on storm drain inlets accompanied by the text: "Do not dump, drains to stream." Appendix E provides further information on the role of education.

Ordinances are most effective in controlling illicit connections, illegal dumping, chemical storage location and covering, erosion from new construction sites, and land use. However, ordinances are ineffective unless they are backed up with adequate enforcement.

Structural Controls

Structural controls are those devices that are designed into a drainage system to remove pollutants from runoff. These devices include swales, filter strips, infiltration basins and trenches, detention facilities, and artificial wetlands. Treatment facilities used for municipal wastewater and industrial wastes are not generally considered viable for stormwater treatment because 1) they are extremely expensive, and 2) they are designed to perform under continuous and fairly uniform loading. Stormwater runoff is highly variable in the frequency and magnitude of both runoff and pollutant concentrations; therefore it does not lend itself to treatment by those practices.

It is commonly but erroneously thought that design storms for sizing water quality controls should be the same as those used for the design of drainage facilities. Design storms for urban drainage systems are large, infrequent storm events ranging from the 5-year storm to the 25-year storm. But the damage done to a receiving water ecosystem by uncontrolled pollutant wash-off in the 25-year storm is inconsequential compared with the hydraulic damage that results naturally to aquatic habitats from such a storm. The same can be said for the five-year storm. Design storms for runoff quality control are small, frequent events smaller than the 1-year storm.

Figure D.9a shows the percentage of annual runoff that will be captured by detention basins of various sizes for six U.S. cities. This information was obtained by analyzing hourly rainfall data over more than 10 years at each city. A 24-hour drawdown time for each basin was used because that is the detention time required for effective pollution removal (Grizzard et al. 1986). For all six cities, less than 1 inch (0.03 million gallons per acre) of detention storage is required to capture 90 percent of the runoff and detain it for 24 hours; for Cincinnati, Detroit, Tucson, and Butte, less than 0.5 inches (0.014 MG/acre) is required. Further study of these curves indicates that the most cost-effective basin size (generally taken at the knee of the curve) varies from 0.7 inches in San Francisco to 0.18 inches in Butte and captures 80 percent and 90 percent of the annual runoff respectively. Thus 80 percent or 90 percent capture of annual runoff would appear to be a good technology-based standard for urban runoff quality control.

Figure D.9b shows how many times per year each basin is expected to fill and overflow. The black dot on Figure D.9b indicates the basin volume required for each city to capture 90 percent of the annual runoff. This figure shows that the overflow frequency of basins sized to capture 90 percent of the runoff volume overflow three to ten times per year, suggesting that the design storm is on the order of about a one-month to a four-month storm.

Structural controls vary from site controls. Site controls attempt to reduce runoff rate and volume at or near the point where the rainfall hits the ground surface. Regional controls (usually detention devices) serve an area of 50 to 100 acres. The following types of structural controls are common:

- minimization of directly-connected impervious areas,
- swales and filter strips,
- porous pavement and parking blocks,
- infiltration devices such as trenches and basins, and
- detention devices.

Minimization of Directly Connected Impervious Areas. Impervious areas are those that do not absorb any rainfall—rooftops, sidewalks, driveways, streets, and parking lots. The minimization of directly connected impervious areas is by far the most effective method of runoff quality control because it delays the concentration of flows into the improved drainage system and maximizes the opportunity for rainfall to infiltrate at or near the point at which it falls.

Swales and Filter Strips. Swales, or grassed waterways, and filter strips are among the oldest stormwater control measures, having been used alongside streets and highways for many years. These devices slow the runoff, giving solids an opportunity to settle or be filtered out by vegetation

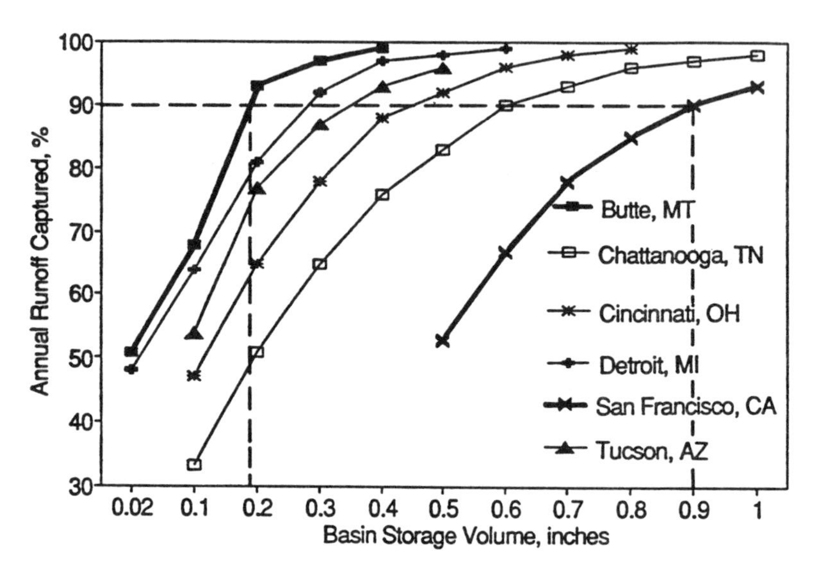

FIGURE D.9a Runoff capture efficiency versus unit storage volume. (Source: Roesner et al. 1991. Reprinted, by permission, from American Society of Civil Engineers, 1991.)

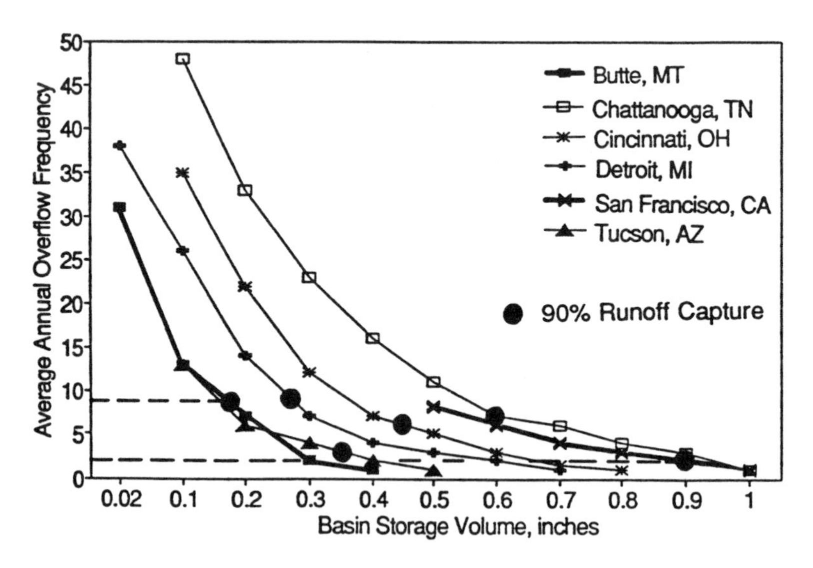

FIGURE D.9b Basin overflow frequency versus unit storage volume. (Source: Roesner et al. 1991. Reprinted, by permission, from American Society of Civil Engineers, 1991.)

and provide an opportunity for the runoff to infiltrate into the ground. To be useful as a treatment device, a swale should have a flat bottom or very flat side slopes so that the water-quality design storm can be conveyed with less than 3 inches of water in the channel. The channel should be planted with vegetation suitable for soil stabilization and nutrient uptake.

A filter strip is simply a mildly sloped strip of land across which stormwater from a street, parking lot, rooftop, or other impervious surface sheet-flows before entering adjacent receiving waters. Minimum length of a filter strip should be 20 feet, and its slope should be such that erosion does not occur, except infrequently during large storms.

Swales and filter strips are widely used for urban runoff treatment in Florida and metropolitan Seattle. They have been used to a limited extent in Virginia and Maryland. They are best suited to non-arid areas.

Porous Pavement and Parking Blocks. Porous pavement has excellent potential for use in parking areas. When properly designed and carefully installed and maintained, porous pavement can have load-bearing strength and longevity similar to conventional pavement. In addition, porous pavement can help to reduce the amount of land needed for stormwater management, preserving the natural water balance at a given site. However, porous pavement is only feasible on sites with permeable soils, fairly flat slopes, and relatively deep water table and bedrock levels. The risk of clogging is high, and once clogging occurs it is difficult and costly to correct.

Another effective site-control device is parking blocks or modular pavement. These are hollow concrete blocks similar to, but smaller than, those used in construction. In parking lots for retail stores, sports arenas, civic theaters, etc., where more than half the parking area is used less than 20 percent of the time, the use of parking blocks in the less-used portions gives them a more attractive appearance and reduces runoff quantity, flow rates, and pollution from these areas.

Infiltration Devices. Infiltration devices are those stormwater quality control measures that completely capture runoff from the water-quality design storm and allow it to infiltrate the ground. They are the most effective stormwater quality control devices that can be implemented because pollutants in the infiltrated flow are removed from the runoff. Advantages of infiltration devices are that they help to maintain the natural water balance of a site and can be integrated into a site's landscaped and open areas. Disadvantages can include a fairly high rate of failure due to unsuitable soils; the need for frequent maintenance; and possible nuisance factors, such as odors, mosquitos, or soggy ground.

An infiltration basin is made by constructing an embankment or by excavating in or down to relatively permeable soils. The basin will temporarily store stormwater until it infiltrates through the bottom and sides of the system. However, the infiltration basin can actually be a landscape depression within open or recreational areas. Infiltration basins generally serve areas ranging from a front yard to areas of five or ten acres in size.

Infiltration trenches, which can be located on the surface of the ground or buried beneath the surface, are usually designed to serve areas of from five to ten acres and are especially appropriate in urban areas where land costs are very high. Stored runoff infiltrates into the surrounding soil.

Detention Devices. Detention basins are widely used throughout the United States to reduce runoff peaks for drainage control (i.e., peak-flow attenuation of the 2-, 10-, 25-, and 100-year storms). They are designed with fixed flow rate outfall pipes. Because these basins are designed for peak shaving of runoff from large storms, the small storms of interest for urban runoff quality control pass through them with little or no detention or pollutant removal. (Peak runoff rates for small storms are much less than the outlet capacity of these detention basins.) However, studies in Virginia and Maryland (Grizzard et al. 1986) and at several other locations (Zariello 1989) indicate that these basins can be effectively retrofitted with adjustable outflow controls to increase detention times and pollutant removal efficiency for small storms. Care must be taken with these retrofit devices not to reduce the flood control effectiveness for which these facilities were designed.

For treatment of urban runoff, detention devices fall into three basic categories: extended detention, detention with filtration, and wet detention. The removal mechanism of extended detention is sedimentation; detention with filtration combines sedimentation with filtration; wet detention incorporates sedimentation and biochemical removal.

Extended detention basins are the most common type of detention basin used around the country. These basins are dry between storms and capture small storms and the first flush from big storms. They then release it slowly. Since stormwater pollutants tend to be associated primarily with very small particles (very fine sands, silts, and clays), relatively long detention times (20 to 40 hours) are required to achieve appreciable removal of suspended pollutants. Even with such extended detention times, however, removal of dissolved stormwater pollutants does not occur. Typical removal efficiencies for extended detention basins are (Hartigan 1989):

TSS: 80% - 90%	TN: 20% - 30%	Pb: 70% - 80%
BOD: 20% - 40%	TP: 20% - 30%	Zn: 40% - 50%

One of the stormwater treatment practices commonly used in Florida is detention coupled with filtration in which stored stormwater is discharged through a filter. Typical filtration systems include bottom or side-bank sand or natural soil filters. Experience in Florida and Texas indicates significant difficulties associated with the design, construction, and especially the maintenance of stormwater filters. It is not a question of if a filter will clog, but when, and who will maintain the filter when clogging becomes a problem. Thus filters should be used when timely maintenance is assured. Livingston et al. (1988) describes design details.

Wet detention basins are characterized by a permanent pool of water and a shallow littoral zone around the perimeter that occupies 30 percent to 50 percent of the pond area. The volume of the pond itself is equal to the runoff from the wettest two weeks of an average year. This volume of water is significant in most locales, and is what distinguishes a *wet pond* from an *extended detention basin* that may have a small permanent pool associated with it. The removal of pollutants in a wet detention system is accomplished by gravity settling and biological uptake of nutrients by aquatic plants and phytoplankton metabolism. A wet pond is the best detention facility for use in locations where nutrients are of concern because they remove two to three times as much phosphorus as extended detention ponds and 1.3 to two times as much total nitrogen, if the plants are harvested.

Artificial Wetlands for Stormwater Quality Enhancement. Wetlands provide water quality enhancement through sedimentation, filtration, absorption, and biological processes. They also provide flood protection through water storage and conveyance. The incorporation of wetlands into a comprehensive stormwater management system achieves wetland preservation and revitalization (Hartigan 1989). While much research has been completed on the ability of wetlands to remove wastewater pollutants (EPA 1985b, Martin 1988), many questions remain. For example, how long can a wetland continue to remove stormwater pollutants effectively? What type of maintenance is required and at what frequency? What is the ultimate fate of pollutants in wetland habitats and how do they affect the wetland ecosystem? Furthermore, treatment wetlands fall under the jurisdiction of the federal wetlands protection law (Clean Water Act, Section 401[k]), which limits the way in which routine maintenance can be done. Design guidance relative to constructed wetlands can be found in Maryland Water Resources Administration (1987) and Livingston (1989).

Retrofitting Structural Controls to Existing Developments. The idea of retrofitting structural controls into an existing setting appears a formidable task at first blush. However, there are a number of ways to retrofit at reasonable cost.

Two devices that are fairly simple to retrofit are oil-water separators and water quality inlets. These can be installed at stormwater inlets in parking lots, service stations, and other areas where oils and greases may be in the runoff. Infiltration trenches have been retrofitted at a number of roadways in Maryland. Also, the city of Orlando, Florida, replaced its entire downtown storm drainage system with infiltration trenches in order to protect its urban lakes from runoff pollution.

Other devices such as street replacement or storm sewer system improvements can be retrofitted as part of infrastructure repairs. For example, porous pavement, modular pavement, or geotextile fabric should be considered as a replacement in parking areas that are in need of upgrading.

There are many ways to retrofit runoff treatment devices to existing development. To do so requires ingenuity and knowledge of how specific treatment devices work to remove pollutants from runoff.

Pollutant Removal Efficiencies of Various Treatment Practices. Table D.29 shows the relative efficiency of various urban runoff quality controls in removing pollutants. This table is based on data collected in the late 1970s and early 1980s. More recent data (Roesner et al. 1989) indicate that wet ponds, filter strips, and swales perform much better than shown, if properly designed. Notice that where suspended sediment removal is good, removal of other pollutants is good. This is because many of the noxious pollutants in urban runoff are attached to particulate matter. As a rule of thumb, if the solids can be removed from the runoff, most of the noxious pollutants will also be removed. This rule does not hold true for nitrogen or bacteria.

Table D.30 shows removal efficiencies for constituents of concern. The table was developed under the assumptions that 1) copper, cadmium, and chromium will behave in the same way as trace metals and lead, 2) PAHs will be associated primarily with suspended solids and behave in the same way as heavy metals, 3) coliforms are an adequate predictor of enterovirus, and 4) oil and grease are 80 percent to 100 percent removed in any properly designed stormwater treatment device.

Rating of Runoff Treatment Practices

The matrix in Table D.31 provides a relative rating of structural controls. This table summarizes the characteristics of the practices described above. It is noteworthy that, in contrast to wastewater treatment processes, most of the practices are not operationally difficult.

Costs for Stormwater Quality Controls

Cost data for source controls are not available. Studies of source control practices (Murray 1989) do not contain information on costs for illicit connection detection or removal. Little information is available on costs of

TABLE D.29 Comparative Pollutant Removal of Urban Runoff Quality Controls (From Schueler 1987. Reprinted, by permission, from Metropolitan Washington Council of Governments, 1987.)

BMP/design	Settleable Solids	Total Phosphorus	Total Nitrogen	Oxygen Demand	Trace Metals	Bacteria	Overall Removal Capability
Extended Detention Pond							
Design 1	◕	⊕	⊕	⊕	◒	⊗	Moderate
Design 2	●	◒	⊕	◒	◕	⊗	Moderate
Design 3	●	◕	◒	◒	◕	⊗	High
Wet Pond							
Design 4	◕	◒	⊕	⊕	⊕	⊗	Moderate
Design 5	◕	◒	⊕	⊕	◕	⊗	Moderate
Design 6	●	◕	◒	◒	◕	⊗	High
Infiltration Trench							
Design 7	◕	◒	◒	◕	◕	◕	Moderate
Design 8	●	◒	◒	◕	●	◕	High
Design 9	●	◕	◕	●	●	●	High
Infiltration Basin							
Design 7	◕	◒	◒	◕	◒	◕	Moderate
Design 8	●	◒	◒	◕	●	◕	High
Design 9	●	◕	◕	●	●	●	High
Porous Pavement							
Design 7	◒	◕	◒	◕	◒	◕	Moderate
Design 8	●	◕	◕	◕	●	●	High
Design 9	●	◕	◕	●	●	●	High
Water Quality Inlet							
Design 10	○	⊗	⊗	⊗	⊗	⊗	Low
Filter Strip							
Design 11	⊕	○	○	○	⊕	⊗	Low
Design 12	●	◒	◒	◕	●	⊗	Moderate
Grassed Swale							
Design 13	○	○	○	○	○	⊗	Low
Design 14	⊕	⊕	⊕	⊕	○	⊗	Low

continued

structural controls other than that collected by the Metropolitan Washington Council of Governments (MWCOG). A MWCOG study (Wiegand et al. 1986) drew together cost data from a survey of engineering estimates and bids for 65 infiltration and detention facilities built since 1982 in the Metropolitan Washington area. Based on these data, regression equations for cost versus volume were developed.

TABLE D.29 *Continued*

Design 1: First-flush runoff volume detained for 6-12 hours.
Design 2: Runoff volume produced by 1.0 inch, detained 26 hours.
Design 3: As in Design 2, but with shallow marsh in bottom stage.
Design 4: Permanent pool equal to 0.5 inch storage per impervious acre.
Design 5: Permanent pool equal to 2.5 (Vr): where Vr = mean storm runoff.
Design 6: Permanent pool equal to 4.0 (Vr): approximately 2 weeks retention.
Design 7: Facility exfiltrates first-flush: 0.5 inch runoff/impervious acre.
Design 8: Facility exfiltrates one inch runoff volume per impervious acre.
Design 9: Facility exfiltrates all runoff, up to the 2 year design storm.
Design 10: 400 cubic feet wet storage per impervious acre.
Design 11: 20 foot wide turf strip.
Design 12: 100 foot wide formated strip with level spreader.
Design 13: High slope swales, with no check dams.
Design 14: Low gradient swales with check dams.

○ - 0% to 20% removal
◔ - 20% to 40% removal
◓ - 40% to 60% removal
◕ - 60% to 80% removal
● - 80% to 100% removal
⊗ - Insufficient knowledge

TABLE D.30 Comparative Removal of Pollutants of Concern by Runoff Treatment Practices

Treatment Practice	TSS	Cu, Cd, Cr	PAHs	ABS	Coliforms, Viruses	Total Nitrogen	Total Phosphorus	Oil and Grease
Grassed Swale (mild slope)	◔	◔	◔	⊗	⊗	◔	◔	◓
Filter Strip (20 feet)	◔	◔	◔	⊗	⊗	◔	◔	◓
Porous Pavement[1]	●	●	●	◕	●	◕	◕	●
Infiltration Basin[1]	●	●	●	●	●	●	●	●
Infiltration Trench[1]	●	●	●	●	●	●	●	●
Extended Detention Basin[1]	●	◕	◕	⊗	⊗	◔	◔	◓
Wet Pond (2 week retention)	●	◕	◕	⊗	◕	◓	◓	◕

[1] Capture first 0.5 inches of runoff from tributary area.

KEY: ○ - 0% to 20% removal
◔ - 20% to 40% removal
◓ - 40% to 60% removal
◕ - 60% to 80% removal
● - 80% to 100% removal
⊗ - Insufficient knowledge

TABLE D.31 Rating of Runoff Treatment Practices

Practice	Reliability	Maintenance	Difficulty of Operation	Land Required	Cost	Demonstrated Effectiveness	Application
Grassed Swales	○	◓	○	○	○	○	Longitudinal slope <2% small drainage area
Filter Strips	○	◓	○	○	○	○	Parking lot runoff, street runoff
Porous Pavement	◓	◓	○	○	○	◓	Parking and sidewalks on pervious soils
Infiltration Basins	◓	◓	○	◓	○	●	Lot and site drainage on pervious soils
Infiltration Trenches	○	●	○	○	●	◓	Lot and site drainage on pervious soils
Extended Detention Basins	●	◓	○	○	○	○	Site and regional facilities
Wet Ponds	●	○	◓	●	◓	●	Site and regional facilities

KEY: ● - High
◓ - Medium
○ - Low

REFERENCES

AMSA (Association of Metropolitan Sewerage Agencies). 1988. Draft CSO Permitting Strategy. Bulletin No. GB88-22. Washington, D.C.: Association of Metropolitan Sewerage Agencies.

AMSA (Association of Metropolitan Sewerage Agencies). 1990. 1989-1990 AMSA Pretreatment Survey Final Report. Washington D.C: Association of Metropolitan Sewerage Agencies.

Arora, M.L. 1992. Water Reuse in California—A Myth of a Reality? In Proceedings from the Water Environment Federation "Urban and Agricultural Water Reuse" Conference, Orlando, Florida, June 28 - July 1, 1992. Alexandria, Virginia: Water Environment Federation.

Asano, T. 1991. Planning and implementation of water reuse projects. Water Science and Technology 24(9):1-10.

Asano, T., L.Y.C. Leong, M. G. Rigby, and R. H. Sakaji. 1992. Evaluation of the California wastewater reclamation criteria using enteric virus monitoring data. Water Science and Technology 26(7-8):1513-1524.

Bannerman, R., K. Baun, M. Bohm, P.E. Hughes, and D.A. Graczyk. 1984. Evaluation of Urban Nonpoint Source Pollution Management in Milwaukee County, Wisconsin. Report No. PB84-114164. Chicago, Illinois: U.S. Environmental Protection Agency, Region V.

Beaulac, M.N., and K.H. Reckhow. 1982. An examination of land use-nutrient export relationships. Water Resources Bulletin 18:1013-1024.

Berube, M.R., and J. Nash. 1991. From Pollution Control to Zero Discharge—Overcoming the Obstacles: the Robbins Company, Attleboro, Massachusetts. Cambridge, Massachusetts: Center for Technology, Policy, and Industrial Development, MIT.

Brombach, H. 1987. Liquid-Solid Separation at Vortex-Storm Overflows. Proceedings Fourth International Conference on Urban Storm Drainage. Lausanne, Switzerland, International Association of Water Pollution Research and Control. Oxford, UK: Pergamon Press Ltd.

Brombach, H., C. Xanthopoulos, H. Hahn, and W. Pisano. 1992. Experience with Vortex Separators for Combined Sewer Overflow Control. Paper read at the International Conference on Sewage into 2000. August 31, 1992. Amsterdam, Holland.

Bruskin, C.A., and K.P. Lindstrom. 1992. Water Conservation. Orange County, California: County Sanitation Districts of Orange County.

Cal-Tech Management Associates and SCADA Systems Inc. 1987. Final Report Waste Reduction for the Printed Circuit Board Industry. Sacramento, California.

Camp Dresser & McKee, Inc. 1991. Proposed Wastewater Management Plan. Prepared for Cincinnati/Hamilton County Metropolitan Sewer District, Cincinnati, Ohio.

CFR (Code of Federal Regulations). Cost-Effectiveness Guidelines. Title 40, Part 35, Subpart E, Appendix A.

Chatel, B. 1992. Pollution Prevention at the Robbins Company. Presentation given at Massachusetts Institute of Technology, Cambridge, Massachusetts.

Chaudhary, R., Y. Shao, J. Crosse, and F. Soroushian. 1991. Evaluation of chemical addition. Water Environment & Technology February:66-71.

CH_2M Hill. 1988. TM-8 Preliminary CSO Control by Sewer Separation. Boston, Massachusetts: CH_2M Hill.

CH_2M Hill. 1989. Best Management Practices Final Report, Combined Sewer Overflow Program. Boston, Massachusetts: CH_2M Hill.

Clinton Bogert Associates. 1991. Combined Sewer Overflow Control Facility Plan. Elizabeth, New Jersey: Clinton Bogert Associates.

CSDOC (County Sanitation Districts of Orange County). 1989. Collection, Treatment, and Disposal Facilities Master Plan. February 1989. Fountain Valley, California: CSDOC.

Drehwing, F. 1979. Disinfection/Treatment of Combined Sewer Overflows. Report No. EPA 600/2-79-134. Syracuse, New York: O'Brien & Gere Inc., and U.S. EPA.

Eganhouse, R.P., and I.R. Kaplan. 1981. Extractable organic matter in urban stormwater runoff: 1. Transport dynamics and mass emission rates. Environmental Science and Technology 15:310-315.

Ellis, J.B. 1986. Pollutional aspects of urban runoff. Pp. 1-38 in Urban Runoff Pollution, H.C. Torna, J. Marsalek, and M. Desbordes, eds. New York: Springer-Verlag New York, Inc.

EPA (U.S. Environmental Protection Agency). 1978. Report to Congress on Control of Combined Sewer Overflow in the United States. Report EPA-430/9-78-006. Washington, D.C.: EPA.

EPA (U.S. Environmental Protection Agency). 1979. Development Document for Existing Source Pretreatment Standards for Electroplating Industry. EPA 440/1-79/003. Washington, D.C.: EPA.

EPA (U.S. Environmental Protection Agency). 1983. Final Report of the Nationwide Urban Runoff Program. Volume I. Washington, D.C.: EPA, Water Planning Division.

EPA (U.S. Environmental Protection Agency). 1985a. Handbook for Estimating Sludge Management Costs. Report No. EPA-625/6-85-010. Cincinnati, Ohio: EPA.

EPA (U.S. Environmental Protection Agency). 1985b. Freshwater Wetlands for Wastewater Management Handbook. Report No. 904/9-85-135. Washington, D.C.: EPA.

EPA (U.S. Environmental Protection Agency). 1986. Report to Congress on the Discharge of Hazardous Waste to Publicly Owned Treatment Works. Washington D.C.: EPA.

EPA (U.S. Environmental Protection Agency). 1989a. Draft Combined Sewer Overflow Guidance Manual. Washington, D.C.: EPA.

EPA (U.S. Environmental Protection Agency). 1989b. National Sewage Sludge Survey. Washington, D.C.: EPA.

EPA (U.S. Environmental Protection Agency). 1989c. Environmental Regulations and Technology: Control of Pathogens in Municipal Wastewater Sludge. EPA/625/10-89/006. Cincinnati, Ohio: EPA.

EPA (U.S. Environmental Protection Agency). 1990. National Water Quality Inventory: 1988 Report to Congress. Report EPA 440/4-90-003. Washington, D.C.: EPA.

EPA (U.S. Environmental Protection Agency). 1991a. Pollution Prevention 1991: Progress on Reducing Industrial Pollutants. Report No. EPA 21P-3003. Washington D.C.: Office of Pollution Prevention, EPA.

EPA (U.S. Environmental Protection Agency). 1991b. Proposed Development and Approved Guidance—State Coastal Nonpoint Pollution Control. Washington, D.C.: EPA.

EPA (U.S. Environmental Protection Agency). 1992a. Facility Pollution Prevention Guide. EPA/600/R-92/088. Washington, D.C.: EPA, Office of Research and Development.

EPA (U.S. Environmental Protection Agency). 1992b. Guidelines for Water Reuse. EPA/625/R-92/004. Washington, D.C.: EPA.

Esmund, S.E., A.C. Petrasek, Jr., H.W. Wolf, and D. Craig. 1980. The Removal of Metals and Viruses in Advanced Wastewater Treatment Sequences. U.S. EPA. Report No. EPA-600/2-80-149. College Station, Texas: Texas A&M University.

FDER (Florida Department of Environmental Regulation). 1991. Guidelines for Presentation of Reuse Feasibility Studies for Applicants Having Responsibility for Wastewater Management. Tallahassee, Florida: Florida Department of Environmental Regulation.

Federal Register. 1989. National Combined Sewer Overflow Control Strategy 54(173).

FHA (Federal Highway Administration). 1985. Management Practices for Mitigation of Highway Stormwater Runoff Pollution. McLean, Virginia: Federal Highway Administration.

Folkard, G.K. 1986. Appropriate Technology for Treatment of Potable Water in Developing Countries: Coagulants Derived from Natural Materials. In Recycling in Chemical Water and Wastewater Treatment. Schriftenriehe ISWW Karlsruhe Bd 50.

Force, J.M., ed. 1991. Primary effluent filtration. Reactor: Wastewater Treatment News from Zimpro Passavant December(72):5.

Forsell, B., and B. Hedstrom. 1975. Lamella sedimentation: A compact separation technique. Journal of the Water Pollution Control Federation 47:834.

GAO (General Accounting Office). 1990. Water Pollution: Greater EPA Leadership Needed to Reduce Nonpoint Source Pollution. Report # GAO/RCED-91-10. Washington, D.C.: General Accounting Office.

Grizzard, T.L., C.W. Randall, B.L. Weand, and K.L. Ellis. 1986. Effectiveness of Extended Detention Ponds. Pp. 323-337 in Urban Runoff Quality—Impact of Quality Enhancement Technology. New York, New York: American Society of Civil Engineers.

Guterstam, B., and J. Todd. 1990. Ecological engineering for wastewater treatment and its application in New England and Sweden. Ambio 19(3).

Hannah, S.A., B. Austern, A. Eralp, and R. Wise. 1986. Comparative removal of toxic pollutants at six wastewater treatment processes. J. Water Pollut. Control Fed. 58(1):27-34.

Hartigan, J.P. 1989. Basis for design of wet detention basin BMPs. Pp. 122-144 in Design Criteria for Urban Runoff Quality Control. New York, New York: American Society of Civil Engineers.

Hauck, J., and S. Masoomian. 1990. Alternate Technologies for Wastewater Treatment. Pollution Engineering May 1990.

Havens & Emerson Engineers. 1987. Value Engineering Investigation for Granite Ave. Relief Sewer. Boston, Massachusetts: Boston Water and Sewer Commission.

Hazen, A. 1904. On sedimentation. Transactions of the American Society of Civil Engineers 53:45.

Heaney, J.P., and W.C. Huber. 1973. Stormwater Management Model: Refinements, Testing and Decision-Making. Gainesville, Florida: University of Florida Department of Environmental Engineering and Science.

Heidman, J. 1990. Sequencing batch reactors. In Proceedings of the U.S.EPA Municipal Wastewater Treatment Technology Forum. Report No. EPA 430/09-90-015. Washington, D.C.: U.S. Environmental Protection Agency.

Heinking, G., and N. Wilcoxon. 1985. Use of a swirl concentrator for combined sewer overflow management. J. Water Pollut. Control Fed. 57(5):398.

Hoffman, E.J. 1985. Urban runoff pollutant inputs to Narragansett Bay: Comparison to point sources. In Perspectives on Nonpoint Pollution: Proceedings of National Conference. Report EPA 440/5-85-001. Washington, D.C.: U.S. Environmental Protection Agency.

Hoffman, E.J., G.L. Mills, J.S. Latimer, and J.G. Quinn. 1983. Annual input of petroleum hydrocarbons to the coastal environment via urban runoff. Can. J. Fish. Aquat. Sci. 40(2):41-53.

Hoffman, E.J., G.L. Mills, J.S. Latimer, and J.G. Quinn. 1984. Urban runoff as a source of polycyclic aromatics to coastal waters. Environmental Science and Technology 18:580-587.

Hudson, D. 1990. Experience in Operation of Real Time Process Control. North East Ohio Regional Sanitary District. Paper presented at Water Pollution Control Federation Real Time Control Workshop, Washington, D.C., October, 1990.

Hunsinger, C. 1987. Combined Sewer Overflow, Operational Study. Decatur, Illinois: Bainbridge, Gee, Milanski Associates and Crawford, Murphy & Tully, Inc.

Jones, M. 1991. Opportunities in water and sewage treatment. EPRI Journal June:42-44.

Kaufman, H., and D. Lai. 1978. Conventional and Advanced Sewerage Design Concepts for Dual Flood and Pollution Control. Report No. EPA 600/2-78-090. Cincinnati, Ohio: U.S. Environmental Protection Agency.

Leblanc, F. 1987. Theory, Technology and Applications of Lamella Settling. Paper presented at the International Symposium of Protection of Marine Environments from Urban Pollution, Marseille, France.

Livingston, E. 1989. The use of wetlands for urban stormwater management. Pp. 467-489 in Design of Urban Runoff Quality Controls, L. Roesner, B. Urbonas, and M. Sonnen, eds. New York: American Society of Civil Engineers.

Livingston, E.H., E. McCarron, J. Cox, and P. Sanzone. 1988. The Florida Department Manual: A Guide to Sound Land and Water Management. Tallahassee, Florida: Florida Department of Environmental Regulation.

Lord, B.N. 1989. Program to Reduce Deicing Chemical Uses. Pp. 421-433 in Design of Urban Runoff Quality Controls, L. Roesner, B. Urbonas, and M. Sonnen, eds. New York: American Society of Civil Engineers.

Marsalek, J. 1978. Pollution Due to Urban Runoff: Unit Loads and Abatement Measures. Windsor, Ontario, Canada: Pollution by Land Use Activities Reference Group of the International Joint Commission.

Marsalek, J. 1986. Toxic contaminants in urban runoff. Pp. 39-57 in Urban Runoff Pollution, H. Torna, J. Marsalek, and M. Desbordes, eds. New York/Berlin: Springer Verlag.

Marshall, T. 1987. Operators Manual—Joint Water Pollution Control Plant. Los Angeles: County Sanitation Districts of Los Angeles County, California.

Martin, E.H. 1988. Effectiveness of an urban runoff detention pond—wetlands system. Journal of the Environmental Engineering Division, Paper No. 22649, July 1988. New York, New York: American Society of Civil Engineers.

Maryland Water Resources Administration. 1987. Guidelines for Constructing Wetland Stormwater Basins. Annapolis, Maryland: Maryland Water Resources Administration.

Matsumoto, M. 1991. Primary Effluent Filtration. Paper presented at 64th Conference of the Water Environment Federation, Toronto, Canada, October, 1991.

McCarty, P.L., M. Reinhard, J. Graydon, J. Schreiner, and K. Sutherland. 1980. Wastewater Contaminant Removal for Groundwater Recharge at Water Factory 21. Report No. EPA-600/2-80-114. Cincinnati, Ohio: U.S. Environmental Protection Agency.

Metcalf and Eddy. 1991. Wastewater Engineering: Treatment, Disposal, Reuse. Third Edition. New York: McGraw-Hill Book Company.

Murcott, S., and D. Harleman. 1992a. Performance and Innovation in Wastewater Treatment. Technical Note # 36. Cambridge, Massachusetts: Massachusetts Institute of Technology.

Murcott, S., and D. Harleman. 1992b. Chitosan, Metal Salt, and Polyacrylamide Jar Tests at the Gloucester Water Pollution Control Facility. Report # MITSG 92-12. Cambridge, Massachusetts: Massachusetts Institute of Technology Sea Grant.

Murray, J.E. 1989. Nonpoint Pollution First Step in Control. Pp. 378-387 in Design of Urban Runoff Quality Controls, L. Roesner, B. Urbonas, and M. Sonnen, eds. New York: American Society of Civil Engineers.

Nieuwstad, T., E. Mulder, A. Havelaar, and M. Van Olphen. 1988. Elimination of microorganisms from wastewater by tertiary precipitation and simultaneous precipitation followed by filtration. Water Research 22(11):1389-1397.

NHSRC (Northeast Hazardous Substances Research Center). 1991. Reader in Preventing Pollution. Cambridge, Massachusetts: Massachusetts Institute of Technology. September 1991.

Novotny, V., and G. Chesters. 1981. Handbook of Nonpoint Pollution: Sources and Management. New York: Van Nostrand Reinhold Co.

Oberts, G.L. 1986. Pollutants associated with sand and salt applied to roads in Minnesota. Water Resources Bulletin 22:479-484.

Olivieri, V., C. Kruse, K. Kawata, and J. Smith. 1977. Microorganisms in Urban Stormwater. Report No. EPA 600/2-77-087. Cincinnati, Ohio: EPA Municipal Environmental Research Laboratory.

Pisano, W.C. 1978. Case Study: Best Management Practice Solution for a Combined Sewer Program, Urban Stormwater Management Workshop Proceedings. Report No. EPA 600/9-78-017. Cincinnati, Ohio: U.S. Environmental Protection Agency.

Pisano, W.C. 1979. Dry Weather Deposition and Flushing for Combined Sewer Overflow Pollution Control. Report No. EPA 600/2-79-133. Cincinnati, Ohio: U.S. Environmental Protection Agency.

Pisano, W.C. 1982a. Inlet Control Program, Ridge Road Sector, Parma, Ohio. Belmont, Massachusetts: Environmental Design & Planning, Inc.

Pisano, W.C. 1982b. An Overview of Four Inlet Control Studies for Mitigating Basement Flooding in Cleveland and Chicago Area. Cleveland, Ohio: American Society of Civil Engineers.

Pisano, W.C., and D. Wolf. 1991. Better Utilization of Storage in Treatment for CSO Control, A Case Study: 14 St. Facility, Saginaw, Michigan. Toronto, Canada: Water Environment Federation.

Pisano, W.C., N. Thibault, and G. Forbes. 1990. The vortex solids separator. Water Environment and Technology 2(5):64-71.

Qualls, R.G., S.F. Ossoff, J.C.H. Chang, M.H. Dorfman, C.M. Dumais, D.C. Lobe, and J.D.

Johnson. 1985. Factors controlling sensitivity in ultraviolet light disinfection of secondary effluents. J. Water Pollut. Control Fed. 57:1006-1007.

Roesner, L., B. Urbonas, and M. Sonnen, eds. 1989. Design of Urban Runoff Quality Controls. New York, New York: American Society of Civil Engineers.

Roesner, L.A., E.H. Burgess, and J.A. Aldrich. 1991. The hydrology of urban runoff quality management. In Proceedings of the 18th National Conference on Water Resources Planning and Management/Symposium on Urban Water Resources. May 20-22, New Orleans, Louisiana. New York, New York: American Society of Civil Engineers.

Rose, J.B., and C.P. Gerba. 1991. Assessing potential health risks from viruses and parasites in reclaimed water in Arizona and Florida, USA. Water Science & Technology 23:2091-2098.

Sartor, J.D., G.B. Boyd, and F.J. Agardy. 1974. Water pollution aspects of street surface contamination. J. Water Pollut. Control Fed. 46:458.

SCCWRP (Southern California Coastal Water Research Project). 1990. Annual Report 1989-90, J. Cross, D. Wiley, eds. Long Beach, California: Southern California Coastal Water Research Project Authority.

Schueler, T. 1987. Controlling Urban Runoff: A Practical Manual for Planning and Designing Urban BMPs. Washington, D.C.: Metropolitan Washington Council of Governments.

Smisson, R. 1981. Storm Sewer Analysis for Implementation of Inlet Control, Euclid, Ohio. U.K.: Hydro Research & Development.

Sobsey, M.D. 1989. Inactivation of health-related microorganisms in water disinfection processes. Water Science & Technology 21(3):179-195.

Sonzogni, W.C. 1980. Pollution from land runoff. Environmental Science and Technology 14:148-153.

Stenstrom, M.K., G.S. Silverman, and T.A. Bursztynsky. 1984. Oil and grease in urban stormwaters. Journal of Environmental Engineering, American Society of Civil Engineers 110(1):58-72.

Sullivan, R.H., J.E. Ure, F. Parkinson, and P. Zielinski. 1982. Design Manual: Swirl and Helical Bend Pollution Control Devices. U.S. EPA Report No. EPA/600/8-82/013. Cincinnati, Ohio: U.S. Environmental Protection Agency.

Tchobanoglous, G., and E.D. Schroeder. 1985. Water Quality. Reading, Massachusetts: Addison Wesley Publishing Company.

Teal, J., and S. Peterson. 1990. The Next Generation of Septage Treatment. Woods Hole, Massachusetts: Woods Hole Oceanographic Institution. April 12, 1990. Unpublished manuscript.

TWA and HRD (Thames Water Authority and Hydro Research & Development). 1987. Interim Engineers Report: Attenuation Storage and Flow Control for Urban Catchments - Conflo '88. September, 1987. U.K.

Uttormark, P.D., J.D. Chapin, and K.M. Green. 1974. Estimating Nutrient Loadings of Lakes from Nonpoint Sources. Report No. EPA-660/3-74-020. Washington, D.C.: EPA, Office of Research and Monitoring.

VSCC (Virginia Soil and Conservation Commission). 1984. Virginia Erosion Control Handbook. Richmond, Virginia: Virginia Soil and Conservation Commission.

Wakeham, S.G. 1977. A characterization of the source of petroleum hydrocarbons in Lake Washington. J. Water Pollut. Control Fed. 49:1680-1687.

Walesh, S. 1985. Stormwater Control Plan for Howard Street. Skokie, Illinois: Donahue Engineers.

WBNSWG (Water Board of the New South Wales Government). 1991. Improved Effluent Quality—Options for Coastal Sewage Treatment Plant Upgrades. Sydney, Australia: WBNSWG.

WEF/ASCE (Water Environment Federation and American Society of Civil Engineers). 1992.

Design of Municipal Wastewater Treatment Plants. Manual of Practice 8. Alexandria, Virginia: Water Environment Federation.

Whipple, W., Jr., J.V. Hunter, and S.L. Yu. 1978. Runoff pollution from multiple family housing. Water Resources Bulletin 14:288-301.

Wiegand, C., T. Schueler, W. Chittenden, D. Jellick. 1986. Cost of urban runoff controls. Pp. 366-380 in Urban Runoff Quality—Impact and Quality Enhancement Technology, B. Urbonas, and L.A. Roesner, eds. New York, New York: American Society of Civil Engineers.

Wisner, P. 1984. Experience with Stormwater Management in Canada. Ottawa, Canada: University of Ottawa.

Wordelman, L. 1984. Swirl Concentrators That Treat Excess Flow at South Toledo, Ohio. Toledo, Ohio: Jones & Henry Engineers.

WPCF (Water Pollution Control Federation). 1986. Wastewater Disinfection. Manual of Practice. FD-10. Alexandria, Virginia: Water Environment Federation.

WPCF (Water Pollution Control Federation). 1990. Natural Systems for Wastewater Treatment. Manual of Practice FD-16. Alexandria, Virginia: Water Environment Federation.

Zariello, P.J. 1989. Simulated water-quality changes in detention basins. Pp. 268-279 in Design of Urban Runoff Quality Controls, L.A. Roesner, B. Urbonas, and M.B. Sonnen, eds. New York, New York: American Society of Civil Engineers.

E

Policy Options and Tools for Controlling Coastal Environmental Water Quality

THE INSTITUTIONAL SETTING

Institutionalized Fragmentation

Three types of fragmentation exist within the institutional setting for wastewater treatment in coastal areas:

Hierarchical—Important responsibilities for coastal water quality reside at every level of government—federal, state, regional, local, and tribal. At each level there may be multiple entities, each with important functions.

Geographic—Bays, sounds, estuaries, and other near-coastal water bodies and their tributary watersheds invariably encompass multiple local and regional jurisdictions. Frequently these jurisdictions have overlapping areas of responsibility. For example, a regional sewage treatment system may have to interact with numerous other local jurisdictions with respect to both sewage and stormwater management.

Functional—Perhaps the greatest challenge to achieving effective water-quality management, as well as accomplishing other environmental goals, is the increasing compartmentalization of functions.

• Water quality considerations themselves are fragmented. Sewage treatment, stormwater management, nonpoint source control, and point source regulatory programs are often conducted in virtual isolation from one another, despite the fact that these sources are in fact interrelated, interacting both in the watersheds and in the marine environment. In most regions, fragmentation of water quality responsibilities is exacerbated by significant gaps in coverage: nonpoint source control and toxicant elimination are two relatively neglected areas.

• Regional land use, growth management, and natural resource planning are usually undertaken in isolation from wastewater planning and discharge permitting programs. In addition, these planning activities are often divorced from implementation activities. With the continuing rapid pace of population growth in many coastal areas, land-use and transportation decisions actually have as much to do with future water quality as the issues usually considered within the scope of water quality programs.

• Water quality is generally evaluated and addressed as if air pollution, solid and hazardous waste, and land-use and water-use decisions were unrelated issues. The converse is also true. This compartmentalization of environmental problems obscures the important ways in which these issues affect each other and hinders effective solutions, especially pollution prevention strategies.

• Funding for water quality and other environmental programs is a patchwork at the federal, state, and local levels. The way money is collected, earmarked, and spent on water quality improvements tends to reinforce fragmentation and the resulting inability to consider and rank priorities on any broad basis.

Jurisdictional Complexity

A typical coastal area includes hundreds of jurisdictions, agencies, and other public bodies on its list of important water-quality actors. Numerous agencies and jurisdictions may exist within a single tributary watershed. Each agency may have several separate bureaucracies with relevant programs, requirements, and responsibilities. A pertinent example is the Puget Sound region of Washington State, where 454 public entities exercise jurisdiction over water quality and related ecosystem management including "6 federal agencies, 5 state agencies, 12 county governments, 14 tribal governments, 100 cities, 40 port districts, 110 water districts, 42 sewer districts, 25 diking districts, 50 drainage districts, 15 flood control districts, 12 soil and water conservation districts, 14 parks and recreation districts, and 9 public utility districts" (PSWQA 1984).

At the federal level, the Environmental Protection Agency (EPA), the National Oceanic and Atmospheric Administration, the U.S. Army Corps of Engineers, the Fish and Wildlife Service, and the Department of Transportation are important actors in any coastal area. In some parts of the country, the U.S. Forest Service, the National Park Service, and others have significant water-quality related programs.

At the state level, separate agencies often are responsible for natural resource management, environmental regulation, community development, land use, and transportation. Local jurisdictions in most states cover very limited geographic areas yet retain major responsibilities for planning, land use, wastewater treatment, stormwater management, water supply, and solid waste. Some states have sub-state regional agencies for some of those responsibilities; all states have a plethora of special purpose districts for sewage treatment, water supply, flood control ports, soil conservation, and other purposes. In some states, tribal governments have direct environmental regulatory programs, sewage treatment responsibilities, and significant jurisdiction over natural resources, including marine fisheries.

Toward Integration of Environmental Decision Making

The foregoing is not intended to represent the details of any specific place, but rather to present the overall institutional challenge: how can environmental decisions in coastal areas, including specific decisions about wastewater management, be made in a large enough context with adequate mechanisms to set and implement priorities? Many past debates over specific wastewater treatment decisions have been couched in terms of priorities but have taken place in institutional situations where implementation of alternate strategies would be very unlikely, even if a particular project or proposal were rejected on the basis that some other project or proposal was more urgent. For example, sewage treatment jurisdictions resisting upgrading to secondary treatment have usually argued that the same amount of money could be more effectively invested in stormwater management or other nonpoint source reduction, even though an equivalent amount of money could not be made available for these needs.

There are signs of progress toward integration of decision-making in coastal areas:

Estuary protection strategies—In some major estuaries, comprehensive plans are being pursued to protect water quality and marine resources, including the sometimes huge tributary watersheds where most of an estuary's problems originate. Such efforts typically demand a specific governmental mechanism to cut through the bureaucratic *turf*, to coordinate multiple agencies, and to oversee plan implementation.

Regional water resource agencies—Some states have created regional water-resource management entities to have a *big picture* perspective on water-related decisions and priorities. Such agencies are becoming major actors in issues regarding wastewater discharges in coastal areas.

Growth management/land-use strategies—Some states are undertaking growth management plans that attempt to integrate land use, natural resource, transportation, and—to some extent—pollution prevention strategies. These strategies can be an important way to link land use and water pollution issues.

Establishing real links among the several dimensions of water quality management, specifically, and resource protection, generally, is an extremely difficult matter, requiring flexibility in approaches to suit the political and institutional realities of each state or region. The integrated state-policy framework being developed in states such as Florida, Maine, Vermont, Georgia, and potentially Washington represents the most comprehensive approach. The Puget Sound estuary management plan is an example of a comprehensive attempt to protect a major water body and its tributaries through the coordinated actions of hundreds of entities at the local, tribal, state, regional, and federal level. Less comprehensive strategies are also possible and can bring about major improvements.

Planning in isolation at all levels—state, regional, and local—has been standard practice in the past, but this resource-destructive pattern is being called into question and actually changed by new integrated growth management systems and estuary management programs. These systems are appearing first in coastal states and thus have a special significance for this study.

Maine and Georgia illustrate the potential for growth management systems to integrate water quality and quantity programs, not only with related resource management programs but also with transportation, land-use, and other relevant systems. In both states, there is a set of state policies (standards and criteria) that frame the requirement of consistency binding the whole system together. All state, regional and local governments are required to develop new plans and to implement programs that are consistent with the state policies and with each other.

In Maine, state agencies are just completing the redrafting of their plans, which will be reviewed for consistency at the state level. Resource management goals and policies are part of the framework in each state, and the existing programs of state resource agencies are being redrawn to fit with each other, with a similar requirement at the regional and local levels.

While these programs are too new to allow more than preliminary judgments about their effectiveness, they show great promise for strengthening the functional links between and among programs in water resources and related resource management, transportation, land-use and economic development programs.

In California, a state growth policy is being fashioned that will set the framework for new regional strategies to better integrate land use and resource management programs. Governor Wilson signaled his commitment to closer coordination by stating, in connection with his proposal for a new umbrella organization called the California Environmental Protection Agency:

"To continue to divide responsibility for the environment between a dozen agencies dilutes accountability for all state environmental programs." The California Environmental Protection Agency, implemented in July 1991, includes the following agencies:

- Air Resources Board,
- Integrated Waste Management Board,
- State Water Quality Resources Control Board (includes the regional Water Quality Control Boards),
- Department of Toxic Substances Control,
- Department of Pesticide Regulation, and
- Office of Environmental Health Hazard Assessment.

When the California EPA effort is coupled with the governor's announced support for a new regional governance framework that will link regional agencies under a policy and fiscal organization capable of assuring the linkage of land use, transportation, and resource management programs, a promising governance framework can be seen. The California experience underscores the importance of leadership, especially at the highest level. The foment of activity in California is a direct result of a new governor with a new agenda that includes, among other things, better management of water resources at the state and regional levels. This new framework unfortunately is being implemented during a time of severe fiscal crisis and only time will tell if the promised results can be achieved.

MANAGEMENT TOOLS

Command-and-Control

The centerpiece of the nation's regulatory system for restoring and protecting the quality of its waters is the Clean Water Act (CWA) (33 U.S.C. 1251 et seq.).[1] This legislation consists primarily of command-and-control measures. Some provisions of the act apply other management tools, however, such as financing/economic instruments (the Construction Grant Program) and limited land-use planning and growth management techniques. In order to simplify exposition, the entire act is discussed in this section.

The modern version of the CWA was first enacted in 1972. There have been four major amendments to the statute since then—in 1977, 1981, 1984, and 1987. The bulk of this section is focused on the CWA; however, there are other important statutes briefly discussed in the concluding paragraphs,

[1]References to United States Code are cited with the title followed by "U.S.C." and the section.

such as the Coastal Zone Management Act (16 U.S.C. 1415 et seq.) and the Marine Protection, Research, and Sanctuaries Act (33 U.S.C. 1401 et seq.).

The purpose of this section is to provide the reader with a general overview of the existing command-and-control regulatory scheme, especially as it relates to the disposal of wastes in urban coastal areas. It will also identify areas where the statutory scheme seems to be inadequate to provide either sufficient protection of aquatic resources or the most effective application of control or management techniques.

The Regulatory System

The CWA regulatory system is complex. Yet, several broad strategies can be identified that form the basic implementation system. Each of these is described briefly below. Certain issues are discussed in greater depth as they particularly relate to the urban setting.

Standards. The CWA seeks to apply standards to those activities that could affect water quality. Beginning in 1972, the focus of this standard system has been upon point source discharges; that is, the pipes releasing effluents of municipalities and manufacturing firms. Standards have been derived from two different perspectives: 1) treatment technology based standards, and 2) water-quality based standards. The technology based standard has generally been considered the minimum acceptable requirement, to be applied universally, while more rigorous water-quality based standards have been implemented where necessary for particular water bodies. For industry, the CWA requires that technology based requirements be applied. The first industry standards were those that were derived from the best practicable treatment technology. Subsequently, a more restrictive standard based on best available technology (BAT) was required.

Permits. Standards are applied to point sources through the National Pollution Discharge Elimination System (NPDES). Under this system, each point source discharger must hold a NPDES permit. The permit contains the numerical standards as well as other requirements which the discharger must meet. Pursuant to the 1987 amendments of the CWA, permits are renewed every ten years, although specific provisions may be modified at any time. Approximately 384,000 dischargers are now covered by NPDES permits.

Compliance. The CWA sets forth a comprehensive system for evaluating whether a permittee is in compliance with its permit requirements, as well as enforcement procedures where non-compliance is found. This system is based on a monitoring system that is carried out and reported upon by the permitted dischargers as well as an independent system carried out by government regulators. Penalties include civil and administrative fines as well as possible criminal prosecution. The government may also seek injunctive relief. There are also substantial provisions for citizen initiated law suits.

Federalism. The federal government has recognized that the states are an essential ingredient in successful implementation of the CWA. Thus, a state that is capable of achieving substantial portions, or all, of the objectives of the federal law may apply for, and receive, delegation of authority (known as primacy) to administer the provisions of the federal program. Where such delegation has occurred, there is a complex system of federal oversight to review the conduct of the state program. As of late 1992, 38 states and 1 territory held primacy over their NPDES programs. Of those, 27 states are authorized to run their own pretreatment programs.

Funding. The federal government has provided financial subsidies to states and municipalities for the purpose of wastewater treatment plant construction and other activities related to water quality since the late 1950s. Starting in 1972, the CWA greatly increased federal funding for eligible projects. Until 1987, this assistance was in the form of federal grants ranging up to 75 percent of the cost of construction of wastewater treatment plants and other central facilities. Through 1991, the total federal expenditure for the Construction Grant Program has been about $50 billion. It is estimated that state and local construction expenditures during the same period total $26 billion. Since that date, the federal role has been to provide assistance to the states for the creation of revolving loan funds. Subsidies to this program have totaled $5 billion; this assistance has been extended by congressional action through 1993.

Nonpoint Sources. Discharges from diffuse sources are subject to varying degrees of regulation under the CWA. Nonpoint source regulation has had only limited effectiveness; these sources of pollution now account for the majority of water quality problems in many of the nation's water bodies. A major category of nonpoint sources is the storm and combined sewer systems of urban centers. While combined sewer overflows (CSOs) have been subject to regulation since 1972, little progress was made until recently. The EPA had no national strategy for CSO control until September 1989 (Federal Register 1989). A second type of nonpoint source is the even more diffuse pollution that comes from general urban development and from agricultural activities. The 1987 amendments to the CWA began the process of regulating these sources of water pollution through the establishment of rudimentary provisions for assessment and state program development. But in 1990, the EPA promulgated final rules requiring all cities and urban counties with populations greater than 100,000 persons to obtain a NPDES permit for their stormwater discharges. These rules require that pollutants in urban runoff be controlled to the "maximum extent practicable" (Federal Register 1990). Agricultural runoff, however, is currently unregulated but is addressed in the final coastal zone management guidance released by NOAA in early 1993.

Industrial Pre-Treatment. Especially in urban areas, the pre-treatment of industrial wastes before discharge to municipal systems is an important component of the CWA's provisions. Although these provisions have existed since 1972, little was done to ensure their implementation until recently. This recent attention is largely due to growing concern about the effect that toxics from industrial sources has on sludges resulting from the municipal treatment process or on the quality of receiving waters. Nationally, about 12,000 individual firms are covered by federal pre-treatment requirements, while as many as 200,000 are not. Of this latter number, many are subject to local limits which are determined on the basis of the NPDES permit for the treatment facility.

The process of implementing a strategy for protecting the nation's water pursuant to the foregoing statutory principles has been difficult. However, an assessment at the end of two decades of effort would have to conclude that substantial progress has been made. Many water bodies have been improved in their quality. The importance of the issues that need to be addressed in the future, such as nonpoint sources, is now more clearly understood. As this report discusses elsewhere, the economic and technical issues associated with their solution remain difficult in many cases. Nowhere is this more true than for the urban centers of the country that discharge to the nation's coastal waters.

Standards

The question of applicable standards for dischargers remains a crucial issue. One focal point for this debate resides with the discharges of municipal wastewater treatment facilities. The CWA requires that all such plants meet, at a minimum, standards equivalent to secondary treatment. The numerical standards for biochemical oxygen demand and total suspended solids, which EPA as adopted as defining secondary treatment, were derived from a study of the effluent characteristics of approximately 80 typical sewage treatment plants (STPs). Thus, the secondary treatment requirement was essentially a technology-derived standard as was required by the statutory language of the 1972 CWA. However, if water quality considerations for the water body to which the STP discharges require it, then even higher levels of treatment (advanced wastewater treatment such as nutrient removal) might be imposed. With approximately 12,000 municipal plants having achieved secondary levels of treatment, the question now remains as to whether water quality requirements will require general application of advanced treatment.

A second focal point of the standards debate has to do with the development of new standards in areas that Congress has determined to be of importance. These include toxics, bottom sediments, and protection of biologic resources. The system for development of these standards is not technology based. Rather, it is driven by the CWA statutory language to "restore and maintain the chemical, physical, and biological integrity of the nation's waters." The process of setting water-quality standards consists of two steps. First, criteria documents that set forth basic numerical and other characteristics are prepared by the EPA for a particular constituent. These criteria are then applied by the states to develop specific standards for particular bodies of water.

To date, criteria have been established by the EPA for about 135 chemical pollutants. Unfortunately, this has not resulted in the wide-spread development of water quality standards. The states have been slow to undertake the standard-setting process. In addition, no criteria documents exist for many toxic or hazardous chemicals, and the criteria that do exist may not be applicable in marine environments or may inadequately assess impacts on living resources.

Specific Urban Issues

Waivers and Variances. There are a number of provisions for variances from the requirements of technology- or water-quality based standards contained in the CWA. These include: 1) economic variances for cases where modified requirements represent maximum use of technology within the economic capability of the owner/operator; 2) water quality variances from BAT limitations for ammonia, chlorine, color, iron, and total phenols; 3) marine discharge variances from secondary treatment requirements; 4) innovative technology variances from BAT/BCT (best conventional pollutant control technology) deadlines for toxic, conventional, or non-conventional pollutants; 5) fundamental difference variances from BAT/BCT/NPDES guidelines where a permittee can demonstrate the existence of a situation that is fundamentally different from factors considered in establishing effluent guidelines; and 6) cooling system variances for thermal discharges.

For coastal urban areas, the marine discharge waiver provision first appeared in the 1977 amendments (section 301(h)). Although waiver applications were accepted for only a short time, this section remains enormously significant for coastal wastewater management. The EPA agreed to grant waivers where the applicant could demonstrate that a discharge to marine waters with less than secondary treatment could meet a number of criteria including preservation of a "balanced and indigenous" population in the aquatic environment. Absent this showing, the municipal system was required to provide secondary treatment. While some 25 percent of all waiver requests were ultimately granted, the fact that several large systems such as Boston and San Diego have been unsuccessful in securing waivers continues to promote controversy.

Complexity. Urban systems, even as traditionally understood, are not simply characterized by STPs treating domestic sewage. They also consist of large combined stormwater and wastewater collection systems—resulting in a series of combined sewer outfalls. Many firms with manufacturing process waste discharge their effluents directly to the municipal sewerage system. Finally, there are separate stormwater collection systems that discharge directly to water bodies.

The response of the regulatory system to each of these problems has been varied. In the case of CSOs, the major issue has been that Congress has not provided funding for controlling the impact that these systems have on receiving waters. Even in the absence of such funding, the polluting impacts of CSOs are increasingly coming under regulatory control (e.g., in Boston, Chicago, and Washington, D.C.). And for stormwater systems, the 1987 CWA amendments provide a separate system of standards and permits that will require control of the quality of the effluent from these sources.

New Issues

Toxic substances and nonpoint sources of pollution remain of serious concern. Accordingly, provisions of the 1987 Amendments established new procedures for evaluating toxic effects in receiving water from both industrial and municipal discharges. In a similar way, concern over nonpoint source discharges resulted in new provisions of the CWA. In both cases, it is too early to know how these provisions will influence local control strategies.

Nonetheless, it is clear that the constellation of environmental challenges facing urban governments has become much greater. The reach of the CWA is far greater and the dimensions of the problems are more severe.

Assessment

The complexity of the pollution sources affecting the marine environment, especially in urban areas, is extraordinary. Necessarily, the statutory scheme which has evolved over the past two decades reflects this complexity. The command-and-control system of the CWA has brought most discrete sources under regulatory control, and this has resulted in a reduction of pollutants discharged to the marine environment. However, it is uncertain that remaining issues such as toxic contaminants and nonpoint sources can be effectively managed through this approach. It is also unclear whether the sheer numbers of sources that may need to be controlled, perhaps as many as one million, can be subjected to a regulatory system of this kind.

The EPA has attempted to address this problem through a fundamental re-examination of its approach to pollution control. The report of its Scientific Advisory Board entitled *Reducing Risks* seeks to re-orient programs so that they "address the significant remaining human and ecological threats" (EPA 1990). The principal recommendations are that 1) the EPA should set priorities on the basis of comparative risks and 2) policies should seek to prevent pollution in the first instance rather than treat it. It remains to be seen whether the complexities of the CWA can be restructured to achieve these goals.

Economic Instruments

Economic instruments are management strategies that provide, for the purpose of environmental improvement, monetary incentives for voluntary, non-coerced actions. Economic instruments do not specify any particular action; they merely encourage certain desired behavior. The central argument for use of economic instruments is that they tend to minimize the cost of achieving any particular level of pollution abatement. This can be illustrated—in an ideal case—by comparison to command-and-control measures. Later sections discuss practical limitations on economic incentives and review their performance with respect to other criteria.

If an effluent standard is applied uniformly to all dischargers of a certain contaminant, each must comply regardless of the cost of doing so. Some dischargers may be able to meet the standard at relatively low cost while others, especially those who have already implemented substantial abatement, may face high unit costs. This situation indicates that the same result could be achieved at lower total cost. Higher removal rates could be required from those who can do so at low cost and relatively less removal from high-cost dischargers.

As low-cost dischargers increase treatment, the marginal cost of abatement (the incremental cost of removing another unit of pollutant) increases. Similarly, as high cost dischargers reduce treatment, their marginal cost falls. So long as some dischargers face marginal abatement costs that are lower than for others, total cost can be reduced by reallocating abatement requirements from high-cost to low-cost dischargers. Total abatement cost is minimized, for any selected environmental target, when each discharger experiences the same marginal cost of abatement.

Under the command-and-control approach, it is possible to modify the uniform discharge standard to reduce total abatement cost. This is done by allocating abatement requirements in inverse relationship to marginal abatement cost. The EPA notes that such regulations "are difficult to design because they require detailed understanding of the costs and benefits of numerous activities" (EPA 1991). Furthermore, it is not clear that the public and dischargers would accept the resulting non-uniform policy as effective and fair.

The same cost-effective result can be achieved by providing dischargers with an economic incentive to reduce pollution, such as a uniform tax that applies to all units of pollutant discharged to the environment (effluent tax). In this case, each discharger is motivated to reduce emissions so long as the cost of discharge (the effluent tax) is greater than the marginal cost of treatment. Cost-minimizing dischargers will increase treatment level until marginal treatment cost rises to equal the effluent tax. Since every discharger equates marginal treatment cost with the same uniform effluent tax, the result should be equal marginal treatment costs, implying least total cost. This is true regardless of the quantity of pollutant removed. If the environmental goal is not met, setting a higher effluent tax causes total removal to increase, although still at least cost. The regulator does not necessarily require any information on abatement cost functions, and the policy is applied uniformly to all dischargers.

In addition to promoting cost minimization, economic incentives offer other advantages over the results of conventional command-and-control regulation. They encourage technological progress by creating "a permanent incentive to further abate pollution" leading to "a permanent inducement to develop more efficient clean-up or preventive technologies." (OECD 1989) Furthermore, economic incentives distribute responsibility for abatement cost fairly, provide needed flexibility in implementation, and exploit opportunities for environmental improvement that may be unknown to regulators or beyond the reach of conventional command-and-control regulation (Boland 1989).

Types of Economic Incentives

At the most elementary level, economic incentives can be divided into those that induce desired behavior by offering a subsidy and those that discourage undesirable behavior by levying a charge. In practice, however, applied and proposed incentives exhibit considerable variety in form and various other useful distinctions. The following paragraphs describe some specific types of incentives, following a framework adapted from the work of Bernstein (1991).

Effluent Charges. The regulator levies fees based on the quantity and/or quality of pollutants discharged to the environment. The fee is levied on each unit of discharge, so that the polluter has an incentive to take any steps that would reduce discharge. The level of the tax may be set according to some measure of incremental environmental damage, or it may be adjusted so as to achieve the desired environmental quality. In the United States, effluent charges have been applied to some solid waste streams. A recent innovation is the use of volume-based charges for municipal trash disposal. These charges can be collected through the sale of mandatory trash bags (Perkasie, Pennsylvania), by selling stickers that must be placed on bags or other trash items (High Bridge, New Jersey), or through yearly subscriptions for collection of a maximum number of trash cans (Seattle, Washington) (EPA 1991).

User Charges. This measure is usually applied by public wastewater treatment systems. Those who discharge waste into the facility are charged on a unit basis for the quantity and/or quality of the waste discharged. The charge is typically set on the basis of capital, operating, and maintenance costs of the system. One survey reports charges to industrial dischargers ranging from $0.22 to $3.56 per 1,000 gallons of water used, with additional charges for suspended solids and/or biochemical oxygen demand in 97 out of 120 utilities surveyed (Ernst and Young 1990). The user charge is equivalent to an effluent charge for individual generators of waste, as well as a revenue source for the system operator.

Product Charges. Where the production and consumption of certain goods are associated with pollutant releases to the environment, the use of such goods may be discouraged by levying a tax on either the final product, or some input used in its production (feedstock tax). Even where such charges are relatively modest, they may be sufficient to encourage consumption of substitute, less polluting goods. The Superfund program for managing abandoned hazardous waste sites is funded by a feedstock tax, levied on a number of chemicals frequently associated with hazardous wastes. A number of states tax the sale of new tires (or motor vehicle transfers) to provide funds for scrap-tire programs (EPA 1991). While both of these programs were enacted as revenue measures, not as economic incentives, charges of this kind can potentially influence the volume of waste generated.

Administrative Charges. Government agencies that perform pollutant-related administrative services (registration of pesticides, permitting of waste discharge points, etc.) may charge fees for services rendered. While usually created for the purpose of recovering administrative costs, these charges also serve to discourage, to some minor degree, the polluting activities.

Tax Differentiation. Where more desirable substitutes exist for polluting products, product charges may be implemented for both goods: a positive tax on the polluting good and a negative tax (subsidy) on the nonpolluting alternative. This policy can be used to create a strong disincentive toward use of the polluting good, while keeping total impact on consumers relatively neutral. Tax credits have been proposed to encourage purchase of equipment for both backyard and centralized yard waste composting and mulching, with the ultimate purpose of reducing community solid waste volume (EPA 1991).

Marketable Permits. Also known as tradable discharge permits, these instruments are similar to discharge permits issued under command-and-control systems: each discharger is required to have a permit for every unit of pollutant released to the environment. In this case, however, the permits may be bought and sold among dischargers. Once a market price for a permit is established, dischargers with low treatment costs will be motivated to sell permits, while high cost dischargers will buy permits rather than invest in further abatement. Marketable permits have been used for some years in the air pollution abatement program. Intrafirm trades have been permitted under the EPA *bubble* program, while the *offset* program has encouraged interfirm trades in nonattainment areas. The Clean Air Act of 1990 expands this practice dramatically by relying mainly on tradable discharge permits to effect the mandated reduction in acid emissions from stationary sources throughout the United States. Further applications of marketable permits have been proposed for greenhouse gases, recycled newsprint, lead, and end uses of volatile organic compounds (EPA 1991).

Liability Insurance. Where dischargers can be liable for damages in the case of some kinds of pollutant discharge, there may be a reason to seek insurance against possible future claims. Insurance carriers, in turn, will be motivated to establish premiums and set conditions for such coverage. It is expected that minimization of insurance costs would provide an incentive for dischargers to reduce releases.

Subsidies. Various kinds of grants in aid of construction, low interest loans, tax allowances, and preferential regulatory treatment (in the case of regulated public utilities) have been used to persuade dischargers to invest in abatement facilities or find other ways to reduce pollutant releases. The CWA Construction Grant Program is an example of this type of measure, although it was not intended to provide abatement incentives (the parallel command-and-control structure governed most behavior).

Deposit-Refund Systems. This method requires the consumer to pay a surcharge on a product that is refunded when the product or container is returned for proper disposal. Nine states and various local governments have implemented deposit-refund systems for beverage containers. Maine and Rhode Island use deposit-refund systems to control the disposal of spent lead-acid batteries. Similar mechanisms have been proposed for pesticide containers and used motor oil. Bills introduced in Congress during the 1990-91 session would have required the EPA to design a national deposit-refund system for motor oil (EPA 1991).

Noncompliance Fees. Although enforcement actions are usually considered to be inherently command-and-control in nature, some features are designed to act as economic incentives rather than pure sanctions. Violations of discharge standards sometimes result in fines. These fines are fees levied on non-compliant polluters under a command-and-control system. Where the fines vary with the intensity and duration of the violation, however, they constitute economic incentives, motivating polluters to avoid serious or lengthy violations.

Performance Bonds. A discharger may be required to make a lump sum payment (or irrevocable promise of such a payment) to a regulatory agency, to be forfeited in the case of violation, or ultimately returned in the absence of violation. This technique is generally employed where the possibility of violation is confined to a short time period (e.g., during the cleanup of a site contaminated with hazardous wastes).

Liability Assignment. Where laws and regulations make it possible to hold dischargers responsible for damages that pollutants impose on others, the possibility of such liability is a potent incentive for pollution abatement.

Critique

While economic incentives share some common features, they may be substantially different in other ways. The following sections discuss economic incentives generally, with some indication of differences. Space does not permit a full analysis of each type of incentive, much less of variants and combinations of incentives.

Effectiveness. Economic incentives are generally capable of producing the desired effect, provided that the incentive can be made large enough. However, it is difficult to know in advance how large the incentive must be to produce any given effect. Where detailed cost data are available for most dischargers, it may be possible to set the proper incentive level at the outset. Otherwise, periodic adjustments are made until the desired result is obtained.

Marketable permits constitute one exception to this uncertainty regarding effectiveness. The initial issuance of permits places an upper bound on total pollutant discharge, in the same way that discharge permits issued under a command-and-control system constrain total discharge. Still, the spatial distribution of discharge cannot always be predicted accurately as the buying and selling of permits can move pollutant releases from place to place within the defined market area.

Economic incentives are expected to be become fully effective shortly after implementation (allowing for any needed installation of equipment). For effectiveness to be sustained in the longer term, it is necessary to insure that relative incentives remain unchanged by factors such as price inflation and possible changes in costs of alternatives. Where the level of incentive is set by the regulator (e.g., effluent taxes), indexing and periodic adjustment may be needed. The market-based methods (e.g., marketable permits) adjust themselves; no action by the regulator is needed.

All economic incentives, no less than command-and-control policies, require the existence of effective monitoring and enforcement programs. Effectiveness is undermined whenever failure to perform—associated with a command-and-control regulation or an economic incentive strategy—is unlikely to be detected or, if detected, unlikely to lead to significant penalty.

Efficiency. In the broadest sense, economic efficiency requires that the *right* level of pollution be identified, then achieved at the least possible overall cost. The right level of pollution is the amount that maximizes net social welfare, considering both environmental damage and the cost of abatement. Economic incentives, as well as other environmental management tools, do not reveal the right pollutant level. Environmental targets must first be set by regulators, then economic incentives can be used to realize them.

Most economic incentives are cost-effective in that they tend to minimize the cost of whatever pollution reduction is achieved. This general result applies to effluent charges, product charges, marketable permits, and others. Cost minimization cannot be assumed, however, in the case of subsidies and enforcement incentives. Individual policies under these categories may or may not promote efficiency; the result depends on the design and conditions of implementation of each application. Only those costs actually borne by dischargers are considered in the course of cost minimization. A municipal wastewater system, for example, has no motivation to minimize abatement costs imposed on others through pretreatment requirements, etc.

Fairness. Economic incentives—with the possible exception of subsidies—may be regarded as fair, since they apply uniformly to all, even if they lead to non-uniform behavior. Some object to these incentives on what may be described as ethical, rather than fairness, grounds. It is argued that measures like effluent taxes permit dischargers to pay for something they should not be allowed to own: the right to pollute. Command-and-control measures, on the other, grant the right to pollute (after abatement targets are met) to dischargers at no cost.

Subsidies are less likely to be perceived as fair. In the first place, they are often applied in a non-uniform manner. Also, they violate a widely held notion of fairness: that the polluter should pay. Subsidies pay the polluter. Abatement actions may be required in order to receive the subsidy, but there is no doubt that the discharger is left better-off than before. On the other hand, the EPA Construction Grant Program, which provided federal subsidies to qualifying local government agencies for the construction of wastewater treatment facilities, was not widely criticized as unfair. This may reflect the fact that subsidies were purely intergovernmental (not available to industrial dischargers) and that the program was large enough and lasted long enough to provide virtually every locality with an opportunity to qualify for one or more grants.

Redistribution of Income. Economic incentives redistribute costs and benefits in various ways. Pollution charges, except for tax differentiation, transfer income from dischargers to regulator (government). Marketable permits may transfer income in the same way or not, depending on how the permits are initially distributed (issued free of charge, auctioned, etc.). To the extent that these permits trade thereafter at their marginal opportunity cost, subsequent transfers of net income are small or nonexistent.

Applicability. Not all economic incentives are suitable for all regulatory tasks. While some are broadly applicable, such as effluent taxes and marketable permits, others are more suited to specific applications (e.g., product charges and deposit-refund systems).

Feasibility. Because economic incentives (except subsidies) have not been widely used, there may be barriers to implementation in many cases. It may be necessary to change federal or state legislation and regulations to implement such measures as effluent taxes or marketable permits. Social acceptability may be an issue for some types of incentives (especially those that appear to reward polluters), while the dischargers themselves may object strongly to others (e.g., effluent taxes). These considerations, in turn, contribute to the presence or absence of the political will to adopt or support a given measure.

Risk. With the partial exception of marketable permits, significant uncertainty exists regarding the effect of economic incentives on the level and spatial distribution of discharges. (Marketable permits result in a specified level of discharge.) This reflects the difficulty of predicting how dischargers will individually and collectively react to any given set of economic incentives. Once a pattern of response is established, reactions to changes in the incentive structure can be predicted with greater confidence. Nevertheless, a transition from complete reliance on command-and-control to a substantial role for economic incentives could involve significant uncertainty regarding the initial outcome.

From the point of view of dischargers, economic incentives are also associated with increased risk. Under any system of regulation, dischargers must invest in treatment works and process adjustments, train employees, seek permits, and make other relatively long-term commitments of resources. Under command-and-control regulation, the nature of this investment is known with some accuracy. In the case of economic incentives, however, the least-cost investment strategy depends on the type and level of economic incentive expected, future abatement costs, and other somewhat uncertain data. Dischargers may overinvest (from their perspective) in treatment, or they may underinvest requiring costly future upgrades.

Incentives for Technology Improvement. Unlike technology-based command-and-control instruments, economic instruments preserve the incentive to improve and refine pollution abatement and production technology. Those who are able to improve the efficiency of a pollution abatement process will benefit in a tangible way: the total cost of compliance will be lowered. Dischargers are motivated to innovate in search of such improvements, and developers of new processes will find a ready market. Also, economic incentives do not distinguish between pollutant reductions achieved by abatement technology and those obtained through production process change. In either case, the incentive for innovation and improvement remains. This can be contrasted to the opposite case, technology-based command-and-control regulation, where the only party interested in more effective treatment technology is the regulator: the new technology could form the basis of more stringent regulations.

Feedback/Adaptive Management/Ease of Evaluation. Economic incentives require the same type of monitoring and reporting that is currently associated with command-and-control policies. Opportunities for feedback and the possibility of adaptive management are, therefore, essentially the same. Program evaluation, however, gains an additional dimension. Data on the payment of taxes or the holding of permits provides information on the economic choices faced by dischargers and on the way in which these choices have been resolved.

Prevention Versus Abatement. Pollution prevention is accomplished by seeking alternative inputs, production processes, or products that result in wastes with a lower volume or lower toxicity. Some economic incentives are specifically intended to influence input or product choice (product charges, tax differentiation, deposit-refund systems). Others provide incentives for pollution reduction that do not discriminate among sources of reduction. The economic incentive to reduce one unit of discharge by changing a production process is exactly the same as the incentive to achieve the reduction through abatement technology.

Discussion

Actual economic instruments, in use in the United States and elsewhere, may or may not resemble the ideal designs of the theoretical literature. Practical considerations may dictate modifications and compromises in the scope and form of these regulatory devices. Design and evaluation of regulatory programs incorporating these incentives should give full attention to factors such as feasibility, risk, and acceptability, rather than assuming the inherent superiority of economic instruments.

When economic incentives are used alone, the results are difficult to predict except after some experience and experimentation. This is true even for marketable permit programs, though the uncertainty may be confined to the spatial distribution of discharge. Also, successful application of economic instruments requires the presence of effective monitoring and enforcement programs.

For these reasons, proposals to abandon command-and-control measures in favor of economic instruments are unlikely to be heard, much less adopted. All U.S. experience so far, primarily in the air quality program, is with economic instruments that have been added to existing command-and-control regulations. In this way, economic instruments are able to provide beneficial improvements at the margin without sacrificing the basic predictability and enforceability of the prior system.

Properly used, economic instruments can claim important advantages:

- They can produce incremental improvements in environmental quality over and above those attainable through direct regulation.
- Some economic instruments provide a revenue source for self-financing pollution control programs.
- Economic instruments preserve incentives for technical progress, with respect to both abatement and prevention technology.
- The total social cost of meeting a given target is reduced; the total cost burden on polluters is also reduced.

• Flexibility in enforcement is enhanced without sacrificing environmental goals.

Growth Management

Growth management strategies, utilizing any of several kinds of planning frameworks, offer the opportunity to integrate environmental decision making. Three types of planning frameworks are relevant to wastewater management in coastal areas:

1. Comprehensive land-use and growth management planning at the regional and state level.
2. Comprehensive planning for regional water bodies such as bays, sounds, and estuaries.
3. Single-purpose planning and implementation efforts specifically directed to effective water resource protection.

Planning approaches can offer significant advantages:

• requiring the integration of land and water resource strategies;
• addressing natural resources and environmental threats on an ecosystem basis;
• employing a watershed or drainage area concept;
• providing a mechanism for overcoming jurisdictional fragmentation;
• providing the context for setting and acting on priorities; and
• focusing on the future, providing the basis for anticipating, and preventing, problems.

Planning approaches are essential for successfully controlling nonpoint and urban stormwater pollution problems, and useful for considering point sources and other water quality issues in the context of broad pollution prevention goals. Planning approaches tend to force attempts to overcome the fragmentation and complexity of the institutional setting.

Comprehensive Land-Use and Growth Management Planning

State and regional land-use and growth management planning efforts are being employed in many but not all coastal states. Typical state strategies include:

• a set of goals and policies, usually adopted by the state legislature, that form a comprehensive planning and implementation system tied together with the concept of consistency (requirements that state, regional, and local actions be consistent with the goals and plans),

• *pay-as-you-go* requirements, often called concurrency, which call for infrastructure, such as stormwater management systems, to be in place and funded (for example through fees or rates) concurrent with new development activity,

• anti-urban sprawl provisions, including incentives, disincentives, and design standards to promote compact, people-friendly, and environmentally sound urban development,

• increased attention to protecting natural resources and environmentally sensitive areas, including farm and forest lands, wetlands, water recharge areas, and wildlife habitat, and

• goals and policies regarding housing and economic development.

Many of these state and regional growth strategies involve a review of existing programs, such as coastal zone planning and management, and efforts to reinforce and integrate such programs with new comprehensive plans and plan implementation.

Successful implementation of these state growth management systems includes the redesign of state, regional, and local land-use and development permitting systems to overcome their inherently limited, case-by-case reactive nature. Comprehensive and integrated land-use planning and plan implementation have significant relevance to water quality efforts, especially nonpoint pollution control and prevention.

Perhaps the most important feature of recent planning strategies, in sharp contrast to past attempts, is that the best of them are enforceable, binding, and mandatory. As a result, they offer consistency in approach both vertically (through each level of government) and horizontally (across programs at each level of government) and thus offer a good mechanism for addressing coastal water-quality problems.

Comprehensive Planning for Bays, Sounds, and Estuaries

In many coastal states, comprehensive strategies are being tried to protect bays, sounds, estuaries, and other near-coastal waters. Some of these plans are being conducted under the auspices of the National Estuary Program, a provision added in 1987 to the federal Clean Water Act.

These planning and implementation programs bear many similarities to comprehensive land-use and growth management strategies, because of the close tie between population pressures and the most significant threats to these water bodies. For example, stormwater management is a significant feature of both types of approaches, and many of the essential features of growth and land-use plans, such as confining urban sprawl and protecting environmentally sensitive areas, have direct benefits to water quality and marine resource protection.

By focusing on a regional water body, the regional ecosystem becomes the integrating factor in the management plan, and the basis for setting and implementing priorities. All of these programs address entire watersheds rather than looking only at activities on or near the marine shoreline.

Like comprehensive land-use planning, the biggest challenge to these estuary and near-coastal management plans is whether they will, in fact, be implemented. Both enforceability and funding are lacking in the National Estuary Program, and states have varying levels of commitment to moving from planning to implementation. Substantial political will, as well as funding, will be required to make even the most well thought out of these plans a reality.

A typical estuary plan addresses these issues:

- rapid and continuing population growth in the basin and the commensurate need to tie land use, transportation, and water quality programs together;
- jurisdictional fragmentation and the need to overcome *turf* concerns;
- significant gaps in scientific research and monitoring related to the regional ecosystem and the effects of pollution;
- the need to broaden the scope of traditional water quality programs, for example, the need to address accumulations of toxic metals and chemicals in sediments underlying the water column;
- the need to address many problems simultaneously in view of the absence of any one dominant threat to water quality;
- the inadequacies of existing water-quality programs, including the NPDES permitting process; and
- the need to sustain political support for implementing complex, long-term plans in the absence of a steady flow of new crises and catastrophes.

It must be recognized that to be successful, these regional water body plans must be living, iterative processes tied to good monitoring and measurements of progress. These are truly long-term projects—in some respects never-ending—in that most of the action strategies involve continuing management responsibilities as much or more than they rely on one-time capital projects.

Institutionally, oversight and coordination responsibility for water-body plans needs to reside somewhere with a strong combination of visibility, objectivity, and political clout.

Improved Water Resource Planning and Management

Even in the absence of comprehensive planning strategies, single-purpose resource agencies can accomplish a more integrated—and thus more effective—system in one functional area. Florida's Regional Water Management Districts and California's Water Quality Control Boards are examples.

Florida's Regional Water Management Districts are organized along watershed lines; they have their own taxing authority, and they possess both planning and regulatory powers extending to both water quantity (primarily) and water quality (increasingly). Much has been accomplished by these agencies as their authority has been expanded to coastal and estuarine waters in recent legislative sessions. Yet there are land-use and other linkages that need to be strengthened.

Florida is discovering that it is difficult to integrate established and powerful systems such as its water management districts into its new comprehensive growth strategy. A renewed effort in that direction has the strong support of Governor Chiles and shows strong evidence of success.

Evaluating the Planning Tool

The planning tool can be held up against evaluation criteria as follows:

Effectiveness—Planning strategies are more effective in the long term than in the short term and are virtually ineffective in the absence of long-term commitments of funding, political will, and institutional capability. With such enforceability and support, planning strategies can be extremely effective and offer the best hope for controlling nonpoint and stormwater pollution problems and the best way to consider the wastewater treatment issue as part of a bigger picture.

Efficiency—Because planning strategies offer an opportunity to set and implement priorities in a broad context, they should result in efficient implementation expenditures. However, they can also be time consuming and slow to develop and implement and can be derailed at any time, risking the loss of initial investments in the planning strategy.

Fairness—Plans tend to be formulated and adopted in public processes where issues of fairness are at least exposed and evaluated and often used as a major criterion for decisions.

Redistribution of income—As with the less objective criterion, *fairness*, planning strategies usually expose issues of costs and benefits as well as who pays and who benefits. The comprehensiveness and complexity of plans, however, can detract from the analytical rigor of this issue as it might apply to any particular project or action.

Applicability—Planning is most applicable to long-range, multi-factor problems and in the water quality area is absolutely necessary for solving nonpoint and stormwater concerns, for setting priorities among water quality choices, and for taking an ecosystem approach to protecting coastal areas.

Feasibility—Planning is usually feasible, whereas plan implementation depends entirely on political will, sustained funding, and long-term institutional capability. These factors are highly variable.

Risk—Planning strategies tend to minimize risk if they are set up with adequate monitoring, evaluation, and course-correction provisions.

Incentives for technology improvement—Planning strategies per se are neutral with respect to technology improvement. The effect of any particular plan could be positive or negative with respect to innovation.

Feedback/adaptive management/ease of evaluation—Planning strategies excel in feedback and adaptability if they are structured as ongoing, iterative processes (as they should be).

Prevention versus abatement—Planning strategies are an excellent—perhaps the best—mechanism for developing and implementing prevention strategies. Abatement strategies are also suitable ingredients of plans but can be accomplished well in other ways.

Integrated versus single medium strategies—Planning approaches are the best hope for achieving integrated strategies and tend away from single medium strategies.

Single versus cumulative impacts—Planning strategies by definition look at the *big picture*, moving beyond case-by-case approaches to consideration, anticipation, and action relative to cumulative impacts.

Conclusion

Planning as a tool is essential for improving water resource management itself and for integrating water quality considerations with other environmental and natural resource objectives. It is also a good mechanism for structuring and implementing decisions and management strategies for regional water bodies in coastal areas. Without a planning approach, nonpoint and stormwater management pollution problems will not be solved, and specific water-quality decisions of any sort, including point source treatment decisions, cannot be made in the context of comprehensive strategies and priorities. There are disadvantages of planning strategies: they are time-consuming and can at times delay needed action; they can result in centralized schemes that hinder creativity and flexibility; they impose costs on some and create windfalls for others; they can raise expectations and then dash hope when plans are not implemented. Implementation is the usual pitfall of most planning strategies, a problem which needs to be addressed at every stage of any such process.

Education

Individual decisions, most of them beyond the direct reach of regulatory strategies, greatly affect water quality. Such decisions are made every day by corporate employees, consumers, householders, and government employees. They include the purchase, use, and disposal of petroleum and other hazardous chemicals; the selection, maintenance, and use of machinery and other equipment; the uses and abuses of urban, suburban, and rural landscapes; and decisions about consumer products ranging from wasteful and resource-intensive ones to essential and resource-conserving ones.

Nonpoint source and pollution and stormwater contamination are caused in large part by myriad, individually insignificant actions. Similarly, discharges into municipal wastewater systems are made up of much more than human sewage: other contaminants, including significant amounts of toxic chemicals, are the product of a multiplicity of everyday decisions by individuals and companies.

Even in industrial facilities subject to point source NPDES permits, individual decisions about equipment maintenance, raw materials, and processes, while ideally made within the parameters of sound permitting requirements, have an important and varying impact on the nature and amount of actual discharges.

Trained environmental professionals are in demand in both the private and public sectors, and the need for basic environmental *literacy* among the public is increasingly apparent, as it is expected that both professionals and the public grasp the relationship of multiple actions and factors and act on that understanding.

Education, therefore, is absolutely necessary as a part of any solution to water quality problems. Education includes a wide range of content, from the most general and simple to the most technical and complex. Similarly, the audiences to be reached range from the general public to students in academic settings to specific employee groups in particular industries. Audiences include children, public officials, treatment plant operators, and water quality regulators. In this broad context, technical assistance, technology transfer, public service messages, *hands-on* stream rehabilitation, and school curricula are all examples of strategies for water quality education.

Education Strategies

Water quality education has, in general, enjoyed little emphasis. Government agencies have gravitated toward the brochure strategy, which in fact is no strategy at all. Effective education is a form of marketing—audiences, messages, targeting, media, and saturation are key concepts in designing a program to modify people's behavior. Effective education is also an essential component of maintaining public support for water quality programs.

Education programs can and do miss the mark. The most helpful and accurate brochure will have no effect if the target audience 1) doesn't get it, 2) doesn't read it, or 3) isn't motivated by it. An in-person training program for technical people in an industry will be a waste of time if 1) the person presenting the information is not credible to the audience; 2) the information isn't tailored to the specific real world of the particular business; or 3) the purpose of the education is to change the policies of management, rather than to change the behavior of the people in the room. Academic programs can exacerbate fragmentation in solving water quality programs by emphasizing information and omitting learning strategies that might broaden context and assist in integration.

On the other hand, education can be extremely effective. Good information, presented at the right time in the right form, can change behavior, avoid battles, empower people, and prevent pollution. Encouraging peer-to-peer education can overcome the credibility problems invariably encountered when government tries to educate business people. Funding citizen involvement programs such as labeling stormdrains can do double duty—addressing a specific water-quality problem while building a more general environmental ethic. Education can also overcome the confines of compartmentalized regulatory programs by integrating environmental information and strategies. Academic curricula can teach both environmental responsibility and technical competence in a *real world* context.

A comprehensive water-quality education strategy would include at least the following:

Technical assistance and technical training—Working through industry and technical/professional associations is especially effective in conveying technical information to targeted audiences. Regulatory programs have generally not proven to be sufficient conduits of technical assistance or training.

Technology transfer—This term refers to methods and approaches as well as to hardware and treatment or manufacturing processes. Most technology transfers occurs informally but can be hastened by conferences and well-thought-out dissemination of information.

Targeted audiences—This concept starts by thinking about the audience rather than the government agency and its program. It asks, "If I owned a dry cleaning establishment or if I were a resident in this watershed or if I were a mayor in this region, what would I need to know to protect water quality or the environment more generally? And how would I learn it?" With this perspective, effective and efficient education strategies can be developed, but only if the educator is up to the challenge of cutting across bureaucratic lines.

General audiences—Messages to general audiences require effective use of mass communication methods, including sufficient saturation to ensure that the messages have an impact. General awareness information (for example the value of marine ecosystems) and information applicable to virtually everyone (what to do with waste oil or paint thinner) require such methods.

Water quality education in the schools—Excellent water quality and other environmental curricula exist for use in schools. They are most effective when adapted to specific local places and issues and teachers are trained in their use. Both of these needs require resources. Basic environmental literacy is an even more important curricula goal than any specific topic; water quality curricula tend to be good for this goal, given the over-arching nature of the question "what affects water quality and how can we protect the water"?

Technical and scientific training in higher education—Integration rather than compartmentalization of technical/scientific education is crucial for the next generation of environmental professionals. Academia needs to address the companion (yet often competing) objectives of producing both *big thinkers* and competent specialists.

Public involvement linked to education—Hands-on projects for volunteers, such as stormdrain stenciling projects, beach clean-ups, restoring streams, and replanting anadromous fish, can educate while simultaneously accomplishing a direct environmental purpose. Such projects are very low cost, and will flourish with some governmental or private seed money.

Pollution prevention programs—Agencies and business associations are increasingly emphasizing "pollution prevention pays" and the technical information to encourage source reduction. Because most regulatory programs focus on the end of the pipe, prevention has largely stayed in the province of education, although, ideally, regulatory pressure and education would work together to achieve prevention.

Examples of Education Strategies

There are examples of efforts to address water quality with effective education strategies. Notable is the Public Involvement and Education (PIE) Fund of the Puget Sound Water Quality Authority in Washington State. Over a period of six years (the program is ongoing), $3 million in small grants have been applied to diverse model education projects sponsored by non-profit organizations, trade associations, local governments, tribes, schools, and others. Projects have ranged from creative (story-telling and song projects) to mundane (manuals and workshops) and have reached a large percentage of the residents of the Puget Sound Basin in some way.

The program has pioneered the concept of *peer-to-peer* education as an alternative to standard agency or academic training or public information programs. An example of this strategy: The Associated General Contractors of Washington were funded to work with their own members to develop a practical handbook and workplace poster addressing the everyday water-quality problems confronted on the *front lines* of their business (how to handle fuels, solvents, and other common toxicants; who to call with a question about hazardous waste; tips on erosion control during construction; etc.). The unique and effective aspect of this project was not the content but the fact that it was presented to the members by a credible source (their own trade association) in a language and format specifically designed around how they work and who makes what decisions on the job. Similar *peer-to-peer* projects have been carried out with the automotive repair industry, the dairy industry, horse owners, and others.

The program has also demonstrated the value of linking hands-on activities, such as cleaning up a stream or beach or taking water quality samples, with broader educational objectives, such as "how can I as a citizen take care of our water resources?" or "what are all the threats to this watershed, and what needs to be done"?

Above all, the PIE Fund program has shown that a small amount of money directed at existing organizations with already established audiences, networks, and programs can have a magnified effect. The largest PIE grants have been under $50,000, and most have been far less.

In academic settings, there are also signs of progress. The EPA has funded programs at the University of New Orleans and Tufts University. The latter is building environmental material and awareness into curricula throughout the university as an alternative to creating a separate specialty in environmental studies. The former incorporates many programs of the PIE-Washington State program into a comprehensive urban wastes control program. Several states, including Washington, have now added environmental education to the basic K-12 curricular requirements.

Challenges and Issues

Accomplishing improved water quality through education—while generally considered non-controversial—in fact is a strategy hindered by significant challenges. For example:

- Where is the line between *information* or *education* and *propaganda*? To motivate behavior and attitude changes, persuasion techniques and value-laden content are essential ingredients. In addition, general audiences need simplified information, which of course can be challenged by experts.
- Where will the funding come from for significantly increased environmental/water quality education programs? While usually afforded *lip service* as important, environmental education programs are rarely afforded high budgetary priority.
- Can higher education science and engineering programs effectively

train both specialists and interdisciplinary scientists and managers? Traditional science and engineering education rewards or even demands intense specialization, but solving environmental problems also demands scientists and engineers who can cross disciplines with facility and competence.

Evaluating Education as a Tool

Effectiveness. Generally speaking, education is more effective in the long term than in the short term, especially education aimed at building an environmental ethic in our society. However, the reverse may be true for specific, targeted messages; there may be only a short-term positive effect in behaviors and attitudes. For example, people may refrain from disposing of oil or other hazardous chemicals into stormdrains during and shortly after an intensive education campaign but may not sustain that behavior over time. Similarly, technical assistance and training can have very immediate effect but must be an ongoing process to address new employees and new problems. Many of the positive effects of education are hard to quantify and evaluate, and thus their effectiveness may be questionable.

Efficiency. Education is not free, but it is inexpensive compared with capital investments. As noted above, effectiveness is sometimes hard to measure, and most water-quality education efforts are not rigorously evaluated with respect to actual changes in behavior or improvements in water quality (although participants in education programs are routinely asked to provide their own evaluation), making assertions about cost-benefit relationships impossible.

Fairness. Water quality education could be judged unfair if funding were distributed inequitably or if education were perceived as a substitute for enforcing pollution control laws. The content of water quality education is sometimes criticized for *singling out* certain polluters or for implying that the individual bears all the responsibility for pollution (as opposed to industry or government). But overall, education is usually applauded as an important, and presumably fair, strategy.

Redistribution of Income. This could become an issue with respect to how water-quality education programs are funded, although the amounts of money (and thus the significance of the issue) are not likely to make this a major concern.

Applicability. Education is uniquely suited to address nonpoint pollution, stormwater contamination, and pollution prevention and to achieve an overall environmental consciousness linking air, water, land, and other environmental issues. Education is an important adjunct to regulatory and planning strategies.

Feasibility. Education is one of the most feasible and least controversial strategies for addressing water quality. However, it is often seen as a frill when budgets are prepared. Education efforts are also often assigned to staff who are not trained or qualified to perform that kind of job. For example, the NPDES permit writer/inspector may be expected to provide technical assistance to the discharger, or the environmental engineer may be expected to train the citizen watershed action team. In schools, the major feasibility issue is the ability to provide teachers the time to learn new curricula.

Risk. Education is a low-risk strategy. The amounts of money are generally not large, and the programs can be modified along the way.

Incentives for Technology Improvement. Technology transfer and pollution prevention programs may be very effective ways of encouraging technology improvement, especially if the education can point to economic and reliability advantages of the improvement.

Feedback/Adaptive Management/Ease of Evaluation. Education programs are often evaluated by the participants and are usually easy to adapt. Exceptions to this are school curricula that become out of date but continue to be used and the difficulty of quantifying the actual effect of education on water quality.

Prevention Versus Abatement. Education is an essential part of pollution prevention efforts and is an adjunct to abatement strategies.

Integrated Versus Single Medium Strategies. Education can apply to single medium strategies but is particularly well-suited to integrating environmental information. For example, a good technical assistance program targeted to a particular industry will take a top-to-bottom approach to the facility, looking at all of the environmental issues and opportunities for pollution prevention.

Single Versus Cumulative Impacts. Education can occur on many levels and is especially important for problems like nonpoint pollution or pouring toxicants down the drain, where cumulative impacts are the issue.

Conclusion

Education is an essential component of water-quality protection strategies. There are excellent pilot and model education projects which should be emulated. Academic programs, advocacy, and education directed toward changing attitudes and behavior are all different and valid components of water quality/environmental education. Its priority has traditionally been very low, however. To deal effectively with nonpoint pollution, stormwater management, toxicants in the municipal treatment system, and pollution prevention strategies, it will be crucial to increase education and public involvement expenditures and other resource commitments. Basic environmental literacy and improved specialized education are also important aspects of an education strategy.

FINANCING MECHANISMS

Introduction

The costs of wastewater management expenditures are borne, in the first instance, by government agencies and private enterprises. In a typical year, the private sector pays directly about two-thirds of the total cost of wastewater management. (In 1988, the business community spent $21.6 billion for water pollution abatement and control, compared with a total expenditure—including regulation, monitoring, and research—of $33.2 billion [Bureau of the Census 1991]). Private sector outlays are financed by increases in product prices and reduced corporate profits. Taken in the aggregate, these changes are small: in 1988, the private sector expenditures for water pollution control were equivalent to less than one percent of total personal consumption expenditures, or about three percent of business profits (Bureau of the Census 1991). Nevertheless, even small changes in prices and profits can potentially influence rates of price inflation, industrial output, employment, and international competitiveness. Examination of these impacts is beyond the scope of this study.

Of more immediate interest, due to its close relationship to the effectiveness of management strategies, is the financing of government expenditures for wastewater management. Most financing decisions are the responsibility of the individual operating agency, whether local or regional in scope. Typically, financing occurs in a decentralized manner, even when compared with other local government programs. Participation of state and federal government is ordinarily limited to the provision of grant and loan programs, which may or may not be available at a particular place and time. Tax and fiscal policies of local and state governments generally have little or no influence on the financing strategies adopted by wastewater agencies.

In developing a financing strategy, wastewater management agencies may consider a number of objectives, including:

- revenue adequacy,
- minimum financing cost,

- minimum total cost,
- acceptable cash flow profile,
- minimum financial risk,
- acceptable incidence of cost on benefitted population,
- flexibility with respect to future financing decisions, and
- public and political acceptability.

At least five major sources of funds may be considered:

- general taxes,
- dedicated taxes,
- user fees and charges,
- intergovernmental transfers, and
- debt financing.

Within each of these categories, there are numerous variants and alternatives. The following paragraphs discuss some of the more common financing instruments, contrasting their positive and negative aspects. Possible economic impacts of financing strategies are also reviewed.

Financing Alternatives

The number of unique financing methods available to wastewater management agencies is very large. Numerous variations of each general type of method are possible; a number of different financing methods can be combined, either in parallel or sequentially, to produce an overall financing strategy. The following paragraphs describe the basic types of methods, without discussing any of the possible variations.

General Tax Revenue

This category includes all revenues derived from non-dedicated taxes and received into a government's general fund. The taxes most often levied by local governments include real property taxes, sales taxes, local income taxes (sometimes consisting of revenue-sharing with a state income tax, called *piggy-backing*), and excise taxes levied on specific commodities or activities (since these taxes are often levied on consumption or sale of alcohol, tobacco, etc., they are called *sin* taxes). In addition, some local governments levy taxes on utilities, business franchises, business inventories, etc. Since these revenues are placed in the government's general fund, they are made available to wastewater management activities by the local legislative body through the normal process of appropriation. This is a straightforward and familiar process with few financing costs. However, local taxes are often politically unpopular and difficult to increase. Wastewater management competes with other local programs for a share of an effectively fixed revenue source.

Dedicated Taxes

It is also possible to levy specific taxes for the purpose of financing wastewater management. The revenues from these taxes do not flow into the general fund but are placed in a trust fund; they can only be withdrawn pursuant to an appropriation to wastewater management activities. These dedicated taxes may be levied on activities thought to contribute to water pollution costs: e.g., on value of manufacturing industry shipments. Where taxes or effluent charges are levied on pollutant discharges, it is common to dedicate the resulting revenues to wastewater management purposes.

The use of dedicated taxes has the advantage of insulating wastewater management from the tax and spending policy controversies surrounding general fund transactions. On the other hand, special legislation is usually needed to create dedicated taxes and their associated trust funds. Since the taxes are often narrowly focused, those who will be taxed may be expected to oppose their adoption.

User Charges

User charges differ from taxes in an important way. Taxes are levied on various kinds of property or activity and cannot be avoided short of disposing of the property or ceasing the activity. User charges are payments required in return for services provided—in this case, by the government. User charges are paid only by those who receive the service; anyone who elects not to receive it is not required to pay. Where possible, user charges are proportionate to the service provided. User charges are, therefore, avoidable.

The most common wastewater application of this financing mechanism is the user charge levied on individual residences, businesses, and institutions for wastewater service. This charge applies only to those actually connected to the collection system and receiving services. In most cases, the charge is based on water use, an approximate measure of the quantity of wastewater services provided. Fees for permits or special services are also user charges.

User charges are generally more acceptable to the public than comparable taxes, and they have advantageous economic characteristics (see below). From the wastewater agency's perspective, user charges involve identifiable financing costs (costs of billing, collecting, etc.) and constitute a less stable source of revenue than taxes (user charge income fluctuates with changes in the use of the related services).

Intergovernmental Transfers

Intergovernmental transfers of funds, or subsidies, may be provided for general purposes, or they can be dedicated to specific activities. Revenue sharing programs are often general purpose in nature, where a state or federal government simply transfers some amount of money from one general fund to another. These funds are appropriated in the same way as other general funds, derived from local taxes. Some revenue transfers may be nonspecific but restricted to a class of purposes, such as federal block grants.

Other transfers are made for a particular purpose, and the funds are restricted to that purpose. State and federal grants for wastewater treatment plant construction constitute intergovernmental transfers of this kind. Grants or fund transfers may also be made to support local or regional regulatory and monitoring programs.

Transfers are attractive to local governments and wastewater agencies, since they involve no local sacrifice and are relatively free of financing cost. For these reasons, they are politically popular at the local level. However, just as they require no local legislative action, they are not subject to local control and may appear and disappear unpredictably, without regard to relative need.

Debt

Another option is the possibility of meeting wastewater management costs with borrowed funds. This has the effect of transferring the financing requirement from the present to the future. A local government or agency may borrow funds on a short-term basis, usually through commercial banks, or for the long term through the issuance of bonds. Long-term borrowings include mortgage bonds (secured by physical assets, not often used by governments), revenue bonds (secured by anticipated future revenue from user charges, fees, and taxes), or general obligation bonds (secured by the full faith and credit of the issuing government). Revenue bonds may be issued by most wastewater utilities organized on an enterprise basis, with some degree of financial autonomy, while general obligation bonds can only be issued by a general-purpose governmental entity (often subject to voter approval).

The use of debt allows an agency to spread the financing of capital outlays over time, producing more uniform and manageable annual cash requirements. Financing costs may be significant, however. Issuing bonds involves one-time underwriting, placement, insurance, and legal costs. Interest must be paid over the life of the bond at rates that may later diverge from the current opportunity cost of capital to the utility. Also, there are market-enforced limitations on the total amount of debt that a particular agency can carry; attempts to increase debt beyond this point will greatly increase the cost of borrowing.

Financing costs are sometimes reduced by subsidies from other levels of government. For example, many states have recently created revolving loan funds with federal assistance. These revolving funds will make loans to local agencies for wastewater treatment plant construction at interest rates and overall financing costs that are less, in general, than those available in the market. The cost savings can be interpreted as an intergovernmental transfer, or subsidy.

Economic Impacts

The choice among financing methods depends upon the relative advantages and disadvantages of each in particular situations. Some of these characteristics, such as the suitability of the cash flow profile or the public and political acceptability of a given strategy, are difficult to generalize. Others derive from the inherent economic impacts of the financing method, as described below.

Revenue Adequacy and Stability

The primary objective of a financing method is that it produce the necessary funds. The requirement is so basic that it seems unnecessary to mention it. Yet it is not always clear that a particular method will, in fact, produce the necessary revenue or that it will do so reliably. Some revenue sources, for example, may increase or decrease in response to external factors. Dedicated taxes may vary with changes in the taxed activity; user charges fluctuate with changes in economic activity, population, or tariff level; intergovernmental transfers may be subject to reduction or discontinuance in times of fiscal stress.

Where variations in the revenue flow are accompanied by changes in financing requirement, as is the case where revenue is derived from user charges and the service area population changes, the indicator of interest is the net revenue (total revenue less total costs). Instability in net revenue can require costly supplemental financing or precipitate a fiscal crisis. Other revenue sources may vary independent of revenue requirements. In any of these cases, the result is financial risk, which must be balanced against other factors in choosing a financing method.

Cost Incidence

Each financing method results in a particular distribution of cost allocation among individuals and organizations and across time. General taxes allocate costs in accordance with the nature of the tax base (e.g., according to real property value). Increases in user charges allocate costs in accordance with water use; debt instruments allocate present costs to future time periods (the nature of the cost incidence in the future depends on the method used to finance the debt service payments). Intergovernmental transfers allocate most costs to residents of other political jurisdictions.

Public and political acceptability of particular methods is influenced by the resulting cost incidence and the perceived fairness of that distribution. Property owners may regard general fund financing as unfair since costs fall on all property owners regardless of use of the wastewater system, and owners of high-value property may pay a relatively large share of the costs. Dedicated taxes, especially those levied on activities or commodities unrelated to wastewater production, may be regarded as unfair. Intergovernmental transfers may be seen as unfair by residents of non-benefiting jurisdictions.

Incentives for Efficient Management

Wastewater management is a service performed for those who generate wastewater as well as those who may benefit from improved quality of the receiving water body. Where the costs of wastewater management are allocated, via the choice of financing mechanisms, to activities and entities who neither contribute to or benefit from the service, there is little incentive for management agencies to improve the efficiency of operation. This is also true when the funds are raised through taxes, even though those who pay the tax may benefit. Their tax liability remains the same regardless of the presence or absence of the benefit.

User charges have a very different characteristic. User charge revenue is the result of voluntary payments for wastewater management services: if the service is not provided, no payment is made; if the service is too expensive, less will be used; if less service is provided, less revenue will be obtained. User charges, alone among the revenue sources discussed, create an incentive for management agencies to provide the service in a cost-effective way. In many cases, the incentive may be weak, especially when compared with the more competitive private-sector industries, but it can promote some degree of efficiency in the operation of wastewater agencies.

Willingness to Pay for Wastewater Services

To the extent that wastewater management provides valuable services to those who either generate waste or have an interest in the quality of receiving waters, those individuals must have a willingness to pay for these services. Willingness to pay is defined as the maximum amount that would be paid for the level of service received rather than forego it altogether (*all or nothing*). Those who pay user charges demonstrate, by doing so, that their willingness to pay is at least as great as the charge paid. In fact, it may be much greater: the user charge merely establishes the lower bound. Estimates of willingness to pay can be performed by various indirect methods, such as econometric demand analysis (for those who pay user charges) or contingent valuation studies (for those who benefit from improved water quality).

Such studies are seldom done for wastewater systems but are potentially important. The results would indicate which groups receive benefits from improved wastewater management and what the approximate magnitude of those benefits may be. These results would also be useful in predicting public acceptance of new financing burdens, especially where large increases in financing requirements are expected. This information could be used to tailor financing strategies to the temporal and spatial distribution of anticipated benefits, thus minimizing the chances of placing unjustified burdens on any sector of the population.

REFERENCES

Bernstein, J.D. 1991. Alternative Approaches to Pollution Control and Waste Management: Regulatory and Economic Instruments, draft report. Washington, D.C.: UNDP/World Bank/UNCHS Urban Management and Environment Program.

Boland, J.J. 1989. Environmental Control Through Economic Incentive: A Survey of Recent Experience. Presented at the Prince Bertil Symposium on Economic Instruments in Environmental Control, Stockholm School of Economics, Stockholm, Sweden, June 12-14.

Bureau of the Census. 1991. Statistical Abstract of the United States, 1991. 111th Edition. Washington, D.C.: U.S. Bureau of the Census.

EPA (U.S. Environmental Protection Agency). 1990. Reducing Risk: Setting Priorities and Strategies for Environmental Protection. SAB-EC-90-021. Washington, D.C.: U.S. Environmental Protection Agency, Science Advisory Board.

EPA (U.S. Environmental Protection Agency). 1991. Economic Incentives: Options for Environmental Protection. Report of the U.S. Environmental Protection Agency Economic Incentives Task Force, March 1991.

Ernst and Young. 1990. 1990 National Water and Wastewater Rate Survey, Ernst & Young's National Environmental Consulting Group.

Federal Register. 1989. September 8. 54(173):37370-37373.

Federal Register. 1990. November 16. 55(222):47990.

OECD (Organization for Economic Co-Operation and Development). 1989. Economic Instruments for Environmental Protection. Paris, France: OECD.

PSWQA (Puget Sound Water Quality Authority). 1984. Annual Report. Olympia, Washington: Puget Sound Water Quality Authority.

F

Biographical Sketches

COMMITTEE AND PANEL MEMBERS

JOHN J. BOLAND, *Chair,* holds a B.E.E. in electrical engineering from Gannon College, an M.A. in governmental administration from George Washington University, and a Ph.D. in environmental economics from Johns Hopkins University. He is a registered professional engineer. His background includes management positions in water and wastewater utilities, teaching, research, and consulting activities at all levels of government and in private industry. He is currently Professor of Geography and Environmental Engineering at Johns Hopkins University. Dr. Boland has published widely on economic aspects of water and resource policy. He is an associate editor of *The Annals of Regional Science* and a member of the Risk Management Technical Advisory Workgroup of the American Water Works Association. He has served on a number of committees and panels of the National Research Council including chair of the Water Science and Technology Board (1985-1988).

BLAKE P. ANDERSON received his B.S. in civil engineering from California State Polytechnic University, Pomona, and has pursued graduate work at California State University, Long Beach, and California State Polytechnic University, Pomona. He is a registered Civil Engineer and a Certified Wastewater Treatment Plant Operator in California. Mr. Anderson is currently the Director of Technical Services for the County Sanitation Districts of Orange County. His responsibilities include directing analytical and technical support for treatment plants, industrial waste control program and research projects; and administering the district's industrial waste; compliance monitoring; ocean monitoring; reclamation, reuse, and conservation programs; and air-quality compliance programs. Mr. Anderson is a member of the American Society of Civil Engineers and the Water Environment Federation. He is chair of the Association of Metropolitan Sewage Agencies Committee on Comprehensive Watershed Management.

TAKASHI ASANO received his B.S. in agricultural chemistry from Hokkaido University in Japan, his M.S. in sanitary engineering from the University of California, Berkeley, and his Ph.D. in Sanitary and Water Resources Engineering from the University of Michigan, Ann Arbor. He is currently an Adjunct Professor in the Department of Civil and Environmental Engineering at the University of California, Davis. He was on the staff of the California State Water Resources Control Board from 1987 to 1992. His consulting activities include many of the United Nations' Agencies. Dr. Asano's research interests include wastewater reclamation and reuse, advanced waste treatment, artificial recharge of ground water, water and wastewater treatment, and planning and regulatory aspects of wastewater reuse. He is Chair of the Specialist Group on Wastewater Reclamation, Recycling and Reuse of the International Association on Water Quality and also serves as a member of the Governing Board. Dr. Asano is a member of the American Society of Civil Engineers, the Water Environment Federation, American Water Works Association, and the American Association of Environmental Engineering Professors.

NORMAN H. BROOKS has an A.B. and M.S. in civil engineering from Harvard University and a Ph.D. in civil engineering and physics from California Institute of Technology. He has conducted research and published extensively in the field of hydraulics with emphasis on ocean waste disposal, sediment transport, turbulent diffusion, stratified flow, and environmental policy. Currently he is the James Irvine Professor of Environmental and Civil Engineering and the Director of the Environmental Quality Laboratory at the California Institute of Technology. He is a member of the National Academy of Engineering and National Academy of Sciences. He has served on many National Research Council committees and is a former member of the Water Science and Technology Board.

RICHARD A. CONWAY (WSTB *ex-officio*) received his B.S. in 1953 from the University of Massachusetts and an S.M. in sanitary engineering from the Massachusetts Institute of Technology in 1957. His expertise is in water treatment, aquatic fate processes, and hazardous waste management. Presently he is with the Central Research and Engineering Technology Department of Chemicals and Plastics Group of Union Carbide Corporation.

Mr. Conway is a member of the National Academy of Engineering and a former member of the Water Science and Technology Board. He has authored or edited several books and technical papers related to pollution control technology.

GLEN DAIGGER received his B.S.C.E., M.S.C.E., and Ph.D. in environmental engineering from Purdue University. Dr. Daigger is Vice President and Director of CH_2M Hill's Wastewater Reclamation Discipline Group. He also serves as a process engineer, project engineer, and project consultant on a variety of municipal and industrial wastewater treatment and reclamation projects. His areas of expertise include biological wastewater treatment and treatment process design. Dr. Daigger has organized major firm-wide technical initiatives in the areas of nutrient control, advanced techniques for wastewater treatment plant analysis, toxics control in wastewater treatment plants, and air emissions from wastewater treatment plants. He is a member of the American Society of Civil Engineers and the Water Pollution Control Federation.

JOHN M. DEGROVE received his B.A. from Rollins College, his M.A. from Emory University, and his Ph.D. from the University of North Carolina. He is the Director of the Joint Center for Environmental and Urban Problems at Florida Atlantic University/Florida International University. The Joint Center's applied research efforts focus on environmental and urban issues affecting governments on a state, regional, or local level—ranging from contractual applied research projects, channeling specific technical information, analyzing legislation, or providing other expertise. Also, he is a Professor in the Department of Political Science at Florida Atlantic University. Dr. DeGrove's current research interests combine urban and environmental areas by focusing on land and growth management.

WILLIAM M. EICHBAUM received his B.A. from Dartmouth College and his LL.B. from Harvard Law School. He specializes in environmental law and public policy and is currently a Vice President on International Environmental Quality of the World Wildlife Fund in Washington, D.C. Prior to his work there, he held posts including Undersecretary, Executive Office of Environmental Affairs, Commonwealth of Massachusetts and Assistant Secretary for Environmental Programs at the Maryland Department of Health and Mental Hygiene. Mr. Eichbaum is a member of the Chesapeake Critical Area Commission, the National Environmental Enforcement Council of the Department of Justice, the Coastal Seas Governance Project, the Patuxent River Commission, and the Environmental Law Institute. He was a member of the National Research Council Committee on Institutional

Considerations in Reducing the Generation of Hazardous Industrial Wastes and the Committee on Marine Environmental Monitoring.

KATHERINE FLETCHER received her B.A. in biology from Harvard University, Radcliffe College in 1971. For five years, Ms. Fletcher was the Chair of the Puget Sound Water Quality Authority, a state agency created to develop and oversee the implementation of a comprehensive plan to clean up and protect the sound. Currently she is on the faculty of the Graduate School of Public Affairs at the University of Washington, where she is teaching courses in environmental policy. She is also a consultant for the Institute for Public Policy and Management at the University of Washington, Seattle.

WAYNE R. GEYER is Associate Scientist of the Department of Applied Ocean Physics and Engineering at Woods Hole Oceanographic Institution. He is also a consultant to Camp, Dresser & McKee and the U.S. Justice Department. Dr. Geyer received his Ph.D. and M.S. in physical oceanography from the University of Washington and his B.A. in Geology from Dartmouth College. His research interests include estuarine and coastal dynamics and transport processes and physical-biological interaction.

LYNN R. GOLDMAN is an environmental epidemiologist who received her B.S. in conservation of natural resources and M.S. in health and medical sciences from the University of California, Berkeley, an M.P.H. from Johns Hopkins University, and an M.D. from the University of California, San Francisco. She is a Board certified pediatrician and currently is Chief of the Division of Environmental and Occupational Disease Control for the California Department of Health Services. Dr. Goldman is a Fellow of the American Academy of Pediatrics for which she is a member of the Committee on Environmental Hazards. She served as a member of the National Research Council Committee to Evaluate the Hazardous Materials Management Program of the Bureau of Land Management and the BEST Committee on Environmental Epidemiology and is a current member of the Water Science and Technology Board.

DONALD R. F. HARLEMAN holds a B.S. from Pennsylvania State University and an S.M. and Sc.D. in civil engineering from Massachusetts Institute of Technology. Professor Harleman's field is environmental engineering, specifically the fate and transport of pollutants in natural water bodies such as lakes, reservoirs, rivers, estuaries, and coastal waters. His research is concerned with the interaction of fluid transport and biogeochemical transformation processes and with innovative wastewater treatment technology involving chemical additives to promote settling through coagulation and flocculation. Currently he is the Ford Professor of Civil Engineering at Massachusetts Institute of Technology, where he has also held the position of Director of the Ralph M. Parsons Laboratory. Professor Harleman is a member of the National Academy of Engineering.

JAMES P. HEANEY received his Ph.D. in civil engineering from Northwestern University in 1968 with an emphasis on water resources engineering, operations research, and urban and regional planning. He is presently Professor and Chair of the Department of Civil, Environmental, and Architectural Engineering at the University of Colorado at Boulder. His research is concerned with developing methods for evaluating cost-effective, multipurpose environmental management systems. Dr. Heaney has developed decision support systems for urban stormwater quality management. He is a member of the American Society of Civil Engineers, American Water Resources Association, and Association of Environmental Engineering Professors. Dr. Heaney was a member of the WSTB from 1986-1990.

ROBERT W. HOWARTH earned his B.A. from Amherst College and his Ph.D. in oceanography from the Massachusetts Institute of Technology and Woods Hole Oceanographic Institute. He is currently a Professor in the Division of Biology and Senior Fellow in the Center for the Environment at Cornell University. Concurrently, he is an adjunct professor at the University of Rhode Island. He also serves as the editor-in-chief of *Biogeochemistry.* Dr. Howarth is a member of the American Society of Limnology and Oceanography, the Ecological Society of America, and the Estuarine Research Federation. His areas of research include environmental management and the effects of pollution on aquatic ecosystems and commercial fisheries, wetland ecosystems, microbial production and activity, sulfur and molybdenum biogeochemistry, and interactions of element cycles in aquatic ecosystems. He was a member of the National Research Council Committee on Petroleum in Marine Environments, and the NRC U.S. National Committee for SCOPE (Scientific Committee on Problems of the Environment). Dr. Howarth currently serves on the NRC Committee on the Coastal Ocean and chairs the new SCOPE Project on Global and Regional Transport of Nitrogen.

ROBERT J. HUGGETT is a Professor in the School of Marine Science of the College of William and Mary and is an Assistant Director of the Virginia Institute of Marine Science (VIMS). He received his M.S. degree in earth science (chemical oceanography) from the Scripps Institution of Oceanography and his Ph.D. in marine science (marine chemistry) from the College of William and Mary. His research interests involve the transport, fate, and effects of toxic chemicals in aqueous systems. Since 1975 he has headed the environmental chemistry research program at VIMS working on such issues as Kepone and its contamination of the James River, Virginia; the transport, fate, and effects of tributlytin (TBT) from antifouling paints in estuarine systems; and the environmental chemistry of polynuclear aromatic hydrocarbons. Professor Huggett is a member of the Executive Committee and the Ecological Processes and Effects Committee of the Science Advisory Board of the U.S. Environmental Protection Agency. He was awarded the Shelton G. Horsley Award for meritorious Fundamental Research by the Virginia Academy of Science in 1980. He received the Izzak Walton League of America Chesapeake Bay Conservation award for fisheries and wildlife in 1989. Dr. Huggett is also a member of the Water Science and Technology Board.

GEORGE A. JACKSON received his B.S. in physics, his M.S. in environmental engineering, and his Ph.D. in environmental engineering science and biology from California Institute of Technology. Recently Dr. Jackson joined the faculty of Texas A & M as a professor of Oceanography. Prior to this, he was an Associate Research Oceanographer at Scripps Institution of Oceanography at the University of California, San Diego. He was Panel Chairman for the National Research Council's Workshop on Land, Sea, and Air Options for the Disposal of Industrial and Domestic Wastes and was also a participant of the NRC's Committee on Ocean Waste Transportation Alternatives. Research interests of Dr. Jackson include physical and chemical properties of aquatic environments, modeling seaweed growth, the role of the seafloor in maintaining deep-ocean chemistry, and internal Kelvin wave propagation in a high drag coastal environment.

THOMAS M. KEINATH acquired his B.S.E., M.S.E., and Ph.D. in civil and water resources engineering from the University of Michigan, Ann Arbor. Currently, he is professor and head of the Department of Environmental Systems Engineering at Clemson University. His research interests include physicochemical water, wastewater and hazardous waste treatment processes, and automation and control of water and wastewater treatment systems. Dr. Keinath has served as an expert science advisor to the Environmental Protection Agency and consults for several environmental engineering firms. He has been an active member of major national and international professional organizations concerned with water quality control. He presently serves as a member of the Governing Board of the International Association of Water Pollution Research and Control and as chairman of the U.S.A. National Committee. He is vice-chair of the Program Committee of the Water Pollution Control Federation and serves the American Society of Civil Engineers as Chair of its Clarifier Research Technical Committee. Dr.

Keinath also is presently serving as President of the Association of Environmental Engineering Professors.

BILLY H. KORNEGAY received his B.S.C.E. in civil engineering from Virginia Military Institute and his M.S.E. in water resources engineering and his Ph.D. in environmental systems engineering from Clemson University. He is currently the Technical Manager of Water and Wastewater Processes at Engineering-Science where he is responsible for providing technical assistance on municipal water and wastewater processes. Dr. Kornegay has a broad range of environmental engineering experience in academia, industry, and the consulting field, including extensive research and pilot studies as well as water and wastewater process design. Wastewater process experience involves physical/chemical and biological processes. He is a member of the Water Pollution Control Federation Research Committee and is a Registered Professional Engineer in Virginia and Georgia.

JAMES F. KREISSL received his B.C.E. from Marquette University and his M.S. from the University of Wisconsin. He is an Environmental Engineer at the U.S. Environmental Protection Agency, Office of Research and Development, Center for Environmental Research Information where he develops design manuals and other technology transfer products to advance the state of the art of environmental control professionals. Research experience of Mr. Kreissl include drinking water and wastewater treatment, wastewater collection, residuals production, treatment and disposal, and regulatory/engineering issues related to the improvement of present practices. He is a 30-year member of the Water Pollution Control Federation.

JOSEPH T. LING received his Ph.D. in sanitary engineering from the University of Minnesota. He was Vice President for 3M from 1974 to 1984, when he retired. His responsibilities at 3M included the coordination and implementation of all civil engineering, environmental engineering, and pollution control programs. Dr. Ling is an internationally recognized expert in environmental management. He was the first person to stress that environmental management must be based on cross-media total environmental impact. In 1974 Dr. Ling initiated the very first Industrial Waste Minimization Program in the United States. He has served on the National Research Council, and is a member of the National Academy of Engineering. Currently, in addition to being Chair of the National Reduction Institute, Dr. Ling is also serving as a Board Director of the World Environmental Center; Board Director of Freshwater Foundation; Board Director of Midwest China Center; Senior Advisor to the Chinese Water and Wastewater Research Institute; and Vice Chair of the Environmental Committee of the U.S. Council of International Business.

ALAN J. MEARNS acquired his B.S. and M.A. from California State University, Long Beach, and his Ph.D. in fisheries from the University of Washington. His research activities include planning and coordinating national and U.S. west coast marine pollution and monitoring programs, developing strategies for marine sewage and sludge disposal, conducting studies of marine pollution in the Southern California Bight and of pollutant flow through marine food webs, and coordinating research on biological effects of oil spills clean-up methods. He is Senior Ecologist with the Hazardous Materials Response and Assessments Division of the National Oceanic and Atmospheric Administration (NOAA) in Seattle, and his BioAssessment Team is currently evaluating recovery of marine ecosystems from the Exxon Valdez oil spill and evaluating the effects and efficacy of bioremediation. Dr. Mearns was Leader of Biology Division of the Southern California Coastal Water Research Project (1973-1980). He was also Ecologist for the NOAA Puget Sound Marine Ecosystems Analysis Program (1980-1984) and for the NOAA National Status and Trends Program (1984-1990). He conducted and participated in five surveys of Prince William Sound, Alaska, following the 1989 Exxon Valdez Oil Spill. He received awards of meritorious service from the San Francisco Bay Regional Water Quality Control Board and the Santa Monica Bay Restoration Project where he is Vice-Chair of the Technical Advisory Committee. Dr. Mearns also served as NOAA liaison to an NRC study of Monitoring in the Southern California Bight.

VLADIMIR NOVOTNY received his Diploma Engineer in sanitary engineering and his Candidate of Science in Sanitary & Water Resources from the Technical University of Brno, Czechoslovakia, and his Ph.D. in environmental engineering from Vanderbilt University. He is a Professor of Civil Engineering at Marquette University. His interests include wet weather (diffuse) urban pollution, water quality impacts, transmission of nonpoint pollution, and modeling of urban runoff. Dr. Novotny is a member of the International Association on Water Pollution Research and Control, the Water Pollution Control Federation, and the American Society of Civil Engineers.

CHARLES R. O'MELIA received a B.C.E. in 1955 from Manhattan College, an M.S.E. in 1956, and a Ph.D. in sanitary engineering in 1963 from the University of Michigan. Currently a professor at the Johns Hopkins University, Dr. O'Melia's professional experience includes positions as assistant engineer for Hazen & Sawyer, Engineers; assistant sanitary engineer, University of Michigan; assistant professor, Georgia Institute of Technology; lecturer, Harvard University; and associate professor of environmental science and engineering at the University of North Carolina, Chapel Hill.

He is an environmental engineer with research interests in aquatic chemistry, environmental fate and transport, predictive modeling of natural systems, and theory of water and wastewater treatment. Dr. O'Melia is a member of the National Academy of Engineering.

WILLIAM C. PISANO earned his B.C.E. from Santa Clara University, his M.S. in sanitary engineering from the University of Arizona, and his Ph.D. from Harvard University. He is Principal of Havens and Emerson, Inc. Prior to this, Dr. Pisano was President of Environmental Design and Planning, Inc., which was an environmental consulting engineering firm in the general area of water pollution and drainage control. Dr. Pisano has 30 years of experience in environmental engineering with particular experience in combined sewer overflow abatement, including research, concept planning, design, and control technologies; stormwater management planning and design; vortex valve technology; water quality and water resources management; systems analysis; water quality modeling; economics; hydrology; and hydraulics. He is a member of the National Water Pollution Control Federation Committee and a registered professional engineer in Massachusetts, Ohio, and Michigan.

DONALD W. PRITCHARD received his B.A. in meteorology from the University of California, Los Angeles, and his M.A. and Ph.D. in oceanography from Scripps Institute of Oceanography. His research interests include dynamics of estuarine circulation and mixing; inshore and coastal oceanography; and turbulent diffusion of natural waters. He is retired from his position at the State University of New York, Stony Brook as the Associate Director of research at the Marine Science Research Center. He is the co-chair of the Technical Advisory Committee of the EPA Long Island Sound Study and a consultant to the Committee on Tidal Hydraulics for the U.S. Army Corps of Engineers. Dr. Pritchard is a member of many professional societies including the International Oceanographic Foundation, the American Association for the Advancement of Science, and the Atlantic Estuarine Research Society. He is also a member of the National Academy of Engineering.

LARRY A. ROESNER received his B.S. in civil engineering from Valparaiso University, his M.S. in hydrology from Colorado State University, and his Ph.D. in sanitary engineering from the University of Washington. He is a registered Professional Engineer in Virginia, Michigan, California, Maryland, and Ohio and a registered Professional Hydrologist. He is currently Senior Vice President and Technical Director of Water Resources and Environmental Sciences for the South Region of Camp Dresser, & McKee. Dr. Roesner has more than 20 years of experience in water resources and water quality engineering and management. He specializes in urban hydrology and nonpoint source pollution control. He has developed and applied sophisticated models for flow-routing urban drainage systems and for analysis of nonpoint source pollution, precipitation and runoff, and combined sewer overflows. Dr. Roesner is a member of the National Academy of Engineering.

JOAN B. ROSE earned her M.S. from the University of Wyoming and her Ph.D. in microbiology from the University of Arizona. She is currently an Assistant Professor of Environmental and Occupational Health at the University of South Florida College of Public Health. Prior to holding that position, she was a research associate at the University of Arizona, Department of Microbiology and Immunology. Her research has focused on methods for detection of pathogens in wastewater and the environment, wastewater treatment for removal of pathogens, wastewater reuse, viruses and parasites in wastewater sludge, and the development and application of risk assessment models to wastewater discharge situations. Dr. Rose has organized and chaired many conferences and meetings on health related microbiology and water borne pathogens. In addition, she is a member of the National Drinking Water Council, serves on the Disinfection Technical Advisory Committee for the City of Portland, and chairs the subcommittee on Waterborne Outbreaks of the American Water Works Association Water Quality Division Committee on Organisms in Water.

JERRY R. SCHUBEL holds a B.S. from Alma College, an M.A.T. from Harvard University, and a Ph.D. in oceanography from Johns Hopkins University. His areas of research include estuarine and shallow water sedimentation, suspended sediment transport, interactions of sediment and organisms, pollution effects, continental shelf sedimentation, marine geophysics, and thermal ecology. Currently, he is the Director of the Marine Sciences Research Center and Dean and Leading Professor of Marine Sciences at SUNY Stony Brook. Dr. Schubel was the senior editor of Coastal Ocean Pollution Assessment and chairman of the Outer Continental Shelf Science Committee for the Department of Interior Mineral Management Service. He is past president of the Estuarine Research Federation. He was a member of the National Research Council's Committee on Marine Environmental Monitoring. Dr. Schubel is currently chair of the Marine Board.

P. AARNE VESILIND earned his B.S. and M.S. in civil engineering from Lehigh University and his M.S. in sanitary engineering and Ph.D. in environmental engineering from the University of North Carolina. Currently he is Professor and Chair of the Department of Civil and Environmental Engineering and is the Director of the Program in Science, Technology and Human Values at Duke University. Some of Dr. Vesilind's research interests include sludge dewatering, standard methods for the examination of water and wastewater, waste management and research, and environmental ethics. He is a member of the Water Pollution Control Federation, the American Association for the Advancement of Sciences, and the American Academy of Environmental Engineering.

COMMITTEE AND PANEL STAFF

PATRICIA L. CICERO received her B.A. in mathematics from Kenyon College. She currently is Senior Project Assistant at the National Research Council's Water Science and Technology Board (WSTB). Ms. Cicero has worked on a variety of studies at the WSTB, including ones on international soil and water research and development, wastewater management in coastal urban areas, techniques for assessing ground water vulnerability, and the environmental effects of the operations at Glen Canyon Dam on the lower Colorado River. She is also Coeditor of the Association for Women in Science Magazine.

SARAH CONNICK earned her A.B. in chemistry from Bryn Mawr College and her M.S. in environmental engineering from Stanford University. She is a Senior Staff Officer with the National Research Council's (NRC) WSTB where she directs studies of wastewater management in coastal urban areas, techniques for assessing ground water vulnerability, and Antarctic policy and science. Prior to joining the WSTB staff, Ms. Connick was a Staff Officer for the NRC's Committee to Provide Interim Oversight of the Department of Energy Nuclear Weapons Complex. Before joining the National Research Council, she served as a Senior Associate at the ILSI-Risk Science Institute where she performed projects related to the use of risk assessment in regulatory decision making. Ms. Connick has also held the position of Research Assistant at ENVIRON Corporation, where she worked on a wide range of environmental science, engineering, and public health policy projects.

LYNN D. KASPER received her B.A. in English from Wesleyan University. She is currently Assistant Editor for the NRC's Commission on Engineering and Technical Systems. Ms. Kasper has done editorial work for a number of the commission's boards, including the Board on Engineering Education and the Board on Natural Disasters.

JACQUELINE MACDONALD served as a research associate for the Committee on Wastewater Management for Coastal Urban Areas. She is now Staff Officer of the Water Science and Technology Board of the National Research Council, where she works on studies including ground water cleanup alternatives, planning and remediation for irrigation-induced water quality problems, and *in situ* bioremediation. Ms. MacDonald holds a master's degree in environmental science in civil engineering from the University of Illinois. She earned a bachelor's degree, *magna cum laude*, in mathematics from Bryn Mawr College.

INTERNS

BETH C. LAMBERT is currently a Senior at Carleton College studying geology. She participated in Carleton's Term in Nepal Program where she studied Nepal's irrigation methods at the Department of Irrigation in Kathmandu, Nepal. Ms. Lambert was a Summer Intern at the National Research Council's WSTB. Her main responsibility at the WSTB was to provide technical support for the Committee on Wastewater Management for Coastal Urban Areas Panel on Sources.

SUSAN E. MURCOTT received a B.A. in English from Wellesley College and an S.B. and S.M. in environmental engineering from the Massachusetts Institute of Technology. Her research interest is in innovative wastewater treatment technologies, including chemically enhanced primary treatment and aerated biofilters. Ms. Murcott is the recipient of the 1990 MIT Sea Grant Dean A. Horn Award for excellence in marine research. She is a member of the Water Environment Federation and the American Society of Civil Engineers.

G

Contributors to the Committee's Effort

Eugene Anderson
City of Tacoma
Washington

John Anderson
New South Wales Public Works Department
Sydney, Australia

Karen Anderson
California Department of Health Services
Emeryville, California

Chester Atkins
U.S. Congress
5th District, Massachusetts

Harold Bailey
Clean Water Program for Greater San Diego California

Joel Baker
Chesapeake Biological Institute
Solomons, Maryland

Paul Baltay
U.S. Environmental Protection Agency
Washington, D.C.

Barbara Bamburger
Sierra Club—San Diego Chapter
California

Mike Barnes
Fairfield-Suisun Sewer District
Suisun City, California

Dan Basta
National Oceanic and Atmospheric Administration
Rockville, Maryland

Robert K. Bastian
U.S. Environmental Protection Agency
Washington, D.C.

Richard Batuk
U.S. Environmental Protection Agency
Annapolis, Maryland

Brock Bernstein
EcoAnalysis, Inc.
Ojai, California

Dollof S. Bishop
U.S. Environmental Protection Agency
Cincinnati, Ohio

Robert Bonnett
Northeast Ohio Regional Sewer District
Cleveland, Ohio

Susan Bradford
County Sanitation Districts of Orange County
Fountain Valley, California

David Breitenstein
City of Eugene
Oregon

Donald R. Brown
Union Sanitary District
Fremont, California

Kenneth Bruland
University of California, Santa Cruz

Caroline Bruskin
County Sanitation Districts of Orange County
Fountain Valley, California

Edward H. Bryan
National Science Foundation
Washington, D.C.

Glen Cannon
National Oceanic and Atmospheric Administration
Seattle, Washington

Gregory D. Cargill
Metropolitan Water Reclamation District of Greater Chicago
Illinois

Robert S. Castle
Marin Municipal Water District
Corte Madera, California

James R. Coe
Central Contra Costa Sanitary District
Martinez, California

Ned Cokelet
National Oceanic and Atmospheric Administration
Seattle, Washington

Mike Conner
Massachusetts Water Resources Authority
Boston, Massachusetts

John Convery
U.S. Environmental Protection Agency
Cincinnati, Ohio

Edmund Cook
Metropolitan Water Reclamation District of Greater Chicago
Lemont, Illinois

Claudia Copeland
Congressional Research Service
Washington, D.C.

Richard Cotugno
Nassau County Department of Public Works
Lawrence, New York

Billy R. Creech
City of Raleigh Public Utilities Department
North Carolina

Jeffrey N. Cross
Southern California Coastal Water Research Project
Long Beach, California

Frank M. Cuffaro
Northeast Ohio Regional Sewer District
Cleveland, Ohio

Herbert Curl
National Oceanic and Atmospheric Administration
Seattle, Washington

Daniel B. Curll
The Boston Harbor Associates
Boston, Massachusetts

Michael A. Dadante
Northeast Ohio Regional Sewer District
Cleveland, Ohio

Tudor T. Davies
U.S. Environmental Protection Agency
Washington, D.C.

Tom Dawes
County Sanitation Districts of Orange County
Fountain Valley, California

Tina de Jesus
City and County of Honolulu
Honolulu, Hawaii

William R. Diamond
U.S. Environmental Protection Agency
Washington, D.C.

Walter Diewald
Transportation Research Board
National Research Council
Washington, D.C.

Dominic DiToro
Hydroqual, Inc.
Mahwah, New Jersey

Eleanor M. Dorsey
Conservation Law Foundation
Boston, Massachusetts

John Dorsey
Hyperion Treatment Plant, City of Los Angeles
Playa Del Rey, California

John Drapp
City of Tampa, Department of Sanitary Sewers
Florida

Michael P. Dunbar
South Coast Water District
Laguna Beach, California

Daniel R. Farrow
National Oceanic and Atmospheric Administration
Rockville, Maryland

Richard Finger
Municipality of Metropolitan Seattle
Renton, Washington

Nick Fisher
State University of New York
Stony Brook, New York

David B. Flaumenbaum
Nassau County Department of Public Works
Wantagh, New York

Virginia Fox-Norse
U.S. Environmental Protection Agency
Washington, D.C.

Jonathan A. French
Camp Dresser & McKee, Inc.
Cambridge, Massachusetts

R. Gallimore
Metropolitan Water Reclamation District of Greater Chicago
Skokie, Illinois

Gina Gartin
The University at Stony Brook
Stony Brook, New York

Sandra Germann
U.S. Environmental Protection Agency
Washington, D.C.

Gerald R. Greczek
Metropolitan Water Reclamation District of Greater Chicago
Des Plaines, Illinois

Bruce Guile
National Academy of Sciences
Washington, D.C.

Frank Hall
U.S. Environmental Protection Agency
Washington, D.C.

Susan Hamilton
Clean Water Program for Greater San Diego
California

Glenn Harvey
Alexandria Sanitation Authority
Alexandria, Virginia

Irwin Haydock
County Sanitation Districts of Orange County
Fountain Valley, California

Margarete Heber
U.S. Environmental Protection Agency
Washington, D.C.

Bruce Henderson
Former City Council Member
San Diego, California

Robert Henken
Congressmember Torricelli's Office
Washington, D.C.

Stephen P. Holcomb
Escambia County Utilities Authority
Pensacola, Florida

Angie Holden
County Sanitation Districts of Orange County
Fountain Valley, California

Tom Holliman
Irvine Ranch Water District
Irvine, California

Nancy Hopps
The World Wildlife Fund
Washington, D.C.

Sarah Horrigan
National Association of State Universities and Land-Grant Colleges
Washington, D.C.

Robert Horvath
Los Angeles County Sanitation Districts
Whittier, California

Richard Houben
Orange County Public Utilities
Orlando, Florida

Joyce Hudson
U.S. Environmental Protection Agency
Washington, D.C.

David Hufford
City of Tacoma
Washington

Norbert Jaworski
U.S. Environmental Protection Agency
Narragansett, Rhode Island

Karen Jones
Camp Dresser & McKee, Inc.
Orlando, Florida

Timothy Kasten
U.S. Environmental Protection Agency
Washington, D.C.

Ed Kerwin
Broward County Office of
Environmental Services
Pompano Beach, Florida

Robert King
U.S. Environmental Protection Agency
Washington, D.C.

Ken Kirk
Association of Metropolitan Sewerage
Agencies
Washington, D.C.

Karen S. Klima
U.S. Environmental Protection Agency
Washington, D.C.

Robert C.Y. Koh
California Institute of Technology
Pasadena

Jessica C. Landman
Natural Resources Defense Council
Washington, D.C.

James F. Langley
City of Los Angeles
California

Alan C. Langworthy
City of San Diego Water Utilities
Department
California

M. Lemore
Hampton Roads Sanitation District
Virginia Beach, Virginia

Howard Levenson
California Integrated Waste
Management Board
Sacramento, California

Charles E. Lewicki
Hampton Roads Sanitation District
Williamsburg, Virginia

Lam Lim
U.S. Environmental Protection Agency
Washington, D.C.

Kris P. Lindstrom
K.P. Lindstrom, Inc.
Pacific Grove, California

Jamison Lowe
National Oceanic and Atmospheric
Administration
Rockville, Maryland

Donald R. Madore
Metropolitan Waste Control
Commission
St. Paul, Minnesota

Judith McDowell
Woods Hole Oceanographic Institution
Massachusetts

Charles McGee
Los Angeles County Sanitation Districts
Whittier, California

Laurie McGilvray
National Oceanic and Atmospheric
Administration
Washington, D.C.

Karen McGlathery
Cornell University
Ithaca, New York

Jack Meade
Unified Sewerage Agency
Forest Grove, Oregon

John Meagher
U.S. Environmental Protection Agency
Washington, D.C.

Eric Mische
San Diego Clean Water Program
California

Marian Mlay
U.S. Environmental Protection Agency
Washington, D.C.

James J. Morgan
California Institute of Technology
Pasadena

David Mullen
Unified Sewerage Agency
Tigard, Oregon

Margaret Nellor
County Sanitation Districts of Orange County
Fountain Valley, California

Ronald A. Neubauer
Metropolitan Water Reclamation District of Greater Chicago
Stickney, Illinois

Harold G. Newhall
South Essex Sewerage District
Salem, Massachusetts

Earl Ng
City and County of Honolulu
Wahiawa, Hawaii

Charles Nichols
County Sanitation Districts of Orange County
Fountain Valley, California

Craig Nishimura
City and County of Honolulu
Kailua, Hawaii

Donald J. O'Connor
Manhattan College
Bronx, New York

Thomas O'Connor
National Oceanic and Atmospheric Administration
Rockville, Maryland

George T. Ohara
City of Los Angeles
California

Dan Olson
Fish and Wildlife Service
Arlington, Virginia

Tony Pait
National Oceanic and Atmospheric Administration
Rockville, Maryland

Paul Pan
U.S. Environmental Protection Agency
Washington, D.C.

Paul Papanek, Jr.
Los Angeles County Department of Health Services
California

Anthony Paulson
National Oceanic and Atmospheric Administration
Seattle, Washington

Kevin Perry
U.S. Environmental Protection Agency
Washington, D.C.

Jack Petralia
Los Angeles County Department of Health Services
California

John Pickelhaupt Jr.
Hampton Roads Sanitation District
Virginia Beach, Virginia

Mark Pisano
Southern California Council of Governments
Los Angeles

Gerald A. Pollock
California Environmental Protection Agency
Sacramento

Martha Prothro
U.S. Environmental Protection Agency
Washington, D.C.

J. Michael Read
Bureau of Environmental Services
Portland, Oregon

Christopher P. Reilly
San Diego County Water Authority
California

Adriana Renescu
County Sanitation Districts of Orange County
Fountain Valley, California

Mary Frances Repko
World Wildlife Fund
Washington, D.C.

Wanda R. Resper
U.S. Environmental Protection Agency
Washington, D.C.

Gordon G. Robeck
Laguna Hills, California

Steve Rohmann
National Oceanic and Atmospheric Administration
Rockville, Maryland

Christine Ruf
U.S. Environmental Protection Agency
Washington, D.C.

James Scanlan
Fort Worth Water Department
Fort Worth, Texas

William Secoy
Orange County Public Utilities
Orlando, Florida

Harry Seraydarian
U.S. Environmental Protection Agency
San Francisco, California

Greg Shaw
Metropolitan Water Reclamation District of Greater Chicago
Hanover Park, Illinois

Kimberly Shea
City of San Diego, Water Utilities Department
California

Bahman Sheikh
City of Los Angeles
California

Peter Shelley
Conservation Law Foundation
Boston, Massachusetts

Mark Shoup
City of Orlando, Environmental Services Department
Florida

David Shulmister
City of St. Petersburg
Florida

Bradley M. Smith
City of Los Angeles Bureau of Engineering
California

Daniel S. Smith
California Department of Health Services
Emeryville

Carlo Spani
Unified Sewerage Agency
Hillsboro, Oregon

James Stahl
Los Angeles County Sanitation District
Whittier, California

Keith Stolzenbach
University of California, Los Angeles

Jan Stull
Los Angeles County Sanitation Districts
Whittier, California

Tim S. Stuart
U.S. Environmental Protection Agency
Washington, D.C.

Robert Summers
Maryland Department of Natural Resources
Baltimore

Richard Sykes
East Bay Municipal Utility District
Oakland, California

James Taft
U.S. Environmental Protection Agency
Washington, D.C.

Keith Takata
U.S. Environmental Protection Agency
San Francisco, California

Martin Talebi
County Sanitation Districts of Orange County
Fountain Valley, California

Mia Tegner
Scripps Institution of Oceanography
La Jolla, California

Chris Toth
City of San Diego Water Utilities Department
San Diego, California

Patricia M. Vainik
City of San Diego Water Utilities Department
California

Michael J. Wallis
East Bay Municipal Utility District
Oakland, California

J. Kris Warren
Anchorage Water & Wastewater Utility
Alaska

Dov Weitman
U.S. Environmental Protection Agency
Washington, D.C.

Wendy Wiltse
U.S. Environmental Protection Agency
San Francisco, California

Lee Marc G. Wolman
American Society of Civil Engineers
Boston, Massachusetts

David W. York
Florida Department of Environmental Regulation
Tallahassee, Florida

Christopher Zarba
U.S. Environmental Protection Agency
Washington, D.C.

Robert Zeller
U.S. Environmental Protection Agency
Washington, D.C.

Gary E. Ziols
Metropolitan Water Reclamation District of Greater Chicago
Schaumburg, Illinois

Index

I

N

O

P

Q

R